Functionality of Cyclodextrins in Encapsulation for Food Applications

Thao M. Ho • Hidefumi Yoshii
Keiji Terao • Bhesh R. Bhandari
Editors

Functionality of Cyclodextrins in Encapsulation for Food Applications

Springer

Editors
Thao M. Ho
Department of Food and Nutrition
University of Helsinki
Helsinki, Finland

Hidefumi Yoshii
Department of Food Science
and Human Nutrition
Setsunan University
Osaka, Japan

Keiji Terao
CycloChem Bio Co., Ltd.
Kobe, Hyogo, Japan

Department of Social/Community
Medicine and Health Science,
Food and Drug Evaluation Science,
Graduate School of Medicine
Kobe University
Kobe, Hyogo, Japan

Bhesh R. Bhandari
School of Agriculture and Food Sciences
University of Queensland
Brisbane, QLD, Australia

ISBN 978-3-030-80055-0 ISBN 978-3-030-80056-7 (eBook)
https://doi.org/10.1007/978-3-030-80056-7

This Springer imprint is published by the registered company Springer Nature Switzerland AG
The registered company address is: Gewerbestrasse 11, 6330 Cham, Switzerland

Preface

Cyclodextrins (CDs), the cyclic oligosaccharides composed of α-(1,4) linked glucopyranose subunits, are typically produced by intramolecular transglycosylation reaction from degradation of starches by glucanotransferase enzymes. The most common types of native CDs are composed of 6, 7, and 8 glucopyranose units, known as α-, β-, and γ-CDs, respectively. To improve the functionality, for example, the solubility, native CDs are modified, and CD derivatives are synthesized. The cyclic molecular structure of CDs has truncated molecular shape with hydrophobic cavity at the centers which can interact (host) with external hydrophobic compounds (guest molecules) and hydrophilic surface. CDs have also been known as generally recognized as safe (GRAS) in the United States, natural products in Japan, and as novel food in Australia, New Zealand, and European countries. Therefore, CDs are being widely used in food production for many purposes, especially to encapsulate hydrophobic compounds including solid, liquid, and gas molecules aiming to solubilize, stabilize, or control release rate of these compounds. There are a large number of studies dedicating to the encapsulation of CDs for various food applications over the last few decades. This book will provide the comprehensive review on the functionality of CDs in the encapsulation for food applications.

The book includes a total of 16 chapters in which Chap. 1 gives general introduction to CD properties and its applications in food processing, and Chaps. 2–16 are about applications of CDs in the encapsulation for many guest compounds. These compounds include gases, flavors, colors, pigments, polyphenols (plant bioactive compounds), essential oils, lipids (cholesterol and polyunsaturated fatty acids), vitamins, and antifungal and antimicrobial compounds. Functionalities of CDs applied to packaging, masking off-flavor and off-taste, and dietary fiber are also described. The book is suitable for both newcomers to encapsulation technology and for those with experiences in the field including academics, undergraduate and postgraduate students, and food industry professionals.

The editors greatly acknowledge with gratitude all authors. Special thanks are also extended to the staff at Springer Nature for their support and highly professional editing of the publication during the course of this book project.

Helsinki, Finland — Thao M. Ho
Osaka, Japan — Hidefumi Yoshii
Kobe, Hyogo, Japan — Keiji Terao
Brisbane, QLD, Australia — Bhesh R. Bhandari

Contents

Contributors

Diana Alves CEB—Centre of Biological Engineering, LIBRO—Laboratório de Investigação em Biofilmes Rosário Oliveira, University of Minho, Braga, Portugal

Hermawan D. Ariyanto Department of Industrial Chemical Engineering, Vocational School of Diponegoro University, Semarang, Indonesia

Bhesh R. Bhandari School of Agriculture and Food Sciences, University of Queensland, Brisbane, QLD, Australia

Keita Chikamoto CycloChem Bio Co., Ltd., Kobe, Hyogo, Japan

Chalida Cholmaitri Division of Postharvest Technology, School of Bioresources and Technology, King Mongkut's University of Technology Thonburi, Bangkok, Thailand

Éva Fenyvesi Cyclodextrin Research & Development Laboratory Ltd., Budapest, Hungary

Takahiro Furune CycloChem Bio Co., Ltd., Kobe, Hyogo, Japan

Francisco García-Carmona Department of Biochemistry and Molecular Biology-A, Faculty of Biology, University of Murcia, Murcia, Spain

Binh T. Ho Department of Food Technology, Faculty of Agriculture and Natural Resources, An Giang University (Vietnam National University Ho Chi Minh City), Long Xuyen City, An Giang, Vietnam

Thao M. Ho Department of Food and Nutrition, University of Helsinki, Helsinki, Finland

Yoshiyuki Ishida CycloChem Bio Co., Ltd., Kobe, Hyogo, Japan

Seid Mahdi Jafari Department of Food Materials and Process Design Engineering, Gorgan University of Agricultural Sciences and Natural Resources, Gorgan, Iran

Nutrition and Bromatology Group, Department of Analytical and Food Chemistry, Faculty of Food Science and Technology, University of Vigo, Ourense, Spain

Natta Laohakunjit Division of Postharvest Technology, School of Bioresources and Technology, King Mongkut's University of Technology Thonburi, Bangkok, Thailand

José Manuel López-Nicolás Department of Biochemistry and Molecular Biology-A, Faculty of Biology, University of Murcia, Murcia, Spain

Adrián Matencio Department of Biochemistry and Molecular Biology-A, Faculty of Biology, University of Murcia, Murcia, Spain

Department of Chemistry, University of Turin, Torino, Italy

Juan-Carlos Mejuto Department of Physical Chemistry, Faculty of Science, University of Vigo, Ourense, Spain

Silvia Navarro-Orcajada Department of Biochemistry and Molecular Biology-A, Faculty of Biology, University of Murcia, Murcia, Spain

Thi Van Anh Nguyen Department of Food Technology, Faculty of Engineering and Food Technology, Hue University of Agriculture and Forestry, Hue University, Phung Hung, Hue, Thua Thien Hue Province, Vietnam

Eva Pinho National Institute for Agrarian and Veterinarian Research (INIAV), Vila do Conde, Portugal

István Puskás Cyclodextrin Research & Development Laboratory Ltd., Budapest, Hungary

Jaruporn Rakmai Biomass and Bio-energy Technology Division, Kasetsart Agricultural and Agro-Industrial Product Improvement Institute (KAPI), Kasetsart University, Bangkok, Thailand

Yaxin Sang College of Food Science and Technology, Hebei Agricultural University, Baoding, China

Jesus Simal-Gandara Nutrition and Bromatology Group, Department of Analytical and Food Chemistry, Faculty of Food Science and Technology, University of Vigo, Ourense, Spain

Tamás Sohajda Cyclodextrin Research & Development Laboratory Ltd., Budapest, Hungary

Afroza Sultana Department of Food Processing and Engineering, Faculty of Food Science and Technology, Chattogram Veterinary and Animal Sciences University, Chattogram, Bangladesh

Kamornrath Sungkaprom School of Agriculture and Food Sciences, University of Queensland, Brisbane, QLD, Australia

Lajos Szente Cyclodextrin Research & Development Laboratory Ltd., Budapest, Hungary

Keiji Terao CycloChem Bio Co., Ltd., Kobe, Hyogo, Japan

Department of Social/Community Medicine and Health Science, Food and Drug Evaluation Science, Graduate School of Medicine, Kobe University, Kobe, Hyogo, Japan

Yukiko Uekaji CycloChem Bio Co., Ltd., Kobe, Hyogo, Japan

Apiradee Uthairatanakij Division of Postharvest Technology, School of Bioresources and Technology, King Mongkut's University of Technology Thonburi, Bangkok, Thailand

Jianbo Xiao Nutrition and Bromatology Group, Department of Analytical and Food Chemistry, Faculty of Food Science and Technology, University of Vigo, Ourense, Spain

Institute of Chinese Medical Sciences, State Key Laboratory of Quality Research in Chinese Medicine, University of Macau, Taipa, Macau

Hidefumi Yoshii Department of Food Science and Human Nutrition, Setsunan University, Osaka, Japan

Chapter 1
Properties of Cyclodextrins and Their Applications in Food Processing

Yoshiyuki Ishida and Thao M. Ho

1.1 Type and Structure of Cyclodextrins

Cyclodextrins (CDs) are non-reducing, chiral cyclic oligosaccharides in which D-(+)-glucopyranose units are linked α-(1,4)-glycosidically into a ring. Depending on the number of D-(+)-glucopyranose units and thus the size of the ring, a distinction is made between α-CD, β-CD, and γ-CD. α-CD consists of six glucose units, β-CD of seven, and γ-CD of eight. Larger CDs consists of more than nine glucose units can be produced, but the utilization of the CDs is limited due to their structures and properties which are different from α-, β-, and γ-CD (Hedges 2009).

CDs are natural conversion compounds of starch, which is a polymer of glucose. In nature, starch is enzymatically decomposed by specific microorganisms into CDs and saccharide chains. Highly active bacteria that can selectively produce oligosaccharide rings have been selected by combining modern and refined techniques (Schmid et al. 1988). The enzyme for producing oligosaccharide rings is called cyclodextrin glucosyl transferase (CGTase). Researchers succeeded in crystallizing the enzyme in 1991 and reported the crystal structure of CGTase (Klein and Schulz 1991).

CDs have a toroid molecular structure with a cavity height of 0.8 nm. The diameter depends on the number of glucopyranose units. The cavity of CDs is hydrophobic due to hydroxyl-group arrangement, and the outer rims are hydrophilic (Fig. 1.1). The primary hydroxyl groups, which are located at the smaller rim, are rotatable in

Y. Ishida (✉)
CycloChem Bio Co., Ltd., Kobe, Hyogo, Japan
e-mail: yoshiyuki.ishida@cyclochem.com

T. M. Ho
Department of Food and Nutrition, University of Helsinki, Helsinki, Finland
e-mail: minh.ho@helsinki.fi

T. M. Ho et al. (eds.), *Functionality of Cyclodextrins in Encapsulation for Food Applications*, https://doi.org/10.1007/978-3-030-80056-7_1

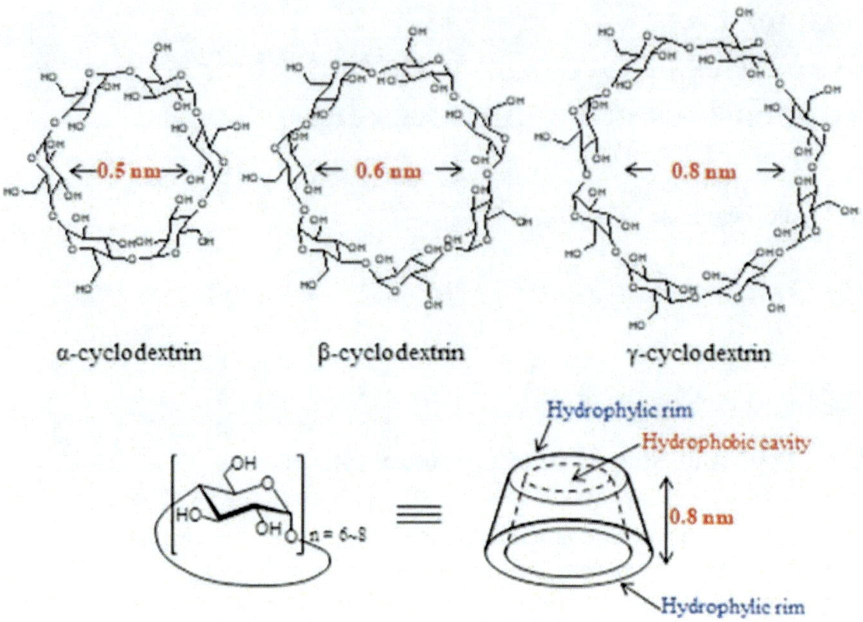

Fig. 1.1 Structures of native α-, β-, and γ-cyclodextrin and their geometric dimensions

such a manner that they can change the size of the rim. Secondary hydroxyl groups are located at the larger rim of the toroid. Intramolecular hydrogen bonding between the secondary hydroxyl groups of each glucopyranose unit plays a major role in the structural flexibility and aqueous solubility of CDs. β-CD has the most rigid structure among all CDs due to the formation of an intramolecular hydrogen bond belt. α-CD is more flexible than β-CD because of truncated formation of the intramolecular hydrogen bond belt due to less glucopyranose units. γ-CD is structurally most flexible and therefore the most soluble of the three CDs (Shieh and Hedges 1996; Davis and Brewster 2004).

A characteristic feature of CD molecules is their ability to enclose lipophilic substances reversibly in their hydrophobic cavity, provided that the guest molecule size and geometric shape fit the cavity. These CD inclusion compounds, which are known as host–guest complexes, are largely held together by van der Waals forces.

Chemically modified CDs, such as hydroxypropylated CD, randomly methylated CD, and sulfobutylated CD, are frequently used for pharmaceutical formulations due to their higher solubilizing power compared to native CDs (Szejtli 1983). In contrast, chemically modified CDs cannot be utilized for food applications from a viewpoint of safety and bioadaptability. Chemically modified CDs generally have a membrane-perturbing effect, resulting in haemolysis and local irritation at higher doses (Szejtli 1984). Therefore, native CDs are the only legally approved CDs to be used for food applications.

1.2 Physical and Chemical Properties of Cyclodextrins

1.2.1 Water Solubility of Cyclodextrins

The water solubility of CDs is lower than that of glucose due to the formation of intramolecular hydrogen bonds between secondary hydroxyl groups on the larger rim of CDs. CDs cannot hydrate enough due to the hydrogen bond belt. The orientation and degree of the hydrogen bonds between the secondary hydroxyl groups of adjacent glucopyranose units are different in each of the CDs. β-CD has the lowest water solubility (1.85 g/100 mL at 25 °C) of the three CDs with the strongest hydrogen bond belt. On the other hand, γ-CD has the highest water solubility (23.2 g/100 mL at 25 °C) of the three CDs due to its weakest hydrogen bond belt. The water solubility of α-CD is the middle of the three CDs (14.5 g/100 mL at 25 °C). The water solubility of CDs increases along with the increase in temperature (Astray et al. 2009).

The acidity of the secondary hydroxyl group of CDs is comparably high. The pK_a of the hydroxyl group is about 12 (Szejtli 1998). Accordingly, CDs are ionized at high pH (pH > 12) and the water solubility of CDs also increases. For example, the water solubility of β-CD reaches 75.0 g/100 mL at pH 12.5 (Hedges 2009).

1.2.2 Thermal Stability of Cyclodextrins

Solid states of α-, β-, and γ-CD are basically stable up to 300 °C. At that temperature, melting of the crystals and thermal decomposition of CDs are observed. The melting of the crystals and decomposition of the structure occur simultaneously and cannot be separated from each other. Therefore, in most cases of food processing, CDs can be applied without any problem (Shieh and Hedges 1996).

1.2.3 Stability of Cyclodextrins in Acids and Bases

The α-1,4 glycosidic bonds of CDs are more resistant to acid hydrolysis than that of starch, because their cyclic structures stabilize them. However, at low pH (pH < 2) and in the presence of strong acids, such as hydrochloric acid or sulfuric acid, α-1,4 glycosidic bonds of CDs are cleaved via hydrolysis yielding a mixture of oligosaccharides and glucose. The rate of hydrolysis increases at lower pH and high temperature (Szejtli and Budai 1976; Szejtli 1988). The α-1,4 glycosidic bonds of CDs are quite stable and not hydrolysed by bases, even at elevated temperatures.

1.2.4 Digestibility of Cyclodextrins in Gut

CDs are digested after oral administration. α-CD and β-CD are not digested by salivary or pancreatic enzymes and are hardly absorbed in small intestines. Approximately, 99% of administered α-CD and β-CD reach the large intestine and they are fermented by the intestinal microflora (Kurkov and Loftsson 2013). The microbial degradation of α-CD results in the generation of short-chain fatty acids (acetate, propionate, butyrate) that provide many health benefits (Nihei et al. 2018; Sakurai et al. 2017). On the other hand, γ-CD differs from α-CD and β-CD in that it is almost completely degraded by salivary and pancreatic α-amylase, similar to digestion of starch and linear dextrins, and therefore γ-CD can be regarded as a slow energy release carbohydrate (Saokham and Loftsson 2017).

1.2.5 Formation of Inclusion Complexes

While CDs are oligosaccharides, their cavities provide a lipophilic environment in aqueous solutions. Various hydrophobic substances can be selectively enclosed in their CDs cavities and form non-covalent reversible inclusion complexes. The formation of inclusion complexes between guest molecules and CDs depends on the polarity and size of guest molecules via driving forces, such as van der Waals forces, hydrophobic interactions, and hydrogen bonds. α-CD forms inclusion complexes with relatively smaller-sized molecules, such as carbon dioxide gas (Neoh et al. 2006) and hydrocarbon chain. β-CD forms inclusion complexes with medium-sized molecules, like monoterpenes, which are typical components of flavor and fragrance (Pollyana et al. 2016), and polyphenols such as hesperidin, which is abundantly found in citrus fruits, and also acts as an antioxidant that contributes to blood-vessel integrity and reduces LDL cholesterol and blood pressure (Monforte et al. 1995; Ohtsuki et al. 2003). γ-CD forms inclusion complexes with larger molecules, such as macrocyclic compounds and lipophilic vitamins, curcumin, coenzyme Q10 (CoQ10), which are used as nutraceutical ingredients in supplements and health foods with various human health benefits (Uekaji and Terao 2019).

1.3 Toxicological Considerations

As for the toxicity of CDs, The FAO/WHO Joint Expert Committee on Food Additives (JECFA) states that they are highly safe because they are natural products of starch.

1.3.1 Toxicity Evaluation of α-CD

Results of acute toxicity studies in rats and mice administered intraperitoneally or intravenously with α-CD have been reported in the range of 500–1000 mg/kg bw as LD_{50}.

In short-term toxicity studies, no significant toxicity was observed in groups of male mice administered orally with α-CD at doses equivalent to 0 or 60 mg/kg bw/day for 15 days. In addition, groups of five Wistar rats of each sex were fed diets equivalent to 0–7500 mg/kg bw/day of α-CD for 4 weeks with no deaths and no significant toxicity observed. Similar results were reported when a group of 20 Wistar rats were fed α-CD at 0–10,000 mg/kg bw/day for 13 weeks, with no deaths and no significant toxicity observed. In addition, groups of four beagle dogs of both sexes were fed diets equivalent to 0–5000 mg/kg bw of α-CD per day for 13 weeks, resulting in no deaths and diarrhoea, but no other toxic effects. There have been some reports of human studies that lack toxicity, but consumption of more than 20 g of α-CD in a single meal may cause gastrointestinal effects.

1.3.2 Toxicity Evaluation of β-CD

In a short-term toxicity study of β-CD, 20 rats of both sexes, aged 6 weeks, were treated with diets containing 0–5% β-CD for 52 weeks. Four deaths were recorded during the study period: three males in the control group, and a female fed a diet containing 1.25% β-CD. None of these deaths were due to β-CD and no other unusual clinical signs were observed. In addition, a group of four beagle dogs of both sexes were treated with diets containing 0–5% β-CD for 52 weeks with no deaths, but elevated urinary protein and potassium levels were observed.

1.3.3 Toxicity Evaluation of γ-CD

In a long-term toxicity study of γ-CD, groups of 20 rats of both sexes were treated with diets containing 0–20% of γ-CD for 12 months, and four deaths were recorded during the study period. No clear toxicity was shown. In a human study, a double-blind crossover study was conducted in two groups of six men and six women to examine gastrointestinal responses to 100 g of plain yogurt containing 0 or 8 g of maltodextrin or γ-CD as a snack within 15 min. There was no statistically significant difference in the incidence of individual symptoms between the maltodextrin group and the γ-CD group, and no toxicity of γ-CD was demonstrated.

Based on the results of the toxicity study of CDs, JECFA evaluates α- and γ-CDs as food additives with no tolerable intake level (Kroes et al. 2006; Abbott 2000), and

Table 1.1 Regulatory status of α-, β-, and γ-CD

Type of CD	Average daily intake (JECFA)	Food approval			
		US	EU	Japan	Australia and New Zealand
α-CD	Not specified	GRAS	Novel food	Available	Novel food
β-CD	5 mg/kg bw/day	GRAS	Food additive	Available	–
γ-CD	Not specified	GRAS	Novel food	Available	Novel food

the acceptable daily intake (ADI) of β-CDs is 5 mg/kg-body-weigh/day based on the data in beagle dogs (Pollit 1996). In the EU, the regulation on the use of CDs is approved as 'Novel Food' for α- and γ-CDs, and as 'Food Additive' for β-CDs. In the USA, CDs are approved by Generally Recognized as Safe (GRAS), and β-CDs are approved to be used as 'Flavor Protectant'. In Japan, α-, β-, and γ-CDs are unconditionally approved for food use. In Australia and New Zealand, α- and γ-CD are approved as Novel Foods (Table 1.1).

1.4 Crystalline and Amorphous Cyclodextrins

CD powders can exist as either crystalline or amorphous structure depending on the arrangement order of molecules in the powders. The commercial CD powders have crystalline form in which molecules are tightly packed in a certain order. In amorphous form, CD molecules are randomly arranged with a loose structure. Theoretically, amorphization of powders can be achieved by quenching of a melt, spray drying, freeze drying, and milling (Einfalt et al. 2013). However, it was found that quenching of a melt and annealing at high temperature (110–150 °C) were not appropriate techniques to produce amorphous CD powders (Kaminski et al. 2012). In the quenching of a melt, CD powders experience extensive chemical degradation (e.g., caramelization) during heating while the annealing approach enables only a small part of the CD structure to undergo phase transformation. Completely amorphous CD powders have been successfully produced by freeze drying (Li et al. 2002), ball milling (Kaminski et al. 2012; Tabary et al. 2011), and spray drying (Ho et al. 2015a, 2017a; Ho 2017; Shrestha et al. 2017; Frieler et al. 2019). During the milling of CD powders, the generation of localized heating effects followed by quenching, or an increase in static disorder adding to the intrinsic dynamic disorder inherent within the lattice up to critical value where structural collapse occurs, results in the amorphous structure of the powder. The degree of amorphization (e.g., partially or completely amorphous) is largely determined by the milling temperature, which must be lower than the glass transition temperature of the powder to induce the amorphization. There is possibly microcrystalline structure remaining in milled CD powders as the mechanical method is unable to completely disrupt all the crystalline structure. Meanwhile, amorphization of CD powders in freeze drying is attributed to rapid sublimation of water vapour from the solid state at reduced pressure and spray drying is attributed to rapid water evaporation, both of which prevent molecules from reorganizing into a crystalline structure.

Like other amorphous solids, amorphous CD powders are in a thermodynamically non-equilibrium state, and thus they are likely to transform to crystalline structure with a thermodynamically stable state during production and storage. The rate of phase transformation from amorphous to crystalline depends on the mobility of molecules in the powders, which is controlled by the moisture content of powders, relative humidity that the materials are exposed to, and powder temperature. The boundary between low (glassy) and high (rubbery) molecular mobility of amorphous materials is defined as the glass transition temperature. Thus, the determination of glass transition temperature and critical moisture content (e.g., moisture content at which phase transformation is initiated) for amorphous CD powders is very important to prevent or initiate the state transformation for a particular application (Ho et al. 2015a). For amorphous α-, β-, and γ-CD powders produced by ball milling, the glass transition temperatures were unable to be determined via conventional DSC scanning, and they were therefore estimated above thermal degradation points, such as ~270 °C (Kaminski et al. 2012; Tabary et al. 2011). In a study by Ho et al. (2015a) on amorphous α-CD powder produced by spray drying, however, the glass transition temperature determined by DSC scanning was about 84 °C. The differences between crystalline α-, β-, and γ-CDs, and amorphous counterparts produced by spray drying under various analytical techniques are illustrated in Table 1.2, and Figs. 1.2 and 1.3 (Ho et al. 2015a; Frieler et al. 2019; Bhandari and Ho 2020). Methods to characterize quantitatively and qualitatively the amorphous and crystalline food powders including CD powders have been well described in a review by Ho et al. (2017a).

Table 1.2 Differences between crystalline and spray-dried amorphous α-CD powders observed under various analytical techniques

Analytical techniques	Amorphous α-CD	Crystalline α-CD
SEM	Spherical-shape particles with a smooth surface and possessing dents and small vacuoles	Irregular shapes with different sizes Many small particles and clefts on their surface Sharp edges of particles
Polarized light microscope	Homogenously dark background	Colored bright crystal granules with different contrast and shape
X-ray	Only two broad peaks	Many sharp and high intensity peaks
FTIR	Low intensity in FTIR peaks	High intensity in FTIR peaks
^{13}C NMR	Broader spectral peaks	Sharp and split spectral peaks
DSC	A big endothermic hump A signal of glass transition	Multiple endothermic peaks

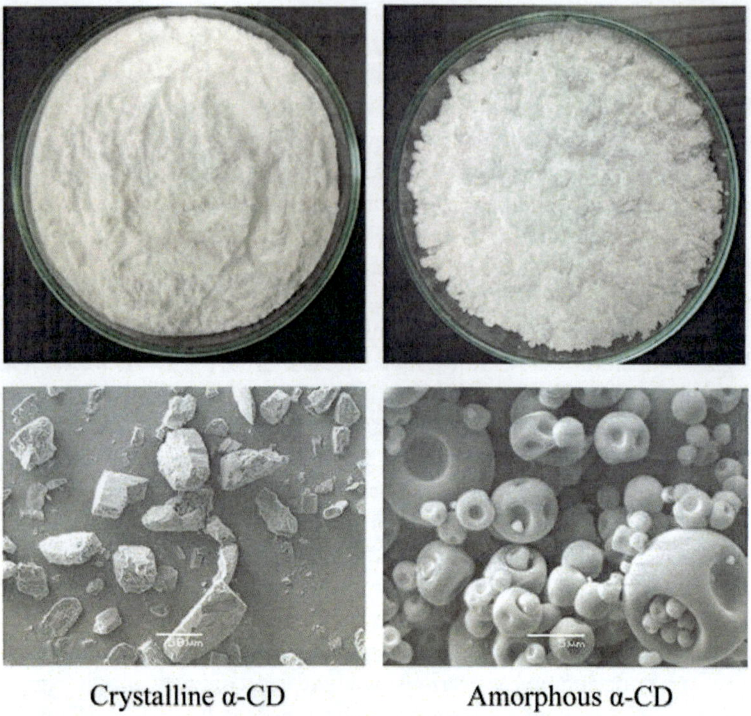

Fig. 1.2 Appearance and SEM images of α-CD powders in crystalline and amorphous structures. (Adapted with permission from Ho et al. 2016b)

When equilibrated at various relative humidities, amorphous α-CD powders crystallized at relative humidities above 65% and absorbed much more water and faster than crystalline α-CD. At equilibrium, the moisture content of amorphous α-CD powder kept at 65% relative humidity was 13.70–14.97 g of moisture per 100 g of dry solids. This relative humidity and moisture content are considered to be critical for the spray-dried amorphous α-CD powder. As shown in Fig. 1.4, a few sharp peaks were observed on the X-ray spectrum of amorphous α-CD powders equilibrated at 65.29% relative humidity, while many sharp peaks were seen on that of amorphous α-CD powders kept at 75.32%, 84.32%, and 97.30% relative humidity (Fig. 1.4a). Some particles of amorphous α-CD powder started to aggregate due to water adsorption when stored at 65.29% relative humidity (Fig. 1.4b), and this phenomenon became clear at 84.32% and 97.30% relative humidity (e.g., powder agglomeration and crystallization).

Recent studies about encapsulation of ethylene and carbon dioxide gases have been performed by directly compressing gases into amorphous solid state α-CD powders, aiming to produce gas complex powders to apply in agriculture and food production (Ho and Bhandari 2016; Ho et al. 2015b). The results demonstrate that the gas encapsulation capacity of crystalline α-CD powder was significantly lower

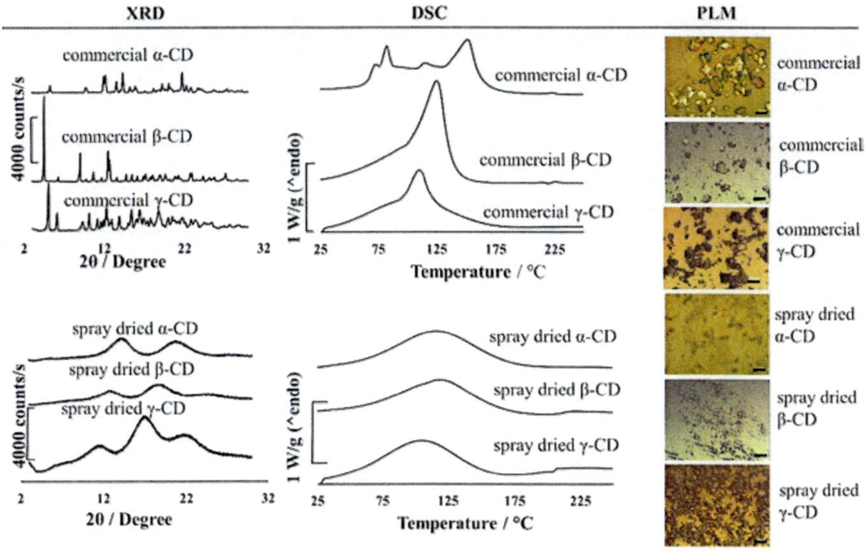

Fig. 1.3 XRD patterns, DSC thermograms, and polarized light microscopy (PLM) of commercial crystalline and spray-dried amorphous α-, β- and γ-CD powders. For images, scale bar = 200μm. (Adapted with permission from Frieler et al. 2019)

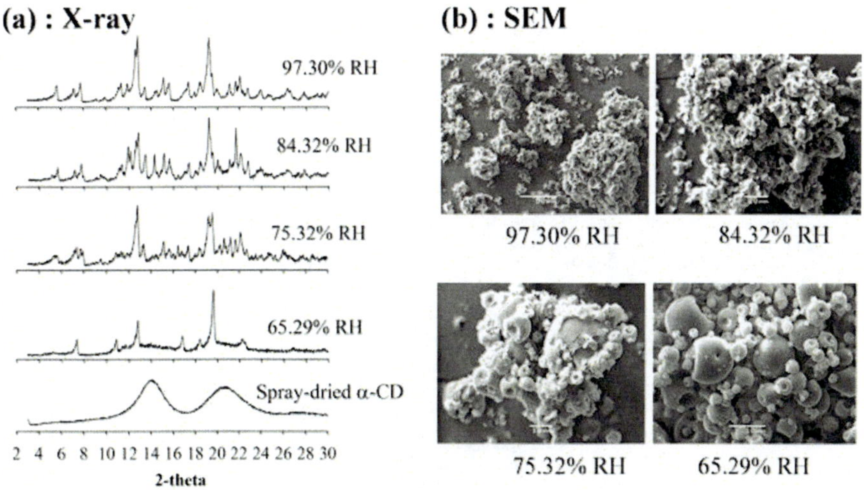

Fig. 1.4 X-ray (**a**) and SEM (**b**) analyses of spray-dried amorphous α-CD powders equilibrated at various relative humidities (RH). (Adapted with permission from Ho et al. 2015a)

than that of amorphous α-CD powder at low pressure and short time (e.g., 0.4–0.8 MPa and 4–24 h). However, due to no long-range order and loose structure of the gas complex powder prepared from amorphous α-CD powder, major part of the encapsulated gas was released upon depressurization. The stability of amorphous gas-α-CD powder can be increased by adding water to a higher than critical value of amorphous α-CD powders (15–17%, w/w) to induce the crystallization of the complex powder during encapsulation (Ho et al. 2016a, b, c). Regarding encapsulation, the advantages of amorphous over crystalline CD powders have been illustrated in studies by Shrestha et al. (2017) and Frieler et al. (2019) for tea tree oil with β-CD powders, and for fish oil with α-, β-, and γ-CD powders, respectively.

1.5 Food Applications of Cyclodextrins

Many reviews describing the applications of CDs have been published (Matencio et al. 2020; Braga 2019; Jansook et al. 2018; Fenyvesi et al. 2016). Indeed, CDs have been frequently applied in the food industry as flavor carriers, protectants of food ingredients, undesired taste masking, and food packaging materials (Szente and Szejtli 2004; Hedges 1998) (Fig. 1.5).

1.5.1 Enhancement Stability of Included Compounds

CD inclusion can stabilize substances that are labile to light (ultraviolet) and heat, as well as unstable guest ingredients that are easily oxidized or hydrolyzed. This effect is applied to prevent flavor volatility, color changing/fading of pigments, and rancidification. CD inclusion can cause changes in color, odor, and taste of fats and oils, and can improve the quality and storage stability of food products (Fenyvesi et al. 2016; Szente and Szejtli 2004; Loftsson 1995). For instance, while allyl

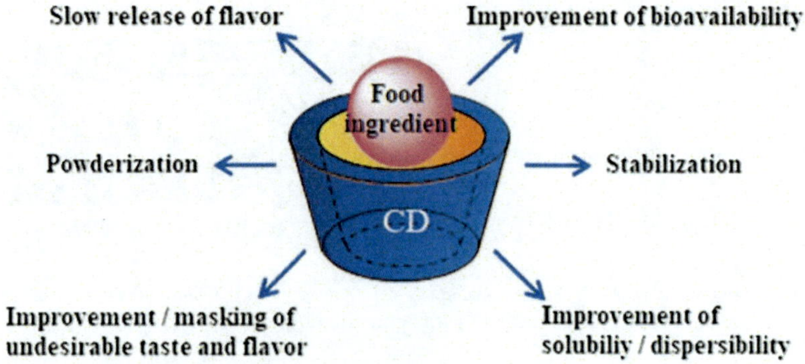

Fig. 1.5 Food applications of CDs by the formation of inclusion complex

isothiocyanate, a pungent ingredient of wasabi and mustard oil, is volatile and easily reacts with some nucleophiles such as water, carotenoids and Vitamin A are readily oxidized and its color fades with light and air. Vitamin D3 is labile under high temperature, and unsaturated fatty acid such as in fish oil is degraded by rancidification due to heat and oxygen. CD can protect from these problems via formation of stable inclusion complexes (Ohta et al. 2000; Sauvant et al. 2012; Szejtli et al. 1980; Yoshii et al. 1997). The inclusion complexes between CD and polyphenols can prevent the color modification of beverages via inhibition of enzymatic browning by polyphenoloxidase (Astray et al. 2009). Stabilization of nutraceutical ingredients by CD inclusion can also improve the absorption of the ingredients into the body after oral administration. R-alpha lipoic acid is a coenzyme in mitochondria and is well known as nutraceutical ingredient. R-alpha lipoic acid is labile in gut and shows poor bioavailability, but CD inclusion can provide sufficient stability to R-alpha lipoic acid and improve its bioavailability (Ikuta et al. 2016).

CD inclusion keeps moisture from entering hygroscopic substances in solid state. Non-solid substances, such as gases and oils, can also be converted into stable powders by encapsulating them with CDs, which makes for easier handling. This effect has been applied to powdered flavor and honey powder, for example. Powdered flavor with CD can easily control the volatile nature of flavor with temperature and moisture (Nguyen and Yoshii 2018), and can be prepared by conventional processes such as by mixing of flavor and CD via kneading and then spray drying or freeze drying (Dziezack 1988; Gibbs et al. 1999; Shahidi and Han 1993). This technique holds great potential for meal flavor enhancement, or for masking of undesirable food flavors.

1.5.2 Enhancement of Solubility of Included Compounds

Water insoluble ingredients cause clouding and sedimentation and may be undesirable in the quality in food products, especially beverages. The formation of inclusion complex by CD can inhibit the aggregation of water-insoluble ingredients, which can improve their water solubility and dispersibility. The solubility enhancement of an ingredient by CD inclusion can be shown in an A-type curve in typical phase-solubility diagram (Higuchi and Connors 1965). As such, CDs are used in processed citrus products, tea drinks, soups, for instance. CDs can also form stable emulsions or whips with oils and some kinds of proteins, and this effect has been used in bakery and confectionery products and milkshakes (Astray et al. 2020). It can also be used in egg-free mayonnaise-like dressings. On the other hand, β-CD has been used as a cholesterol-removing agent from egg yolk and daily products via formation of a water-insoluble complex with cholesterol (Fenyvesi et al. 2016).

CD-inclusion enhancement of water solubility of guest ingredients can improve the absorption of those ingredients into the body after oral administration. It is well known that absorbability of the ingredient is important in the nutraceutical and pharmaceutical fields. It has recently been shown that γ-CD inclusion can uniquely increase the efficiency of micelle formation by bile acids secreted in the intestine and greatly improve the absorption of the guest ingredient into the body with improved water solubility, especially in the case of highly fat-soluble guest ingredients (Uekaji et al. 2013). The above technique has been used as a γ-CD inclusion body to enhance the absorption of supplement ingredients, such as coenzyme Q10 and curcumin, a component of turmeric (Terao et al. 2006; Purpura et al. 2018).

1.5.3 Improvement of Taste and Flavor

The sense of taste and flavor of food is one of the principal factors to determine the quality and consumer acceptability, as well as shelf-life of food. CDs are useful to control taste and flavor in food industry because of the specific or selective nature of inclusion between CD and the taste and flavor components. Since the inclusion phenomenon of CD reduces the volatile nature of flavor and inhibits the interaction between taste components and their receptors on taste buds, many researchers have reported controlling, reducing, or masking the sensation of taste and flavor by CD (Szejtli and Szente 2005). α-CD has been used to reduce the bitter taste of amino acid and soy protein hydrolysate (Linde et al. 2009). β-CD has been used to reduce undesirable flavors, such as beany flavor from soy protein and soy milk (Lee et al. 2020; Arora and Damodaran 2010), unsaturated aldehyde from watermelon (Yang et al. 2020), and the goaty flavor of several branched chain fatty acids from goat milk (Wang et al. 2018). β-CD has also been used to reduce the bitter tastes of limonin and naringin of citrus juice (Hedges 1998). γ-CD has been applied to reduce the bitter tastes of relatively larger compounds, such as triterpenoid saponin of red ginseng (Tamamoto et al. 2010).

1.6 Conclusion and Future Perspectives

This chapter introduced the basic properties, safety, and inclusion effects of CDs in the food industry. CDs are a very attractive material that has a unique structure and the simple function of inclusion shows a variety of effects. CDs have long been used for foods. CDs are very useful, not only to improve and stabilize the quality of food, including physical properties and flavor, but also to improve the bioavailability of guest ingredients. In addition, CDs are used as a packaging material for controlling ripening and antibacterial treatment. Furthermore, the effect of supplementation of α-CD as a dietary fibre has recently been attracting attention. CD incorporation in foods is expected to become more widely used as a key method in food development to maintain and improve the quality of life and health.

References

Abbott PJ (2000) JEFCA 55th meeting. Safety evaluation of certain food additives and contaminants (γ-cyclodextrin). WHO Food Addit Ser 44:969

Arora A, Damodaran S (2010) Competitive binding of off-flavor compounds with soy protein and β-cyclodextrin in a ternary system: a model study. J Am Oil Chem Soc 87:673–679

Astray G, Gonzalez-Barreiro C, Mejuto JC et al (2009) A review on the use of cyclodextrins in foods. Food Hydrocoll 23:1631–1640

Astray G, Mejuto JC, Simal-Gandara J (2020) Latest developments in the application of cyclodextrin host-guest complexes in beverage technology processes. Food Hydrocoll 106:105882

Bhandari B, Ho TM (2020) Stickiness and caking in food preservation. In: Rahman MS (ed) Handbook of food preservation, 3rd edn. CRC Press, Boca Raton, pp 413–426

Braga SS (2019) Cyclodextrins: emerging medicines of the new millennium. Biomol Ther 9(12):801

Davis ME, Brewster ME (2004) Cyclodextrin-based pharmaceutics: past, present and future. Nat Rev Drug Discov 3:1023–1035

Dziezack JD (1988) Microencapsulation and encapsulated ingredients. Food Technol 42:136–151

Einfalt T, Planinšek O, Hrovat K (2013) Methods of amorphization and investigation of the amorphous state. Acta Pharma 63(3):305–334

Fenyvesi E, Vikmon M, Szente L (2016) Cyclodextrins in food technology and human nutrition: benefits and limitations. Crit Rev Food Sci Nutr 56:1981–2004

Frieler L, Ho TM, Anthony A et al (2019) Crystallisation properties of amorphous cyclodextrin powders and their complexation with fish oil. J Food Sci Technol 56(3):1519–1529

Gibbs BF, Kermasha S, Alli I et al (1999) Encapsulation in the food industry: a review. Int J Food Sci Nutr 50:213–224

Hedges AR (1998) Industrial application of cyclodextrins. Chem Rev 98:2035–2044

Hedges AR (2009) Cyclodextrins: properties and applications. In: Starch, 3rd edn. Academic Press, Cambridge, MA, pp 833–851

Higuchi T, Connors KA (1965) Phase-solubility techniques. Adv Anal Chem Instrum 4:117–212

Ho TM (2017) Studies of the encapsulation and release of carbon dioxide from amorphous and crystalline alpha-cyclodextrin powders and its application in food systems. PhD thesis, University of Queensland, Australia

Ho BT, Bhandari BR (2016) Novel solid encapsulation of ethylene gas using amorphous α-cyclodextrin and the release characteristics. J Agric Food Chem 64(17):3318–3323

Ho TM, Howes T, Bhandari BR (2015a) Characterization of crystalline and spray-dried amorphous α-cyclodextrin powders. Powder Technol 284:585–594

Ho TM, Howes T, Bhandari BR (2015b) Encapsulation of CO_2 into amorphous and crystalline α-cyclodextrin powders and the characterization of the complexes formed. Food Chem 187:407–415

Ho TM, Howes T, Bhandari BR (2016a) Encapsulation of CO_2 into amorphous alpha-cyclodextrin powder at different moisture contents—part 1: encapsulation capacity and stability of inclusion complexes. Food Chem 203:348–355

Ho TM, Howes T, Jack KS et al (2016b) Encapsulation of CO_2 into amorphous alpha-cyclodextrin powder at different moisture contents—part 2: characterization of complexed powders and determination of crystalline structure. Food Chem 206:92–101

Ho TM, Howes T, Bhandari BR (2016c) Dehydration of CO_2-α-cyclodextrin complex powder by desiccant adsorption method and its release properties. J Microencapsul 33(8):763–772

Ho TM, Truong T, Bhandari BR (2017a) Chapter 5: Spray-drying and non-equilibrium states/glass transition. In: Bhandari BR, Roos YH (eds) Non-equilibrium states and glass transitions in foods—processing effects and product-specific implications. Elsevier, Duxford, pp 111–136

Ho TM, Truong T, Bhandari BR (2017a) Methods to characterize the structure of food powders–a review. Biosci Biotechnol Biochem 81(4):651–671

Ikuta N, Okamoto H, Furune T et al (2016) Bioavailability of an R-α-lipoic acid/γ-cyclodextrin complex in healthy volunteers. Int J Mol Sci 17:949
Jansook P, Ogawa N, Loftsson T (2018) Cyclodextrins: structure, physicochemical properties and pharmaceutical applications. Int J Pharm 535:272–284
Kaminski K, Adrjanowicz K, Kaminska E et al (2012) Impact of water on molecular dynamics of amorphous α-, β-, and γ-cyclodextrins studied by dielectric spectroscopy. Phys Rev E 86(3):031506
Klein C, Schulz GE (1991) Structure of cyclodextrin glycosyltransferase refined at 2.0A resolution. J Mol Biol 217(4):737–750
Kroes R, Verger P, Larsen JC (2006) α-Cyclodextrin. WHO Food Addit Ser 54:3–15
Kurkov SV, Loftsson T (2013) Cyclodextrins. Int J Pharm 453:167–180
Lee EJ, Kim H, Lee JY et al (2020) β-Cyclodextrin-mediated beany flavor masking and textural modification of an isolated soy protein-based yuba film. Foods 9(6):818
Li J, Guo Y, Zografi G (2002) The solid-state stability of amorphous quinapril in the presence of β-cyclodextrins. J Pharm Sci 91(1):229–243
Linde GA, Junior AL, de Faria EV (2009) Taste modification of amino acids and protein hydrolysate by α-cyclodextrin. Food Res Int 42:814–818
Loftsson T (1995) Effects of cyclodextrins on the chemical stability of drugs in aqueous solutions. Drug Stab 1:22–33
Matencio A, Caldera F, Cecone C et al (2020) Cyclic oligosaccharides as active drugs, an updated review. Pharmaceuticals 13(10):281
Monforte MT, Trovato A, Kirjavainen S et al (1995) Biological effects of hesperidin, a citrus flavonoid. (noteII): hypolipidemic activity on experimental hypercholesterolemia in rat. IL Farmaco 50(9):595–599
Neoh TL, Yoshii H, Furuta T (2006) Encapsulation and release characteristics of carbon dioxide in α-cyclodextrin. J Incl Phenom Macrocycl Chem 56:125–133
Nguyen TVA, Yoshii H (2018) Release behavior of allyl sulfide from cyclodextrin inclusion complex of allyl sulfide under different storage conditions. Biosci Biotechnol Biochem 82(5):848–855
Nihei N, Okamoto H, Furune F et al (2018) Dietary α-cyclodextrin modifies gut microbiota and reduces fat accumulation in high-fat-diet-fed obese mice. Biofactors 44(4):336–347
Ohta Y, Takatani K, Kawakishi S (2000) Kinetic and thermodynamic analyses of the cyclodextrin-allyl isothiocyanate inclusion complex in an aqueous solution. Biosci Biotechnol Biochem 64:190–193
Ohtsuki K, Abe A, Mitsuzumi H et al (2003) Glucosyl hesperidin improves serum cholesterol composition and inhibits hypertrophy in vasculature. J Nutr Sci Vitaminol 49(6):447–450
Pollit FD (1996) Safety evaluation of certain food additives (β-cyclodextrin). WHO Food Addit Ser 35:257–268
Pollyana SSL, Angélica ML, Heitor GAF et al (2016) Inclusion of terpenes in cyclodextrins: preparation, characterization and pharmacological approaches. Carbohydr Polym 151:965–987
Purpura M, Lowery RP, Wilson JM et al (2018) Analysis of different innovative formulations of curcumin for improved relative oral bioavailability in human subjects. Eur J Nutr 57:929–938
Sakurai T, Sakurai A, Chen Y et al (2017) Dietary α-cyclodextrin reduces atherosclerosis and modifies gut flora in apolipoprotein E-deficient mice. Mol Nutr Food Res 61(8):1600804
Saokham P, Loftsson T (2017) γ-Cyclodextrin. Int J Pharm 516:278–292
Sauvant P, Cansell M, Sassi AH et al (2012) Vitamin A enrichment: caution with encapsulation strategies used for food applications. Food Res Int 46(2):469–479
Schmid G, Huber OS, Eberle HJ (1988) Selective complexing agents for the production of γ-cyclodextrin. In: Proceedings of the 4th international cyclodextrin symposium. F.R.G., Munich, pp 87–92
Shahidi F, Han XQ (1993) Encapsulation of food ingredients. Crit Rev Food Sci Nutr 33:501–547
Shieh WJ, Hedges AR (1996) Properties and applications of cyclodextrins. JMS-Pure Appl Chem A33(5):673–683

Shrestha M, Ho TM, Bhandari BR (2017) Encapsulation of tea tree oil by amorphous beta-cyclodextrin powder. Food Chem 221:1474–1483
Szejtli J (1983) Dimethyl-β-cyclodextrin as parenteral drug carrier. J Incl Phenom Macrocycl Chem 1(2):135–150
Szejtli J (1984) Highly soluble β-cyclodextrin derivatives. Starch/Stärke 36(12):429–432
Szejtli J (1988) Cyclodextrin technology. Kluwer Academic Publishers, Dordrecht
Szejtli J (1998) Introduction and general overview of cyclodextrin chemistry. Chem Rev 98:1743–1753
Szejtli J, Budai ZS (1976) Acid hydrolysis of β-cyclodextrin. Acta Chim Acad Sci Hung 91(1):73–80
Szejtli J, Szente L (2005) Elimination of bitter, disgusting tastes of drugs and foods by cyclodextrins. Eur J Pharm Biopharm 61:115–125
Szejtli J, Bolla E, Szabo P (1980) Enhancement of stability and biological effect of cholecalciferol by β-cyclodextrin complexation. Parmazie 35:779–787
Szente L, Szejtli J (2004) Cyclodextrins as food ingredients. Trends Food Sci Technol 15:137–142
Tabary N, Mahieu A, Willart J-F et al (2011) Characterization of the hidden glass transition of amorphous cyclomaltoheptaose. Carbohydr Res 346(14):2193–2199
Tamamoto LC, Schmidt SJ, Lee SY (2010) Sensory properties of ginseng solutions modified by masking agents. Food Sci 75:S341–S347
Terao K, Nakata D, Fukumi H et al (2006) Enhancement of oral bioavailability of coenzyme Q10 by complexation with γ-cyclodextrin in healthy adults. Nutr Res 26:503–508
Uekaji Y, Terao K (2019) Bioavailability enhancement of hydrophobic nutraceuticals using γ-cyclodextrin. Incl Phenom Macrocycl Chem 93:3–15
Uekaji Y, Jo A, Urano A et al (2013) Application of γ-cyclodextrin in nanomedicinal foods and cosmetics. In: Bio-nanotechnology: a revolution in food, biomedical and health sciences. Wiley & Sons, Chichester, pp 179–211
Wang C, Wang C, Gao F (2018) Effects of polymerized whey protein on goaty flavor and texture properties of fermented goat milk in comparison with β-cyclodextrin. J Dairy Res 85:465–471
Yang X, Yang F, Liu Y (2020) Off-flavor removal from thermal-treated watermelon juice by adsorbent treatment with β-cyclodextrin, xanthan gum, carboxymethyl cellulose sodium, and sugar/acid. LWT 131:109775
Yoshii H, Furuta T, Kawasaki K et al (1997) Oxidative stability of powdery tridocosazhexaenoin included in cyclodextrin and its application to fish meal paste. Biosci Biotechnol Biochem 61(8):1376–1378

Chapter 2
Solid Encapsulation Method: Ethylene Gas Encapsulation into Amorphous Alpha-Cyclodextrin Powder

Thao M. Ho, Kamornrath Sungkaprom, Binh T. Ho, and Bhesh R. Bhandari

2.1 Introduction

Ethylene (C_2H_4) gas is a well-known olefin with a simple structure of two carbons symmetrically linked to each other by a double bond (CH_2=CH_2) (Sundaram et al. 1991). It is an odorless and colorless gas, which occurs naturally and can also be created by man-made sources from industrial activities such as car exhausts from internal combustion engines and gas pipes (Wills et al. 2007). It is a very reactive molecule which can be involved in various chemical reactions, such as polymerization addition, oxidation, and hydration. Biologically, C_2H_4 is a gaseous hormone and is involved in many physiological processes during plant development. The largest producers of C_2H_4 compound are plant and plant products, for example, fruits, vegetables, and floral products, which produce it within their tissues and release it into the surrounding atmosphere. Fruits, vegetables, and flowers contain receptors which serve as binding sites to absorb free atmospheric C_2H_4 molecules (Saltveit 1999; Sundaram et al. 1991). For commercial uses, C_2H_4 gas is usually stored in pressurized cylinders. However, safety is a significant issue in its handling, transport, and storage due to its flammability.

T. M. Ho
Department of Food and Nutrition, University of Helsinki, Helsinki, Finland
e-mail: minh.ho@helsinki.fi

K. Sungkaprom · B. R. Bhandari (✉)
School of Agriculture and Food Sciences, University of Queensland, Brisbane, QLD, Australia
e-mail: kamornrath.sungkaprom@uq.net.au; b.bhandari@uq.edu.au

B. T. Ho
Department of Food Technology, Faculty of Agriculture and Natural Resources, An Giang University (Vietnam National University Ho Chi Minh City), Long Xuyen City, An Giang, Vietnam

T. M. Ho et al. (eds.), *Functionality of Cyclodextrins in Encapsulation for Food Applications*, https://doi.org/10.1007/978-3-030-80056-7_2

Table 2.1 The beneficial and detrimental effects of C_2H_4 gas on fresh products (adapted with permission from Saltveit 1999)

Beneficial effects	Detrimental effects
• Promotes color development in fruits • Stimulates ripening of climacteric fruits • Promotes de-greening of citrus • Stimulates dehiscence in nuts • Alters sex expression in the cucurbitaceae • Promotes flowering in bromeliaceae (e.g. pineapple) • Reduces lodging of cereals by inhibiting stem elongation	• Accelerates senescence • Stimulates chlorophyll loss and yellowing • Enhances excessive softening of fruits • Stimulates sprouting of potatoes • Promotes abscission of leaves and flowers • Stimulates phenylpropanoid metabolism • Promotes discoloration (e.g. browning) • Hastens toughening of vegetables

C_2H_4 gas is a naturally produced gaseous plant hormone that plays a significant role in plants' physiology, development, and metabolism. It accelerates senescence, stimulates yellowing, enhances excessive softening, and other undesirable effects on fruits and vegetables. Despite its deleterious effects, it can be used to improve fruit quality by promoting uniform ripening of batches of fruits (Saltveit 1999). Table 2.1 summarizes the beneficial and the detrimental effects of C_2H_4 on quality of fresh fruits and vegetables.

There are many ways that C_2H_4 gas can be used for ripening and other applications in postharvest technology. For instance, the utilization of C_2H_4 generators where a liquid, comprising of ethanol and catalyzing agents, is heated to produce C_2H_4 gas, in the addition of measured quantities of C_2H_4 for gas cylinders and the use of C_2H_4-releasing chemicals such as Ethephon (Barry and Giovannoni 2007). However, the utilization of C_2H_4 in postharvest and plant growth regulation is limited due to its high volatility, safety issues, and the needs for strict handling protocols (Ho et al. 2011a; Ho 2013).

Cyclodextrin (CD) powders have been increasingly employed in food technology as encapsulating agents to form inclusion complexes with a wide variety of hydrophobic guest molecules. Among three common types of CDs, α-CD has the smallest cavity dimension comparing with other CDs, being able to complex with guest molecules having five carbons or less (including gases) more easily than β- and γ-CDs (Dodziuk 2006). The encapsulation of various gases using α-CD has been well described in Chap. 3 of this book. C_2H_4 gas is an appropriate structure, with its apolar characteristic and small size (C=C and C–H lengths of 132 and 108.2 pm, respectively) (Djedaïni-Pilard and Bonnet 2007) to form complexes with α-CD.

The studies have shown that C_2H_4 gas can be converted into a powder form of C_2H_4-α-CD inclusion complex (Bhandari and Ho 2014). Powder form produced by encapsulation of C_2H_4 gas into α-CD could provide several benefits which are easy and safe in handling as compared to gaseous form of C_2H_4, and being able to be delivered in smallest portion as required. Moreover, the product is convenient to transport, which could also be in favor of choice over gaseous state. This has also been effective in in-transit ripening of fruits in trucks or fruit-transport vehicles (Ho et al. 2016a). Thus, the usage of powder form of C_2H_4-α-CD inclusion complex as the novel product is relatively simple way of managing fruit ripening and other C_2H_4 regulated plant processes such as seed germination (Ho et al. 2011a; Ho 2013). The powdered form of C_2H_4-α-CD is usually produced by following liquid encapsulation method, whereby the complexation takes place under the pressure of the gas in a saturated solution of the α-CD. In this liquid encapsulation, precipitation or crystallization of the inclusion complex occurs slowly, with the recuperation of the crystallized complexes being done in a batch in several days. Thus, this is a time consuming and low yielding process because the crystallization and diffusion processes are time dependent. In addition, there will be a significant amount of residual uncomplexed CD remaining in the liquid. The yield of complexed C_2H_4-α-CD is normally around 50% (Ho 2013). An innovative approach for encapsulating C_2H_4 gas into solid α-CD directly with expectation of shortened process and 100% yield, would have potential practical application and demand. Commercial α-CD exists in crystalline structure. The molecular structure of the crystalline form is tightly packed in a long-range order. That would be relatively difficult for C_2H_4 gas to be included into the α-CD molecule in the crystalline solid, as the molecules are tightly packed, making the cavity less accessible, resulting in the difficulty of C_2H_4 gas penetration (Dodziuk 1994). Amorphous structure of α-CD molecule is more porous, anarchial and tangled (Bhandari and Hartel 2005). The CD cavities in the amorphous structure would potentially be more accessible to interact with external molecules than would be the case for the crystalline structure. Solid encapsulation was used to encapsulate tea tree oil (TTO) by Shrestha et al. (2017). They added ethanol to the mixture of TTO and amorphous β-CD at 1:2 and 1:3 TTO:ethanol ratios. This resulted in the inclusion of 94.3 and 98.45 mg of TTO/g β-CD, respectively, which was similar to that of TTO encapsulated in the conventional paste method (95.56 mg TTO/g β-CD), suggesting an effective amorphous solid encapsulation method. Solid complexation by direct compression of carbon dioxide (CO_2) gas into amorphous and crystalline α-CD in solid state was also reported by Ho et al. (2015b) and Ho (2017). The results indicated that amorphous α-CD powder was more effective in the complexation with CO_2 at low pressure and short time than crystalline α-CD powder. For instant, complexation at 0.4 MPa for 4 h, amorphous α-CD powder encapsulated about 0.60 mol CO_2/mol α-CD whereas crystalline α-CD powder entrapped only 0.05 mol CO_2/mol α-CD.

In this chapter, we report the encapsulation of C_2H_4 gas in amorphous α-CD structure to increase the yield of entrapped C_2H_4 gas in the complexes, and qualitative characterization of the amorphous C_2H_4-α-CD complexes. Amorphous CD powders can be successfully produced by freeze drying (Li et al. 2002), ball milling (Kaminski et al. 2012; Tabary et al. 2011), and spray drying (Ho et al. 2015a, 2017; Ho 2017; Shrestha et al. 2017; Frieler et al. 2019). Details about production of amorphous CD powders and their properties can be found in Sect. 1.4, Chap. 1 of this book.

2.2 Complexation of C_2H_4 and Amorphous α-CD Powder

The encapsulation of C_2H_4 gas into α-CD allows for the conversion of C_2H_4 gas into a powder form that is easier to handle and use. This encapsulation technique relates to the incorporation of guest molecule physically into the core of a carrier molecule (Ho et al. 2011a). The C_2H_4-α-CD inclusion complex prepared by liquid encapsulation are in crystalline form. There are two different patterns of crystalline CD inclusion complex form, the cage and the tunnel type. The types of crystal formation can be responsible for further locking up of the guest molecule and release property of complexed gas (Saenger et al. 1998). Thus, C_2H_4 gas, olefin with two carbons symmetrically linked to each other by double bond, can be easily incorporated into α-CD. Although Ho et al. (2011a) had accomplished encapsulation of C_2H_4 gas into α-CDs to produce solid form inclusion complex crystals, there was some problems from crystalline molecule such as yield of complexed and uncomplexed C_2H_4 gas in crystallization technique. There is also little research of encapsulation of C_2H_4 gas using α-CDs into amorphous form.

Encapsulation of C_2H_4 gas into amorphous α-CD can be simply done in a pressure vessel. Amorphous α-CD powder is added into a multi-layered plastic container which is then placed inside the chamber of the pressure vessel. C_2H_4 gas is flushed into the vessel at set pressures and times. In the encapsulation process, the amorphous powder is exposed to C_2H_4 gas under the pressure without agitation. The complexation between α-CD and C_2H_4 occurs at the interface. In this case, the complexation is dependent on the penetration of the C_2H_4 gas into the CD molecular cavity. Figure 2.1 showed that an increase in encapsulation time results in significantly higher concentration of C_2H_4 in the amorphous C_2H_4-α-CD inclusion complexes. Meanwhile, in liquid encapsulation where C_2H_4-α-CD inclusion complex crystallises due to the low of solubility of complex compound, increasing pressure and time of encapsulation does not cause a profound increase (as compared to data

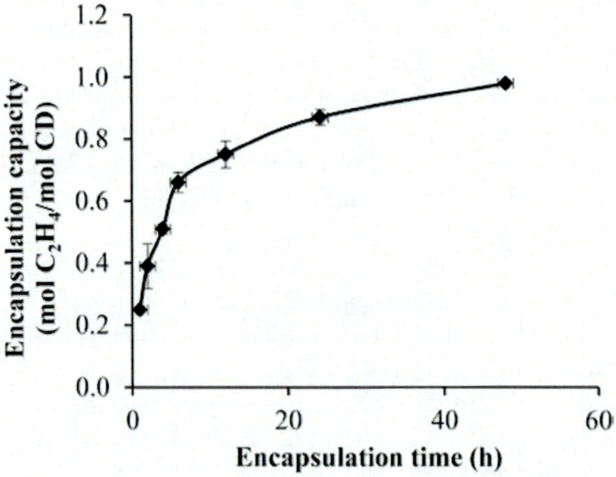

Fig. 2.1 C_2H_4 concentration (mol C_2H_4/mol α-CD) in amorphous C_2H_4-α-CD inclusion complexes encapsulated at 1.5 MPa at various time points

in Fig. 2.1) in encapsulated C_2H_4 concentration (Ho et al. 2011a). This result indicated that the time of encapsulation strongly influenced the C_2H_4 concentration in solid encapsulation technique.

The longer time for encapsulation will provide the deeper penetration of C_2H_4 gas into the amorphous α-CD molecule through the loose packing of amorphous molecules. At a similar encapsulation conditions of pressure, crystalline inclusion complex prepared by liquid encapsulation and amorphous inclusion complexes prepared by direct compression of C_2H_4 to amorphous α-CD reached the equilibrium of encapsulated C_2H_4 content (e.g. 0.98 mol/mol CD) at the different time of encapsulation. Crystalline inclusion complexes took 96 h while amorphous inclusion complex took only 48 h. The yield of amorphous complex will obviously be 100% while for crystalline complex less than 50% yield has been reported (Ho et al. 2011a).

The encapsulation time of amorphous inclusion complexation was faster than crystalline inclusion complexation because of the differences in structural properties of amorphous and crystalline α-CD molecules. The arrangement of amorphous molecules will be random and loosely packed in a short-range periodic structure which accommodates the penetration of C_2H_4 faster in time. The arrangement of crystalline molecules is a lattice pattern exhibiting long-range order and in tightly packed structure. SEM and XRD results shown in Fig. 2.2 confirm amorphous structure of C_2H_4-α-CD inclusion complex.

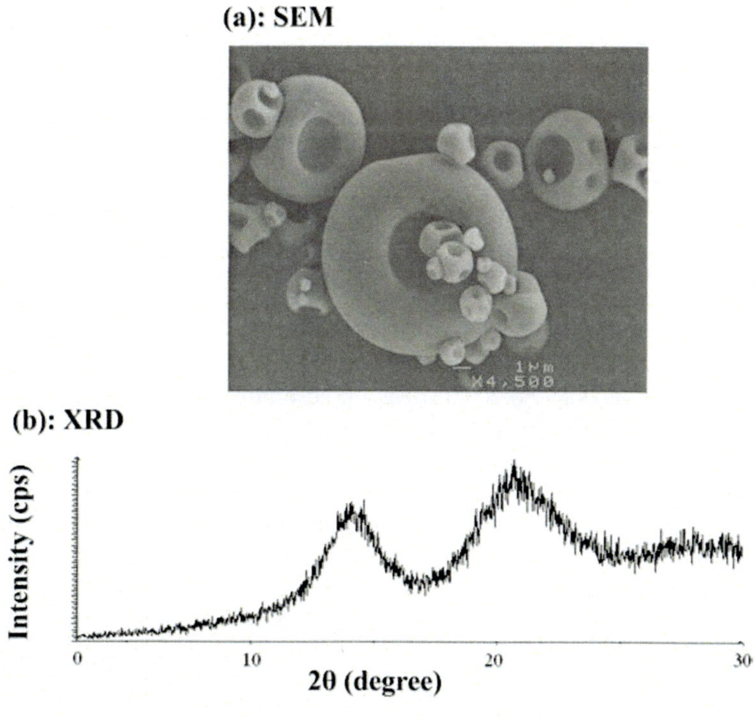

Fig. 2.2 SEM (**a**) and X-ray diffractogram (**b**) of amorphous C_2H_4-α-CD inclusion complex. (Adapted with permission from Ho and Bhandari 2016)

2.3 Release Properties of C_2H_4 Gas from Amorphous C_2H_4-α-CD Inclusion Complex Powder

2.3.1 Release Property at Various Relative Humidities

A release system shown in Fig. 2.3, in combination with gas chromatography (GC) analysis was used to investigated release property of C_2H_4 gas from amorphous C_2H_4-α-CD inclusion complex powder at 11.15%, 32.73%, and 75.32% relative humidities (RH), and room temperature, over 120 h. The RH level inside the system was controlled at 11.15%, 32.73% and 75.32%, using saturated sodium chloride, magnesium chloride, and lithium chloride solutions, respectively (Rahman 2009). The saturated salt solutions were prepared and equilibrated in the system for at least 24 h before the experiment is conducted. The system was integrated with a fan system to distribute the released C_2H_4 gas throughout inside the system. The system was tested and verified with the release properties of CO_2 gas from CO_2-α-CD inclusion complex powder (Ho et al. 2016b). C_2H_4 in the inclusion complex powders was quantified using GC using headspace analysis. About 1 g of C_2H_4-α-CD complex powder was weighed out approximately into a plastic petri dish and transferred into the system fitted with airtight cover. The headspace of the system was subsequently sampled for the GC analysis.

Figure 2.4 shows the release time course of C_2H_4 gas from C_2H_4-α-CD inclusion complex powder. Release of C_2H_4 depended markedly upon RH. C_2H_4 started to release right after taking out of the encapsulator (pressure vessel). RH of 75.32% had the most rapid release as compared to 32.73% and 11.15% RH. C_2H_4 release at 75.32% RH attained equilibrium after 4 h, while 32.73% and 11.15% RH showed a slower rate of release. The results suggested that release rate of C_2H_4 can be controlled by its surrounding RH. Thus, increased storage RH accelerated the release of C_2H_4 from the complex. Reineccius et al. (2002) reported that water was essential to release guests from CD inclusion complex. It might be the evidence that water absorption by the complex powder resulted in replacement of the C_2H_4 molecule by a water molecule in the CD cavity. Similar effects of humidity on release of CO_2 (Neoh et al. 2006) and allyl isothiocyanate (Li et al. 2007) have also been reported. CDs are hygroscopic and absorb moisture when exposed to high RH conditions (Blanchard and Proniuk 1999). RH involves the adsorption of water, which might be the cause of break down of the inclusion complexes (Cevallos et al. 2010). Moreover, Ho et al. (2011b) found that the dissolution of complex powder in water led to complete release of C_2H_4 from the complex powder. This phenomenon could be explained by the adsorption of water vapor occurring on the surface of the amorphous α-CD molecule. This occurred due to more water vapor being adsorbed at the surface and condensation (Salameh et al. 2006). As a result, the powder started to absorb more liquid water and released C_2H_4 gas from the amorphous complexes over time. Also, Ho et al. (2011b) showed that increased storage humidity

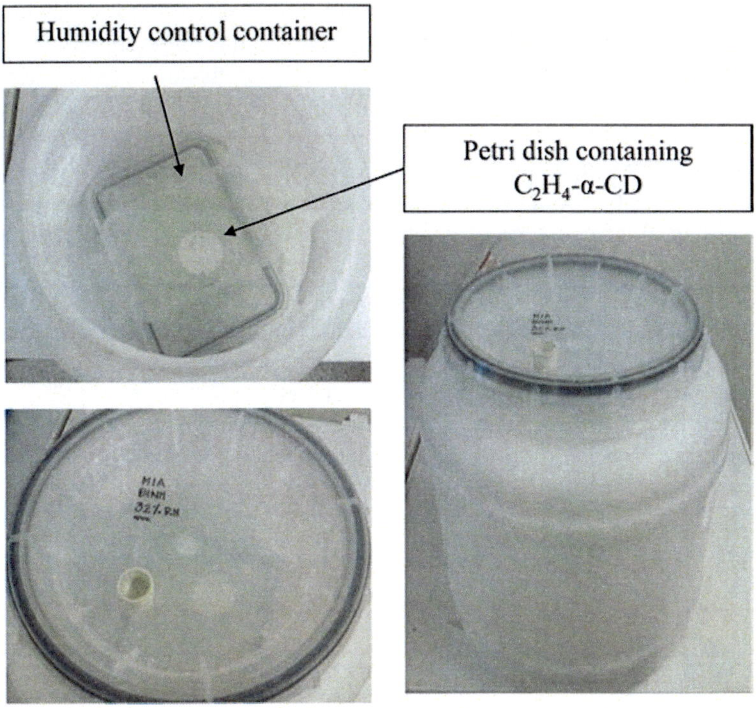

Fig. 2.3 A system used to investigate the release rate of C_2H_4 gas from C_2H_4-α-CD inclusion complex powder at different relative humidities

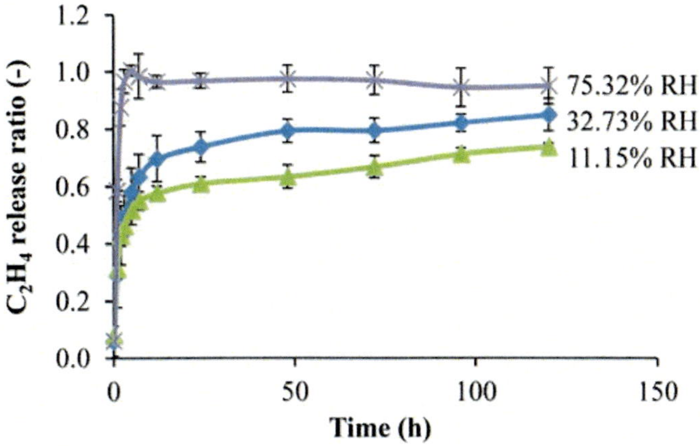

Fig. 2.4 The release of C_2H_4 gas from C_2H_4-α-CD inclusion complex powder at 11.15%, 32.73% and 75.32% RH over 120 h (25 °C). C_2H_4 release ratio as the ratio of the released to initial concentration of the C_2H_4 from the C_2H_4-α-CD inclusion complex powder expresses the relative release of C_2H_4 gas over time

accelerated the release of C_2H_4 from the complex crystals, but the inclusion complex was stable at 52.29% RH. This could be explained that under certain RH conditions, amorphous compounds could undergo a glass transition event. Amorphous materials lacked the three-dimensional long-range order characteristic of the crystalline state; instead consisting of a more random arrangement of molecules, somewhat like the liquid state, leading to physical properties that differ from those of crystalline materials. Generally, amorphous solids draw up considerably more water than their crystalline counterparts below deliquescence point of an individual crystalline ingredient (RH_o), since amorphous substances could also absorb water into their bulk while crystalline solids below RH_o only adsorb moisture at the surface. Moisture sorption upon exposure of amorphous compounds to high RH results in plasticization and a decrease in the glass transition temperature (T_g) for the solid. Following the transition from the glassy to the rubbery state, greater molecular mobility, increased free volume, and decreased viscosity have been reported, allowing reactions, especially diffusion-controlled reactions to occur more readily (Zografi 1988; Labuza and Hyman 1998; Fennema 1996).

2.3.2 Pressure Determination of Released C_2H_4 from Amorphous C_2H_4-α-CD Inclusion Complex Powder

Pressure of C_2H_4 gas released from C_2H_4-α-CD complex powder was measured using digital pressure gauge (Fig. 2.5). Amorphous α-CDs powder obtained from encapsulation at 1.5 MPa for 48 h was used. Forty grams of complexes were weighed out approximately into a 40 mL glass bottle. The pressure was quantified using pressure gauge headspace analysis with an airtight cover.

Figure 2.6 shows the pressure dramatically increased from 0 to 12 h before gradually decreasing towards 120 h. The 12 h indicated the highest pressure of 0.153 MPa. The pressure slightly decreased from 12 to 120 h, and the pressure was 0.074 MPa at 120 h. The reason for pressure drop could be attributable to negligible leakage during measurement of pressure and/or internal factors of spontaneous reverse of gas penetration into the amorphous α-CD complexes. There were little studies of pressure release from the amorphous α-CD complexes. Spontaneous reverse of gas penetration into amorphous α-CD complexes in the system occurs when pressure is high enough to push the gas back into the complexes. However, further study is required on pressure release for a better understanding and development of a product packaging.

It was calculated to have an expected pressure of 1.35 MPa, given that all C_2H_4 gas has been released. A pressure of 1.5 MPa was used to encapsulate, and it was seen that the highest pressure released was almost equivalent to the pressure used to encapsulate. From the results, the headspace pressure was determined as the highest

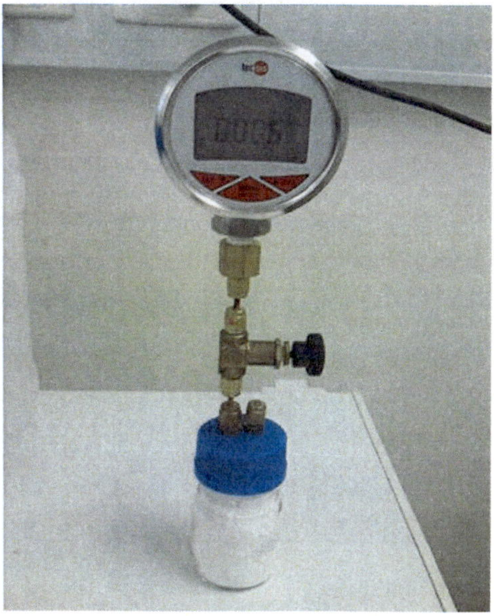

Fig. 2.5 Digital pressure gauge connected with glass bottle cap to measure pressure created by the released C_2H_4 from the complex powder

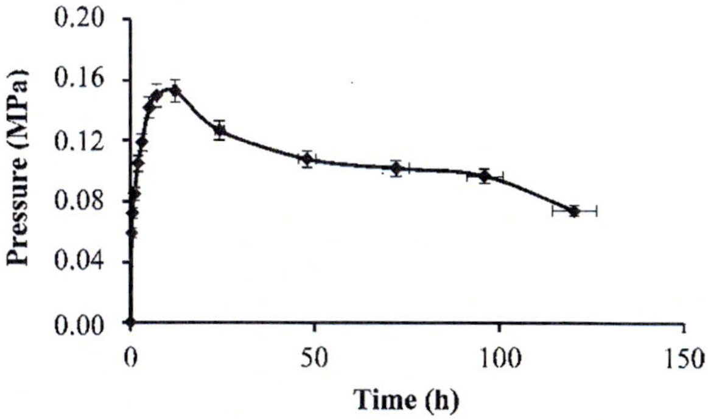

Fig. 2.6 The pressure (MPa) created by the release of C_2H_4 from amorphous C_2H_4-α-CD inclusion complexes at time intervals

pressure was 0.153 MPa at 12 h. The pressure of 0.153 MPa was used to calculate the percentage of the C_2H_4 released to the headspace. The percentage showed a proportion in the headspace was 11.33%, indicating that the C_2H_4 gas inside the amorphous α-CD complexes was approximately 90%. This showed that C_2H_4 gas was released gradually, and inversely increased the pressure. This study would contribute towards controlling the pressure of C_2H_4 gas released when packed into product packages. The study also showed the proportion of complexed ethylene gas being released which could be useful in future applications.

2.4 Conclusion and Further Perspectives

Ethylene gas can be encapsulated into amorphous α-CD powder by solid encapsulation method with a 100% yield. It is a simple and efficient method of encapsulation as compared to conventional solution method of encapsulation. The maximum load of the C_2H_4 and the optimized time in which the C_2H_4 could be encapsulated under given conditions was 0.98 mol C_2H_4/mol α-CD and 48 h, respectively. This is similar to the conventional method complexation capacity. The relative humidity for release of C_2H_4 from amorphous C_2H_4-α-CD complex powders significantly ($p < 0.05$) affected C_2H_4 release rate ratio. At 75.32% RH, the C_2H_4 release rate was the highest, followed by 32.73% and 11.15% RH, respectively.

Future work could explore the study of re-crystallisation of amorphous complexes to control the premature release of the gas. Re-crystallisation of amorphous complex molecules could lock the C_2H_4 gas molecules inside the CD cavity in an effective way.

References

Barry CS, Giovannoni JJ (2007) Ethylene and fruit ripening. J Plant Growth Regul 26:143–159

Bhandari BR, Hartel RW (2005) Phase transition during food powder production and powder stability. In: Onwulata C (ed) Encapsulated and powdered foods. Dekker/CRC Press, Boca Raton, pp 262–290

Bhandari B, Ho BT (2014) Encapsulation of gases within cyclodextrins. Patent No. WO 2014/056035 A1

Blanchard J, Proniuk S (1999) Some important considerations in the use of cyclodextrins. Pharm Res 16:1796–1798

Cevallos PAP, Buera MP, Elizalde BE (2010) Encapsulation of cinnamon and thyme essential oils components (cinnamaldehyde and thymol) in β-cyclodextrin: effect of interactions with water on complex stability. J Food Eng 99:70–75

Djedaïni-pilard F, Bonnet V (2007) Cyclodextrins and their complexes. In: Dodziuk H (ed) Chemistry, analytical methods, applications. Angew Chem Int Ed 46:2352–2353

Dodziuk H (1994) Structure of cyclodextrins and their complexes: part 2. Do cyclodextrins have a rigid truncated-cone structure? J Mol Struct Theochem 304:61–68

Dodziuk H (2006) Cyclodextrins and their complexes: chemistry, analytical methods, applications, 1st edn. Wiley-VCH, Hoboken

Fennema OR (1996) Food chemistry. Marcel Dekker, New York

Frieler L, Ho TM, Anthony A et al (2019) Crystallisation properties of amorphous cyclodextrin powders and their complexation with fish oil. J Food Sci Technol 56(3):1519–1529

Ho BT (2013) Production of ethylene powder by encapsulation of ethylene gas into α-cyclodextrin and its application for the ripening of fruit. PhD thesis, University of Queensland, Australia

Ho TM (2017) Studies of the encapsulation and release of carbon dioxide from amorphous and crystalline alpha-cyclodextrin powders and its application in food systems. PhD thesis, University of Queensland, Australia

Ho BT, Bhandari BR (2016) Novel solid encapsulation of ethylene gas using amorphous α-cyclodextrin and the release characteristics. J Agric Food Chem 64(17):3318–3323

Ho BT, Joyce DC, Bhandari BR (2011a) Encapsulation of ethylene gas into α-cyclodextrin and characterisation of the inclusion complexes. Food Chem 127:572–580

Ho BT, Joyce DC, Bhandari BR (2011b) Release kinetics of ethylene gas from ethylene-α-cyclodextrin inclusion complexes. Food Chem 129(2):259–266

Ho TM, Howes T, Bhandari BR (2015a) Characterization of crystalline and spray-dried amorphous α-cyclodextrin powders. Powder Technol 284:585–594

Ho TM, Howes T, Bhandari BR (2015b) Encapsulation of CO_2 into amorphous and crystalline α-cyclodextrin powders and the characterization of the complexes formed. Food Chem 187:407–415

Ho BT, Hofman P, Joyce DC, Bhandari BR (2016a) Uses of an innovative ethylene-α-cyclodextrin inclusion complex powder for ripening of mango fruit. Postharvest Biol Technol 113:77–86

Ho TM, Tuyen T, Howes T, Bhandari BR (2016b) Method of measurement of CO_2 adsorbed into α-cyclodextrin by infra-red CO_2 probe. Int J Food Prop 19(8):1696–1707

Ho TM, Truong T, Bhandari BR (2017) Chapter 5: Spray-drying and non-equilibrium states/glass transition. In: Bhandari BR, Roos YH (eds) Non-equilibrium states and glass transitions in foods—processing effects and product-specific implications. Elsevier, Duxford, pp 111–136

Kaminski K, Adrjanowicz K, Kaminska E et al (2012) Impact of water on molecular dynamics of amorphous α-, β-, and γ-cyclodextrins studied by dielectric spectroscopy. Phys Rev E 86(3):031506

Labuza TP, Hyman CR (1998) Moisture migration and control in multi-domain foods. Trends Food Sci Technol 9:47–55

Li J, Guo Y, Zografi G (2002) The soli-state stability of amorphous quinapril in the presence of β-cyclodextrins. J Pharm Sci 91(1):229–243

Li X, Jin Z, Wang J (2007) Complexation of allyl isothiocyanate by α- and β-cyclodextrin and its controlled release characteristics. Food Chem 103(2):461–466

Neoh TL, Yoshi H, Furuta T (2006) Encapsulation and release characteristics of carbon dioxide in α-cyclodextrin. J Incl Phenom Macrocycl Chem 56:125–133

Rahman SM (2009) Food properties handbook. CRC Press/Taylor & Francis Group, Boca Raton

Reineccius TA, Reineccius GA, Peppard TL (2002) Encapsulation of flavors using cyclodextrins: comparison of flavor retention in alpha, beta, and gamma types. J Food Sci 67(9):3271–3279

Saenger WR, Jacob J, Gessler K et al (1998) Structures of the common cyclodextrins and their larger analogues—beyond the doughnut. Chem Rev 98:1787–1802

Salameh AK, Mauer LJ, Taylor LS (2006) Deliquescence lowering in food ingredient mixtures. J Food Sci 71:E10–E16

Saltveit ME (1999) Effect of ethylene on quality of fresh fruits and vegetables. Postharvest Biol Technol 15:279–292

Shrestha M, Ho TM, Bhandari BR (2017) Encapsulation of tea tree oil by amorphous beta-cyclodextrin powder. Food Chem 221:1474–1483

Sundaram KM, Shreehan MM, Olszewski EF (1991) Ethylene. In: Kirk RE (ed) Kirk-Othmer encyclopedia of chemical technology. Wiley-Blackwell, New York, pp 593–631

Tabary N, Mahieu A, Willart J-F et al (2011) Characterization of the hidden glass transition of amorphous cyclomaltoheptaose. Carbohydr Res 346(14):2193–2199

Wills RBH, McGlasson B, Graham D et al (2007) Postharvest: an introduction to the physiology and handling of fruit, vegetables and ornamentals, 5th edn. University of New South Wales Press, Sydney

Zografi G (1988) States of water associated with solids. Drug Dev Ind Pharm 14:1905–1926

Chapter 3
Encapsulation of Gases

Thao M. Ho and Bhesh R. Bhandari

3.1 Introduction

Many kinds of gases have been used in a wide range of industries such as biology, medicine, science, technology, agriculture, and food (Rudkevich 2004). Methane (CH_4) is considered as a possible source of clean energy (Adisa et al. 2011), while hydrogen (H_2) is used in the design of energy-rich fuel-cell devices (Dincer 2002). Oxygen (O_2) plays an important role in medicine and steel making. Nitrogen (N_2) is the most commonly used inert gas in biological and food applications, space aircraft, and ammonia production. Nitrogen, argon (Ar), and helium (He) are extensively used for the operation of analytical equipment, especially gas chromatography (Hall 1978; Rudkevich 2004). Nitrogen and carbon dioxide (CO_2) are also used to modify the atmosphere in packaging headspace to control the respiration rate of many agricultural products, which extends their shelf life and retards the growth of undesirable organisms during storage (Arvanitoyannis 2012). Nitrous oxide (N_2O) is extensively used in anaesthesia, while nitric oxide (NO) serves as an important messenger in signal transduction processes in smooth muscle cells and neurons and has antithrombotic effects (McKinlay et al. 2008; Wheatley et al. 2006). Nitrous oxide is also used in canister sprays, such as cooking vegetable oil and whipped cream (Hanson and Smith 1993). Carbon dioxide is used as a bubble-creating agent in many kinds of beverages to enhance their sensory properties (Zeller and Kim

T. M. Ho (✉)
Department of Food and Nutrition, University of Helsinki, Helsinki, Finland
e-mail: minh.ho@helsinki.fi

B. R. Bhandari
School of Agriculture and Food Sciences, University of Queensland,
Brisbane, QLD, Australia
e-mail: b.bhandari@uq.edu.au

T. M. Ho et al. (eds.), *Functionality of Cyclodextrins in Encapsulation for Food Applications*, https://doi.org/10.1007/978-3-030-80056-7_3

2008) and is used to extend shelf life and to improve the quality of dairy products and fruit juices (Hotchkiss et al. 2006; Shomer et al. 1994). Carbon dioxide also has many benefits in blood circulation improvement, blood-vessel dilation, and activation of gastrointestinal movement (Neoh et al. 2006). Ethylene (C_2H_4) is recognized as a useful phytohormone to trigger ripening processes, enhance color development of some types of fruits, de-green citrus fruits, and promote germination of many non-dormant seeds such as mung bean sprouts (Saltveit 1999; Khang et al. 2013). 1-Methylcyclopropene (1-MCP), an anti-ethylene gas, is used for extending shelf life and quality of agricultural products (Blankenship and Dole 2003). Hydrogen sulphide (H_2S) is useful for modulating cellular functions (Allan 2012). Chlorine dioxide (ClO_2) is used to disinfect objects contaminated by microorganisms (Wood et al. 2010). Ozone (O_3) is used in water treatment and sanitization of raw fruits and vegetable products (Guzel-Seydim et al. 2004). The applications of various gases are summarized in Table 3.1 (Ho et al. 2014).

Typically, gases are stored and transported in a high-pressure compressed form in metal cylinders or are alternatively obtained from gas-producing processes, such as CO_2 gas from alcohol fermentation. The utilization and transportation of gas

Table 3.1 Applications of various gases in different fields (adapted with permission from Ho et al. 2014)

Gases	Applications
CO_2	Disinfestation and fumigation of grain; modified atmosphere in food storage; carbonating water and soft drinks; refrigerated transport maintaining the cold chain; bubble-creating and leavening agent; extraction of oils, aromas, and volatile compounds; extension of shelf life of dairy products and fruit juice; pH control; cryocleaning of conveyor belts and moulds
N_2	Deep-freezing and individually quick-frozen (IQF) freezing; cryogrinding; modified atmosphere in food storage; space aircraft; production of ammonia; inertization and blanketing of storage tanks; stripping of oxygen from drinkable liquids; deodorizing edible oils; pneumatic transportation of oxidizable ingredients; cryogenic freeze-drying
O_2	Oxygenation of fish farms; biofermentation; modified atmosphere in food storage; biological treatment of wastewaters; medicine; steel making
CO	Modified atmosphere packaging for meat products; gasotransmitter and therapeutics
H_2	Energy-rich fuel-cell devices; hydrogenation process for fats
CH_4	Clean energy production
C_2H_4	Fruit ripening and promotion of seed germination
N_2O	Canister sprays (non-stick cooking oil, whipped cream); anaesthesia
SO_2	Anti-browning of fruits and vegetables
1-MCP	Delay ripening of fruits and vegetables
O_3	Water treatment; sanitization of raw fruits and vegetables
He, Ar, N_2	Wine-making industry and in food packaging (Ar), analytical equipment (e.g. gas chromatography)
ClO_2	Surface disinfection, sanitization
NO, H_2S	Gasotransmitter and therapeutics (e.g. platelet aggregation inhibition and antibacterial activity, modulating cellular functions)

from these sources requires special care to prevent explosions and unknown leaks. In addition, due to the inherent properties of gases (low solubility in water and low diffusion in solid materials), it is difficult to maintain them at high concentration in solution or solid matrices in desired conditions as they are intended to be mixed or solubilized into food products to extend their shelf life. Entrapment of gases in solid matrices to produce the powder form of gases can minimize these shortcomings, as this can offer safer methods to store gases for further use. In addition, the gas powders are quite useful in many cases as a minimal amount of gas is required.

Many solid matrices have been employed to encapsulate or adsorb gases for various applications in clean energy production, environmental protection (emission control and removal of hazardous gases or vapours), nanodevice production, food and agriculture production, and pharmacy. Common solid matrices include activated carbons, carbon nanotubes, zeolites, metal-organic frameworks, and cyclodextrins (Ho et al. 2014). These matrices exist in powder form with a very high surface area and have many binding sites on the particle surface or in the cavity (or both), which can form physical or chemical (or both) interactions with gas molecules. In physical interactions, gas molecules are loosely held by physical forces (e.g. dipole-dipole, apolar, electrostatic, hydrophobic associations, Van der Waals) with a low sorption energy (~8–41 kJ/mol-gas) and are thus generally reversible. In contrast, in chemical interactions, chemical bonds are formed between the gas molecules and the solid surface with high sorption energy range (~62–418 kJ/mol-gas), thus extra energy is required to remove adsorbed gas molecules and the interactions are possibly irreversible (Masel 1996; Allan 2012). Gas encapsulation/adsorption into solid matrices is governed by temperature and pressure at which gas encapsulation or adsorption is accomplished. In most cases, increasing encapsulation pressure enhances the adsorption capacity because high pressure facilitates contact of gas molecules at binding sites on the surface or pores of solid matrices. The effects of temperature on adsorption capacity depend greatly on the structure of solid matrices. For example, the gas adsorption capacity of H_2 gas into carbon nanotubes was higher at lower temperatures, while H_2 gas adsorption capacity on zeolites increased at elevated temperatures (Anson et al. 2004; Yoon and Heo 1992). Thus, by manipulation of temperature and pressure, gas encapsulation capacity into solid matrices can be optimized.

As solid matrices differ in construction compositions and properties, the final application of gas-solid matrix complexes is a prerequisite for selecting a solid matrix to entrap gas. For gas storage, a high gas adsorption capacity of solid matrices is required. For food, pharmaceutical, or agriculture systems, controlled release of the adsorbed gas and non-toxic, biodegradable, and biocompatible solid matrices are more important. Among reported solid matrices, cyclodextrin (CD), also known as cyclic oligosaccharide, is the most suitable matrix to entrap gases for food application. This is because CD is non-toxic and is generally recognized as safe (GRAS) in the USA, a natural product in Japan, and a novel food in Australia, New Zealand, and European countries (Irie and Uekama 1997; Loftsson and Duchêne 2007). In this chapter, techniques to encapsulate gas molecules using CD powders and applications of the complexes in food production are described.

3.2 Preparation and Characteristics of Gas-CD Complexes

3.2.1 Techniques of Gas-CD Complex Preparation

Two common techniques to encapsulate gas molecules into CD powders, namely liquid encapsulation and solid encapsulation, have been reported. Their main characteristics are shown in Table 3.2 and Fig. 3.1 (Ho 2017).

3.2.1.1 Liquid Encapsulation

The first step in the liquid encapsulation is to completely dissolve CD powder into water such that the initial crystalline structure of CD molecules is entirely disrupted. Typically, the saturated CD solution is preferred to maximize the obtained yield of gas-CD complexes. To obtain saturated CD solution, an oversaturated CD solution is prepared and then filtration is applied to remove any insoluble particles. The CD solution is then pressurized with an encapsulating gas at a certain pressure level. In aqueous solution, the CD hydrophobic cavities are always occupied by water molecules via energetically unfavourable hydrophobic-hydrophilic interactions, and these water molecules are easily displaced by less hydrophilic gas molecules during encapsulation (Szejtli 1989). The main driving force for the gas-CD complex formation is the substitution of the high-enthalpy water molecules in the CD cavities

Table 3.2 Main characteristics of liquid and solid approaches for gas encapsulation using CD powder

	Gas encapsulation methods	
Characteristics	Liquid method	Solid method
Mechanisms	Direct compression of gases into aqueous solution of CD, followed by crystallization, collection, and dehydration of complexes	Direct exposure of CD powder in solid state to gas at high pressure
Yield	Low (<50%)	High (100%)
Encapsulation time	Long (several days)	Short (several hours)
Complex structure	Crystalline form	Depends on initial structure of CD powder (e.g. amorphous and crystalline CD powder results in amorphous and crystalline complex powder, respectively) and moisture content during encapsulation (e.g. high moisture content induces amorphous-crystalline structure transformation)
Stability at normal conditions	High (several months)	Low (from several hours to several days)

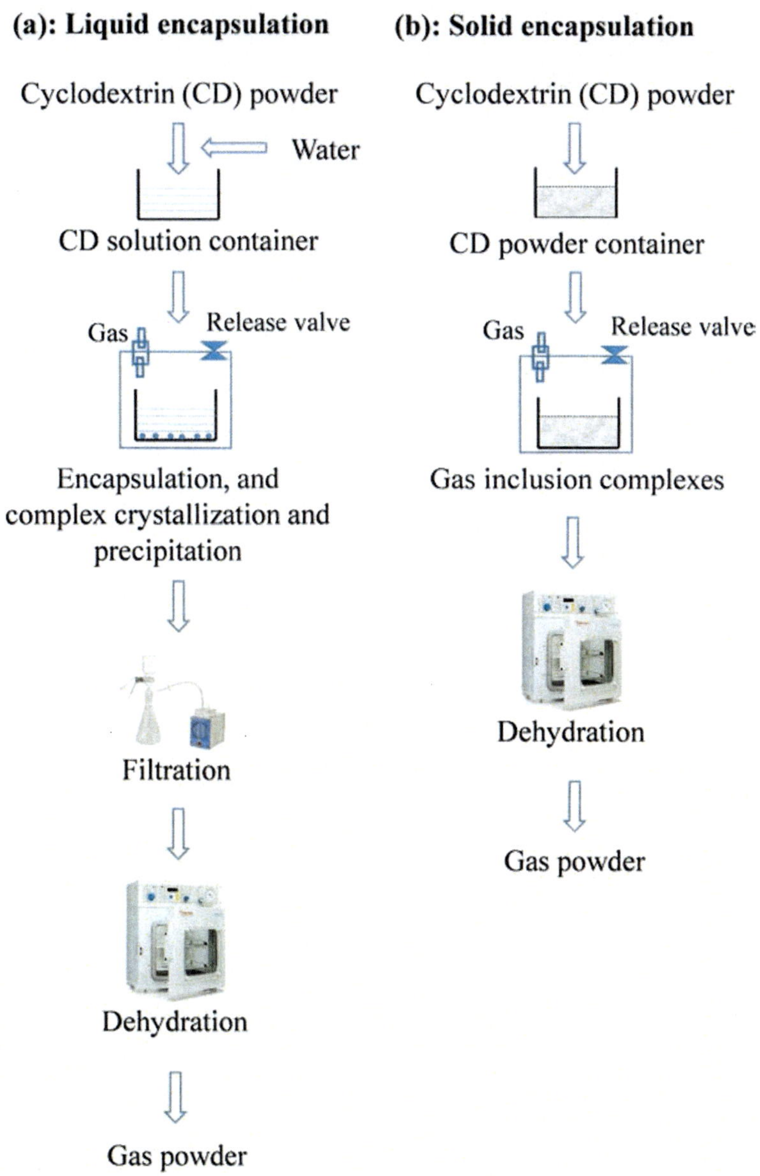

Fig. 3.1 Approaches used for gas encapsulation using cyclodextrin powder (**a**) liquid encapsulation and (**b**) solid encapsulation

by hydrophobic gas molecules (Astray et al. 2009). The physical interactions (dipole-dipole bonds, hydrophobic-hydrophilic interactions, hydrophobic associations, and Van der Waals) are responsible for the complexation between CD cavities and gas molecules with no evidence of forming or breaking covalent bonds (Schneiderman and Stalcup 2000). Following complexation, gas-CD complexes can

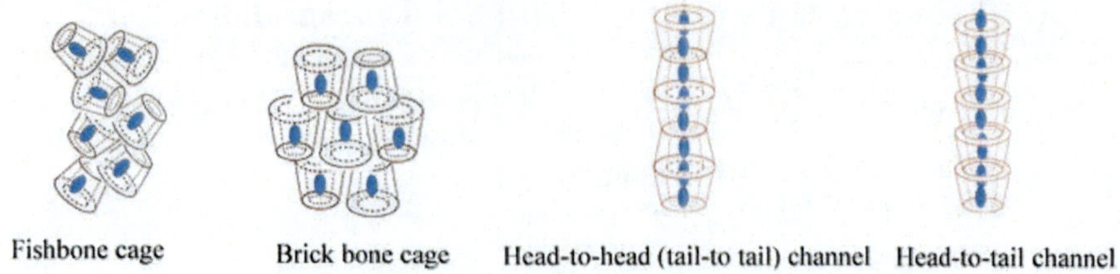

Fig. 3.2 Possible arrangement schemes of crystalline gas-CD complexes. (Adapted with permission from Saenger et al. 1998)

crystallize in two basically different patterns, known as the cage and the channel (tunnel) types, as illustrated in Fig. 3.2 (Saenger 1985). The cage type results when CD molecules are packed crosswise (fishbone) or side-by-side (brick bone) in layers, with adjacent layers being displaced by approximately one-half a molecule. In each of these molecular arrangements, the internal cavity of a CD is locked on both sides by neighbouring CD molecules further entrapping the gas molecules in isolated cavities. In contrast, channel types are formed when CD molecules are stacked like coins in a roll (head-to-head, tail-to-tail, and head-to-tail patterns) and therefore the CD cavities line up to produce long channels holding gas molecules in a tunnel-like cavity. Crystal packaging of gas-CD complexes depends on the size and nature of gas molecules. The cage type structures are usually formed with small molecular gases, while long molecular and ionic guests induce channel- or tunnel-type structures (Saenger et al. 1998). The crystals of gas-CD complexes are collected by filtration after precipitation and are then dehydrated in open air at room temperature to a moisture content level safe for long-term storage (~5.0% moisture content).

Gas-CD complexes are very stable at normal conditions in the crystalline state. However, a low yield of gas-CD complex powder (less than 50%), together with a time-consuming process (several days), are major drawbacks that prevent this method from commercial application. However, this is the most widely-used technique in laboratory scale for studies on gas encapsulation using CD powder (Ho et al. 2011a, b, 2013, 2015a; Neoh et al. 2006; Szejtli 1988; Zeller and Kim 2008; Pereva et al. 2015; Bazzano et al. 2016).

3.2.1.2 Solid Encapsulation

In solid encapsulation, CD powder in solid state is directly exposed to an encapsulating gas at high pressure such that the yield of the gas-CD complex powder is very high (~100%) and the time for encapsulation is much shorter (Ho 2013, 2017; Ho and Bhandari 2016; Neoh et al. 2006). However, the encapsulation capacity and stability of the complex are quite low, and thus improvement of the technique is required. In the solid encapsulation approach, encapsulation capacity is highly dependent on the initial moisture content and structure of CD powder (Ho 2017). Commercial CD powders exist in a crystalline state with highly packed structure,

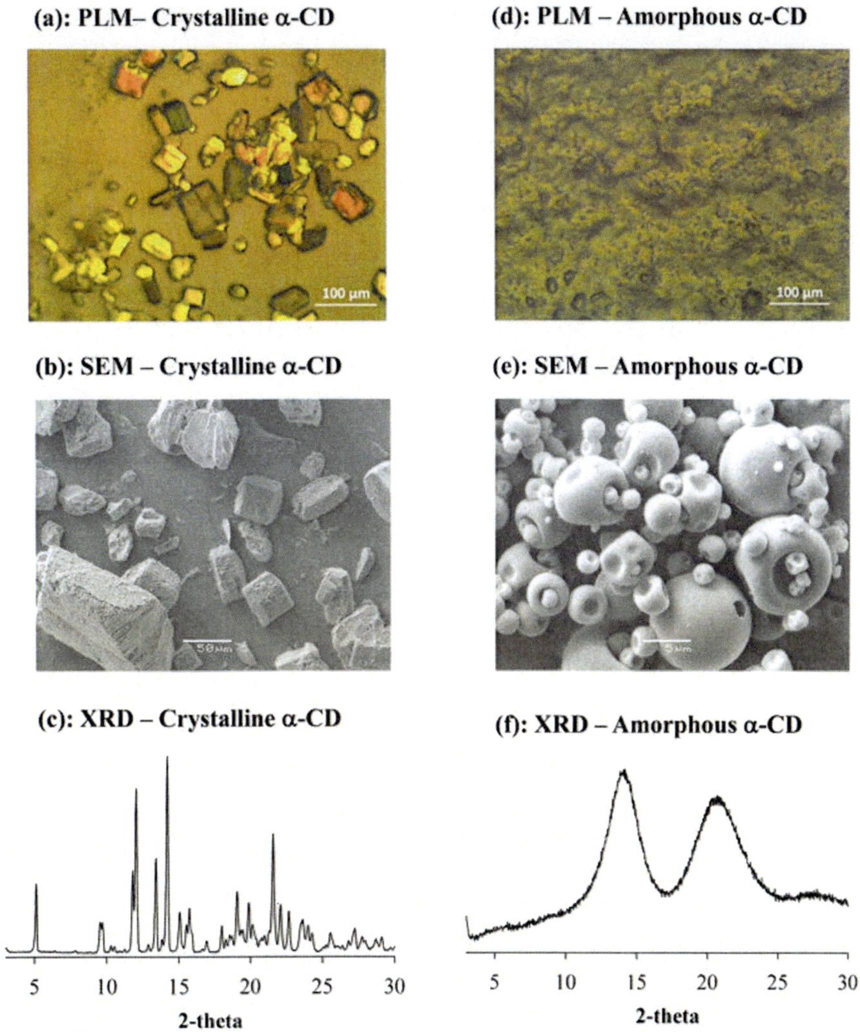

Fig. 3.3 Polarized microscope (PLM), scanning electron microscope (SEM), and X-ray diffraction (XRD) of commercial crystalline α-CD powder (**a**–**c**) and spray-dried amorphous α-CD powder (**d**–**f**). (Adapted with permission from Ho et al. 2015b)

while spray drying can produce CD powder with a completely amorphous structure (Ho et al. 2017). As an example, the differences in morphology as determined by polarized microscope (PLM) and scanning electron microscope (SEM) and the structure measured by X-ray diffraction (XRD) between commercial (crystalline) and spray-dried (amorphous) α-CD powders are illustrated in Fig. 3.3 (Ho et al. 2015b). Crystalline α-CD powders are characterized by many sharp peaks in XRD diffractogram, colored bright crystal granules with different shapes under PLM, and several shapes (including cubic, tetragonal, or triclinic) with different sizes and many small particles and clefts on their surface under SEM. Amorphous counterparts exhibit broad peaks (halos) in XRD diffractogram, a homogenously dark background under PLM, and a spherical shape under SEM. Amorphous α-CD powder encapsulated much more CO_2 gas molecules than crystalline CD powder at low pressure (<0.5 MPa) and with a short encapsulation time (~4 h). This is because amorphous α-CD powder, with no long-range order and a loose structure, allows CO_2 to diffuse easily into the molecular cavities where interfacial interactions between CO_2 and α-CD molecules are formed. In contrast, due to the highly packed structure of crystalline α-CD powders that restrict CO_2 diffusion to their cavities, the time required to reach near equilibrium encapsulation capacity of crystalline α-CD powder is typically twice that of amorphous α-CD powder.

Regarding stability of gas-α-CD complexes, those produced by the solid method are not stable, especially those produced from amorphous CD powder in which the encapsulated gas is rapidly lost during the depressurization step. This is because the penetration of gases into the CD cavities in the solid state do not cause any changes in the structure and morphology of the gas-CD complexes that can assist stabilization of the encapsulated gas. If the initial structure of CD powders used to encapsulate gases is in an amorphous or crystalline form, the structure of formed gas-CD complexes remains in the same form. Therefore, there is no mechanism to prevent the encapsulated gas from escaping under depressurization. For instance, as indicated in Fig. 3.4a–d, which exhibit the morphology (SEM) and structure (XRD) of CO_2-α-CD powder prepared from amorphous and crystalline α-CD powders, there

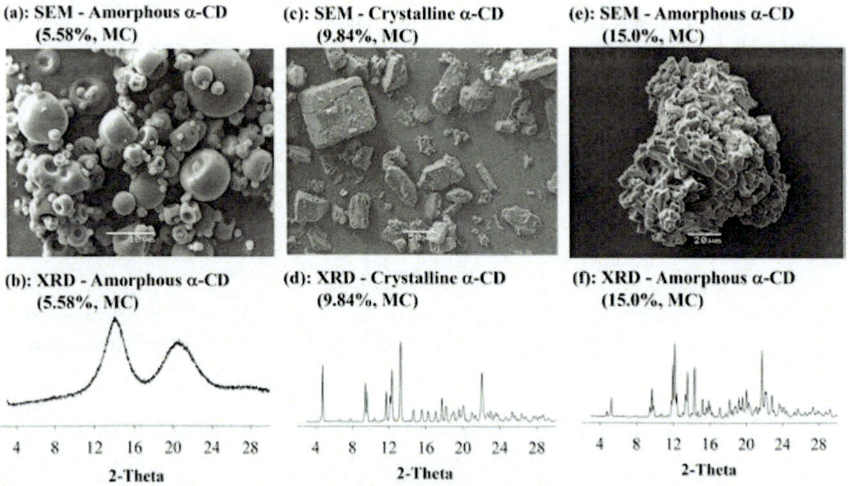

Fig. 3.4 SEM and XRD of inclusion complex prepared from amorphous α-CD powder at 5.58% moisture content (**a**, **b**), crystalline α-CD powder at 9.84% moisture content (**c**, **d**), and amorphous α-CD powder at 15.0% moisture content (**e**, **f**). For (**e**) and (**f**), due to crystallization of amorphous α-CD powder at high moisture content, SEM and XRD results exhibited the characteristics of crystalline powders that are similar to those in (**c**) and (**d**). (Adapted with permission from Ho et al. 2016e)

is no change in the morphology and structure of α-CD powder during encapsulation (Ho et al. 2015c). A lower stability of CO_2-α-CD powder produced by solid approach than liquid approach was also reported (Neoh et al. 2006). Therefore, further steps are required to improve the stability of gas-CD complexes produced by solid approach.

Amorphous CD powders are thermodynamically in a non-equilibrium state and tend to convert to crystalline (thermodynamically stable) structure at a moisture content that is higher than the crystallization-induced level, also known as critical moisture content. This suggests that during solid encapsulation, if amorphous CD powders are crystallized, the crystalline structure of gas-CD complexes might assist with the stabilization of the gases in the CD cavities, by which the encapsulation capacity of amorphous CD powder is increased and the stability of its complex powder is improved. Ho et al. (2015b) reported that the critical moisture content of amorphous α-CD powder is 13.70–14.97% (w/w). In addition, the effects of water-induced crystallization of amorphous α-CD powder during CO_2 encapsulation were also reported (Ho et al. 2016c). The addition of water to increase the initial moisture content of amorphous α-CD powder from 5.58% (w/w) to a level close to or higher than the critical level of crystallization (e.g. 13%, 15%, and 17%) during CO_2 encapsulation process (1.0–1.6 MPa and 48 h) markedly increased its encapsulation capacity up to 1.1–1.2 mol CO_2/mol α-CD. This was higher than that of amorphous α-CD powder without adding water (~0.9 mol CO_2/mol α-CD) and of crystalline α-CD powder under the same conditions (1.0–1.1 mol CO_2/mol α-CD). These differences were much more apparent as the encapsulation process was performed at lower pressure (~0.4 MPa) and shorter time (~4 h). The simultaneous crystallization of amorphous α-CD powder during CO_2 encapsulation is the reason for improvement of encapsulation capacity, especially as the encapsulation process is performed at low pressure and for a short time. The morphology and structure of CO_2-α-CD complexes produced from amorphous α-CD powder added with 15% moisture content (Fig. 3.4e, f) exhibited the characteristics of crystalline powders.

However, due to water exclusion during crystallization and encapsulation, the quantity of water on the CO_2-α-CD particle surface is too high to make the complex powder sufficiently stable for actual application. The surface of a CO_2-α-CD complex powder particle is nearly saturated by water (e.g. water activity ~0.9) and the half-life time of gas release at normal conditions (e.g. 75% RH, 25 °C) of the most stable CO_2-α-CD complex powder was only 30 min. Therefore, the reduction of water on the surface of CO_2-α-CD complex powder particles could be a potential means to improve its stability. Regarding dehydration of CO_2-α-CD complex powder, desiccant adsorption (25 °C) by storage of CO_2-α-CD complex powder and a small bag of silica gel or $CaCl_2$ at 1:1 ratio in an airtight package was performed. The results indicated that $CaCl_2$ adsorbed water on the surface of the CO_2-α-CD complex powder much faster than silica gel, especially at the initial time of desiccation. However, after being equilibrated for 60 h, water activity of CO_2-α-CD complex powder treated with both silica gel and $CaCl_2$ had nearly equilibrated at 0.247 and 0.225, respectively. Although the removal of the surface water by silica gel and $CaCl_2$ did not affect the physical properties of complex powder, desiccation with

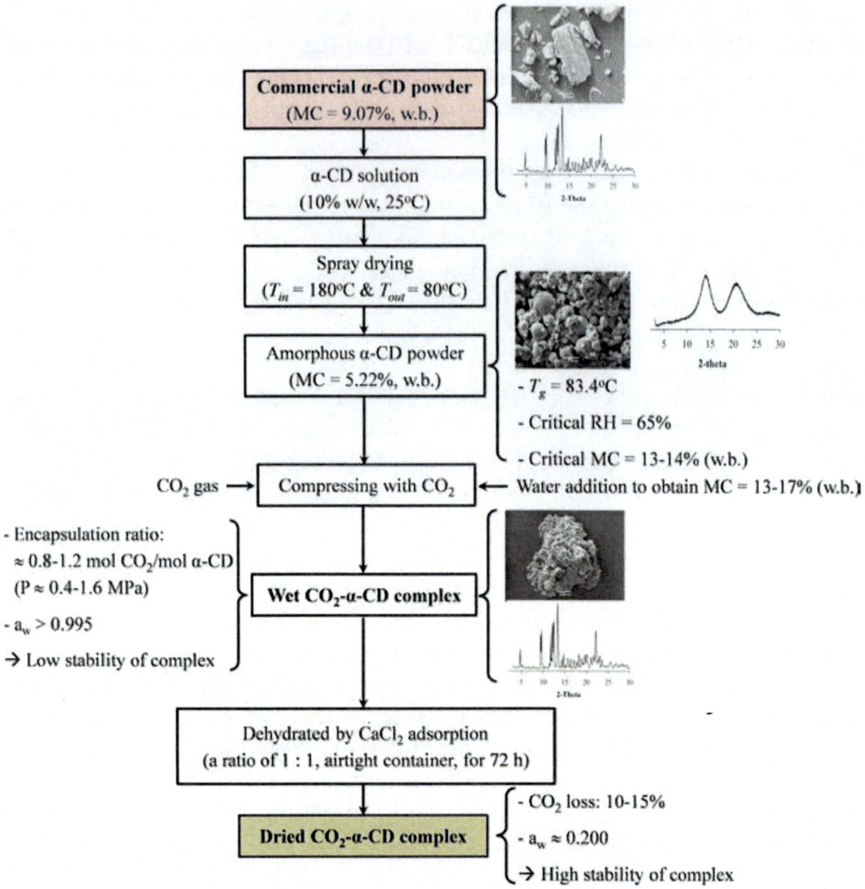

Fig. 3.5 Procedural details of production of a stable CO_2-α-CD complex powder by solid encapsulation approach. *MC* moisture content, *w.b.* wet basis, *RH* relative humidity, a_w water activity, T_{in} and T_{out} spray drying temperature for inlet and outlet

silica gel resulted in loss of about half of encapsulated CO_2, which was three times greater than the CO_2 loss caused by the use of $CaCl_2$. Dehydration of the complex powder by desiccant adsorption significantly improved its stability, as a result of which the CO_2 concentration of the desiccated complex powder remained unchanged during a 1-month storage period when kept in an airtight package with normal and vacuum sealing (Ho et al. 2016b). Details of the procedure to produce a stable CO_2-α-CD complex powder are presented in Fig. 3.5 (Ho 2017).

By following a method reported by Caira (2001), the arrangement of α-CD molecules in the crystals of CO_2-α-CD complex powders was also determined. This method is based on matching the experimental powder XRDs of CO_2-α-CD complex powders with the reference XRDs of α-CD complexes with other guest molecules in which the crystal arrangement (cage or channel type) is known. The reference XRDs were constructed and reported by extracting the full single crystal XRD data files from the Cambridge Structural Database (https://www.ccdc.cam.ac.uk/solutions/csd-system/components/csd/) and then using other software (e.g. Lazy Pulverix) to compute the data files of the respective XRD patterns. These were shown in the study by Caira (2001). The analytical results revealed that during crystallization, α-CD molecules were arranged into a cage-type structure in which an α-CD molecule is blocked at both ends by surrounding α-CD molecules by which CO_2 was entrapped in isolated cavities (Ho et al. 2016d).

3.2.2 *Encapsulation Capacity*

Among three common forms of CD, α-CDs have the smallest interior cavity (0.47–0.53 nm) and are most suitable for forming complexes with organic guests with less than five carbon atoms in their molecules. This is because the smaller cavity will offer more interaction and better binding force between the guest molecules and walls of cavities (Hedges et al. 1995). Most common gases are of low molecular weight and small size and can easily fit into the cavities of α-CDs (Trotta et al. 2011). Therefore, most gas complexes are produced using α-CD powder. It is likely that parent β- and γ-CD molecules with a large cavity (0.60–0.6 and 0.75–0.83 nm, respectively) are not suitable to form a complex with most gases in the way that the parent α-CD does. It was reported among N_2, N_2O, CO_2, and Ar gases that β- and γ-CD molecules can only form complexes with N_2 while α-CD molecules can complex with all of these gases (Pereva et al. 2015). However, by cross-linking with carbonildiimidazole (known as CD nanosponges), β- and γ-CD molecules enable encapsulation of some types of gases due to changes in channels between CD molecules or the formation of porous networks during modification process, which can serve as places to occupy gases (Trotta et al. 2012). This has been confirmed by Cavalli et al. (2010), who showed that α-, β-, and γ-CD cross-linking with carbonildiimidazole at 353–373 K for 4 h can also become an oxygen-delivery system due to their high oxygen storage ability. Similarly, Trotta et al. (2011) also reported that β-CD nanosponges prepared by a similar method can store large amounts of CO_2 and 1-MCP. The encapsulation capacity of CD molecules for different gas molecules is summarized in Table 3.3 (Ho 2017). Interestingly, it appears that Ar gas could not be encapsulated by α-, β-, and γ-CD (Zeller and Kim 2008; Pereva et al. 2015). Various gases such CH_4, C_2H_6, C_3H_8, Kr, 1-MCP, Xe, CO_2, and C_2H_4 were bubbled into an α-CD solution at controlled pressure (0.7–12 MPa) and temperature (293.15 K), which resulted in encapsulated gas content ranging from 1.04–10.27 g/100 g in crystallized complexes collected after 5–7 days of encapsulation (Szejtli 1988; Cramer and Henglein 1956). By following a similar technique (e.g. liquid encapsulation), the encapsulation capacity of these gases and others (such as Ar, N_2, and N_2O) at different pressure and time levels was reported (Ho et al. 2011a; Neoh et al. 2006, 2007; Zeller and Kim 2008; Pereva et al. 2015) (Table 3.3). Recently, encapsulation of ClO_2 gas using α-CD solution was also

Table 3.3 Summary of research studies on gas encapsulation using CDs (adapted with permission from Ho et al. 2014)

Gases	CD type	Encapsulation conditions	Encapsulation capacity (g/100 g)	References
C_2H_4	α-CD	P = 0.2–1.5 MPa T = 298.15 K	2.74–2.88	Ho et al. (2011a)
		P = 0.2–1.0 MPa T = 298.15 K	2.23–2.35	Bazzano et al. (2016)
		P = 0.5–1.5 MPa T = 298.15 K	α-CD (4.1% MC): 1.29–2.50	Ho and Bhandari (2016)
			α-CD (7.4% MC): 1.21–1.56	
CO_2		P = 1–3 MPa T = 298.15 K	4.20–4.66	Neoh et al. (2006)
		P = 0.4–1.6 MPa T = 298.15 K	3.20–4.56	Ho (2017)
CO_2	β-CD nanosponges[a]	P = 0.10 MPa T = 293.15 K T = 393.15 K	293.15 K = 0.40 393.15 K = 0.60	Trotta et al. (2011)
1-MCP		P = 0.101 MPa T = 298.15 K	8.10–9.10	
Cl_2	α-CD	P = 0.7–12 MPa T = 293.15 K	2.14	Szejtli (1988); Cramer and Henglein (1956)
Kr			2.84	
Xe			10.27	
O_2			1.04	
CO_2			5.87	
C_2H_4			1.81	
CH_4			1.62	
C_3H_8			4.43	
C_4H_{10}			6.69	
1-MCP	α-CD	P = 0.101 MPa T = 288.15 K	5.02	Neoh et al. (2007)
CO_2	α-CD	$P_{(CO_2)}$ = 2.2 MPa $P_{(N_2O)}$ = 1.8 MPa $P_{(Ar, N_2)}$ = 3.4 MPa T = 298.15 K	CO_2 = 3.59	Zeller and Kim (2008)
N_2O			N_2O = 3.59	
Ar			No complex formed	
N_2				
CO_2	β-CD	P = 3.1 and 3.4 MPa T = 298.15 K	No complex formed	
Ar				
CO_2	γ-CD			
Ar				

(continued)

Table 3.3 (continued)

Gases	CD type	Encapsulation conditions	Encapsulation capacity (g/100 g)	References
N_2	α-CD	P = 0.5–4.0 MPa T = 278.15–293.15 K	0.50–0.62	Pereva et al. (2015)
N_2O			1.36–1.96	
CO_2			0.97–1.35	
Ar			Very low yield	
N_2	β-CD γ-CD	P = 0.5–4.0 MPa T = 278.15–293.15 K	Only formed complexes with N_2: 0.62–1.84	
N_2O				
CO_2				
Ar				
CO_2	β-CD derivative with NaOH	P = 0.35 MPa T = 298–403 K	1.98–3.96	Guo et al. (2017a)
	β-CD		278 K: Almost no adsorbed CO_2 334–403 K: 1.76–5.06	
	β-CD derivative with HCl	P = 0.30 MPa T = 298 and 403 K	2.86–7.48	Guo et al. (2017b)
ClO_2	α-CD	Co-precipitation method at normal pressure and 4 °C	6.44	Chen et al. (2020)

MC moisture content
[a]Cross-linking of β-CD with carbonyldiimidazole

reported (Chen et al. 2020). ClO_2 is green-yellow gas under ambient conditions and is used as a washing antimicrobial solution for industrial practices. However, because of its explosiveness, reactivity, and corrosiveness, it is not kept as is and is normally generated on-site from its stabilized form. By using the co-precipitation encapsulation method in which ClO_2 gas (generated from acidification of chlorite) was brought to mix with α-CD solution in an airtight system, encapsulation capacity reached to 6.44% (which is very close to the theoretical maximum loading of 6.45%).

Regarding the solid encapsulation method, Neoh et al. (2006) attempted to complex CO_2 into crystalline α-CD in solid state with different initial moisture content (2%, 10%, and 30%) at a pressure of 1 or 3 MPa. The results showed that an increase of encapsulating pressure from 1 to 3 MPa significantly increased the inclusion ratio while the effects of initial moisture content of α-CD on the inclusion ratio was not consistent. At 3 MPa, the highest inclusion ratio was observed in a sample with 10% moisture content (5.92 g gas/100 g), which was higher than at 2% and 30% moisture content (5.06 and 4.88 g gas/100 g, respectively). A similar result in CO_2 encapsulation capacity of crystalline α-CD powder in solid state at 13%, 15%, and 17% moisture content at 0.4–1.6 MPa was also reported (Ho et al. 2016c). Both studies indicated that in solid encapsulation, increasing initial moisture content of crystalline α-CD powder reduced its CO_2 encapsulation capacity. The accumulation of water molecules on the surface of crystalline powder particles inhibits the substitution of the high-enthalpy water molecules in the α-CD cavity by hydrophobic CO_2

molecules, which is known as the main driving force of the complexation process. Likewise, an increase of moisture content of amorphous α-CD from 4.1% to 7.4% reduced its maximum C_2H_4 encapsulation capacity from 2.50% to 1.56% (Ho and Bhandari 2016). However, for solid encapsulation of amorphous CDs at moisture content higher than the critical moisture content level, gas encapsulation capacity behaviour was different. Moisture content levels of 13%, 15%, and 17% were close to the critical moisture content of amorphous α-CD powder (~14%) and thus were sufficient to initiate state transformation from amorphous to crystalline. The initiation of crystallization by adding water helped the crystallization of the amorphous CO_2-α-CD complex powder, which assisted in locking the CO_2 in the cavity of α-CD during the encapsulation process. Therefore, for the same encapsulation conditions, amorphous α-CD powders at 13%, 15%, and 17% moisture content encapsulated much more CO_2 gas than the original amorphous α-CD powder at 5.22% moisture content. However, due to the rapid crystallization at high moisture content, part of the α-CD possibly becomes crystallized without first complexing with the CO_2. Subsequently, the crystalline structure could act as a barrier for gas penetration. Thus, among the three levels of moisture content investigated (13%, 15%, and 17%), amorphous α-CD powder at 17% moisture content had the lowest encapsulation capacity (Ho 2017). As a result, for solid encapsulation of amorphous CD powder, the amount of water added into the powder must be optimized to not only improve encapsulation capacity, but also to enhance the stability of the formed complexes.

By using a dynamic adsorption system (e.g. gravimetric-IGA system), Guo et al. (2017a) and Guo et al. (2017b) reported that native β-CD powder in solid state almost did not adsorb CO_2 gas at normal temperature (289 K) regardless of adsorption pressure (up to 0.35 MPa). However, after increasing the adsorption temperature to 403 K, the CO_2 adsorption capacity of β-CD powder increased up to 5.06%. In addition, modifying the structure of β-CD powder by sodium hydroxide and sulfuric acid markedly enhanced CO_2 adsorption capacity, especially at low adsorption temperature.

3.2.3 Release Properties of Gas-CD Complex

The interactions of guest molecules within the host CD are not permanent but rather are a dynamic equilibrium (Singh et al. 2002). This explains why encapsulated gases are easily released from gas-CD complexes. The important environmental factors that are effective in controlling the release rate of gases are humidity and temperature. An increase in humidity and temperature will accelerate the release rate of gases from gas-CD complexes, and gases are almost immediately released upon dissolving the complexes in water (Ho et al. 2016e). Due to very low moisture content, a gas-CD complex powder can act as hygroscopic material, thereby higher storage humidity results in a greater amount of water being adsorbed by a gas-CD complex powder. The replacement of encapsulated gas by adsorbed water in the cavity of gas-CD complex powders could be the main driving force for the release of encapsulated gas under the effects of RH (Neoh et al. 2006; Zeller and Kim 2008; Ho et al. 2011a). In an aqueous environment, the solubility of CD in water causes the weak physical bindings between guests and the CD cavity to break, and consequently the encapsulated gases escape easily from gas-CD complexes.

The solubility of CD in an aqueous solution is strongly affected by inorganic salts. The presence of chloride (Cl^-) and nitrate (NO^{3-}) salts of groups II (Ca^{2+} or Mg^{2+}) and III (Fe^{3+} or Al^{3+}) metals in water results in an increase in CD solubility. In contrast, the converse trend was observed for sulphate (SO_4^{2-}) salts of group I (Na^+ or K^+) metals (Coleman and Nicolis 1993; Eddaoudi et al. 1996). Therefore, mixing the CD complexes with various inorganic salts can modify the release rate of the encapsulated gases. Most inorganic salts can absorb moisture from the surrounding environment to liquefy and form a saturated salt solution. The phase transformation only occurs at a particular relative humidity (RH) known as the deliquescent RH, which is specific for each type of salt and depends on temperature (Lipasek et al. 2012). Therefore, the release of encapsulated gases (C_2H_4 or 1-MCP) from CD complexes can be controlled by blending gas-CD complexes with deliquescent salts (e.g. $CaCl_2$ and $MgCl_2$) followed by placing this mixture in an high RH environment (higher than deliquescent RH of salts) (Ho 2013; Kostansek 2002). The rate of gas release is dependent on the type of deliquescent salt and the RH in which the deliquescent salts and gas-CD complexes are exposed. Ho et al. (2015a) reported that in the mixture of C_2H_4-α-CD complexes prepared by liquid method and deliquescent salts ($CaCl_2$ and $MgCl_2$), C_2H_4 gas was not released at 11.2% RH even up to 168 h. However, at 52.9%, 75.5%, and 93.6% RH, C_2H_4 gas commenced from C_2H_4-α-CD powder after 2 h and markedly increased to maximum at 24 h. At 32.7% RH, the ethylene release ratio reached only 0.83 for C_2H_4-α-CD powder mixed with $MgCl_2$ and 0.87 for C_2H_4-α-CD powder mixed with $CaCl_2$ at 168 h.

The kinetics for gas release from CD complexes has been studied and reported. Neoh et al. (2006) observed that the CO_2 release rate from CO_2-α-CD complexes under various RH levels (5%, 33%, and 75%) was well described by Avrami's equation (Weibull equation). This was despite the fact that this model was initially developed to describe the crystallization of polymers (Avrami 1940) and that the CO_2 release mechanism was diffusion-controlled and the rate was significantly enhanced by high RH. Similarly, in a study on ethylene release kinetics from C_2H_4-α-CD complexes under various RH levels (52.9%, 75.5%, and 93.6%) and temperature (318, 338, 358, and 378 K), Ho (2013) claimed that both Avrami's and power law models can be used to describe ethylene release rate for all conditions, and the former was a better fit to the experimental data than the latter and the amount of energy required to activate ethylene release in inclusion complexes was 65.9 kJ/mol. Similar effects of RH and temperature on release rate of 1-MPC and ClO_2 gas from 1-MPC-α-CD and ClO_2-α-CD complex powder were also reported by Ariyanto and Yoshii (2019) and Chen et al. (2020), respectively. However, as the ClO_2-α-CD complex powder was kept in plastic tubes in desiccators at −20, 4, and 23 °C, after 3 months there was no loss at −20 °C whereas there was approximately 8% loss at 4 °C and 80% loss at 23 °C. In addition, the release of 1-MPC from its complexes with α-CD powder at high temperature (~120 °C) can be also investigated by thermal analyses, such as thermogravimetry and differential scanning calorimetry (Neoh et al. 2008).

For CO_2-α-CD complexes prepared by solid encapsulation followed by desiccant dehydration with $CaCl_2$ salt (as described in Fig. 3.5), release properties at different RH levels, different packaging conditions, and different aqueous medium (oil and water) were also reported and are illustrated in Fig. 3.6 (Ho et al. 2016b). A rapid release of CO_2 was observed for complex powders kept at 75% and 97% RHs, with more than 98% of the encapsulated CO_2 being released after only 360 and 120 h of storage, respectively. At 32% and 52% RHs, the release fraction measured after a 4-week period was approximately 0.72 and 0.92, respectively (Fig. 3.6a). Similar effects of RH on the release rate of C_2H_2 and CO_2 from C_2H_2-α-CD and CO_2-α-CD complex powders prepared by liquid encapsulation were also reported by Ho et al. (2011b) and Neoh et al. (2006), respectively. In addition, as indicated in Fig. 3.6a, the removal of a large proportion of the air from the package headspace by normal or vacuum sealing, which prevents moisture exchange between the complex powder and surrounding air, has a great effect on the stability of the dry CO_2 powder. In sealed packages, the CO_2 concentration in the complex powder was observed to be almost unchanged during a 4-week period of storage, while there was no difference in CO_2 concentration in packages sealed using normal or vacuum conditions.

In liquids, especially in water, the erosion of CO_2 complex powder is caused by its dissolution followed by solubilization of the released CO_2 into solution. The released CO_2 then diffuses through the solution into the air. Thus, diffusion, which is affected by solubility of the complex powder, viscosity of solution, or agitation of the mixture, has a great impact on the release rate of gas. Under the same conditions, the release rate of CO_2 in an oil environment was significantly slower than in a water environment. The amount of oil added into the complex powder highly influenced the release rate of CO_2. As more oil is added, the release rate of CO_2 decreases (Fig. 3.6b, c). This is because oil has a higher viscosity than water, which prevents the released CO_2 molecules from escaping into the headspace. In addition, the structure of α-CD molecules with a hydrophilic surface makes its complex powder less soluble in apolar liquids like oil. This retards disruption of the coating material of complex powders and therefore the release of encapsulated CO_2 is restricted when complex powders are dispersed into oil. Complex powders are completely submerged into oil once mixed with oil; therefore the released CO_2 molecules require a long time to travel through the highly viscous oil. The release kinetics of gases from CD inclusion complexes at different conditions such as temperatures, relative humidities and solvents could be investigated by using the convenient and inexpensive method developed by Ho et al. (2016d) in which gas probe is utilized to measure the gas content.

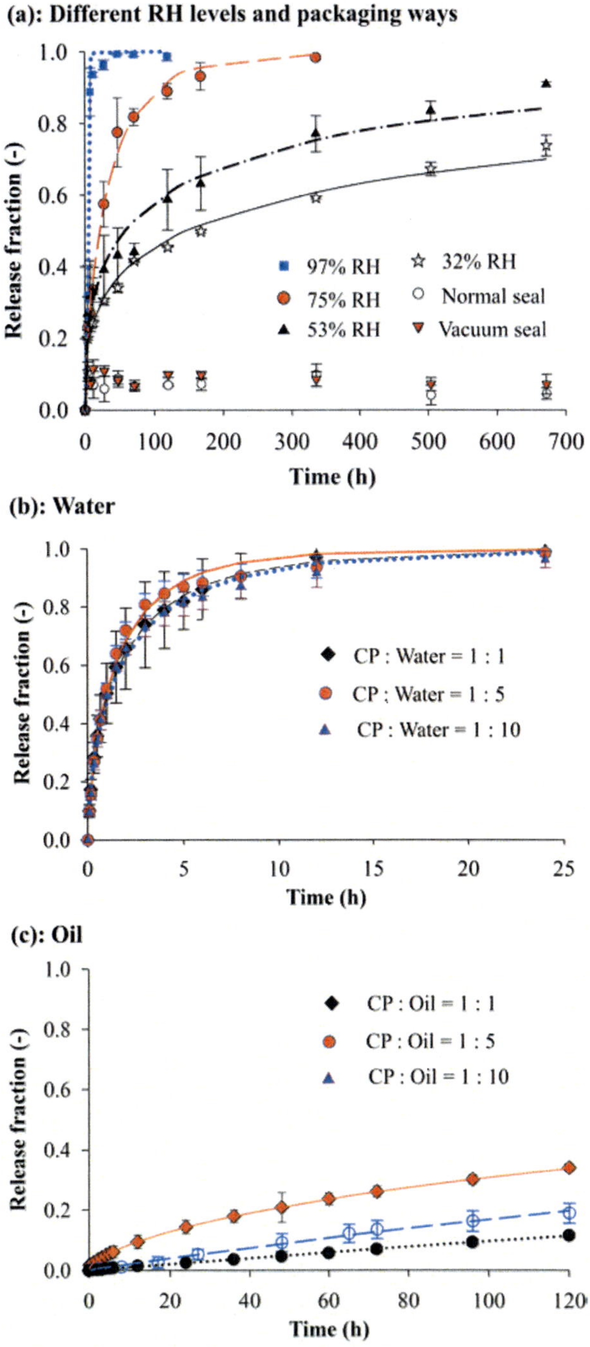

Fig. 3.6 Release properties of CO_2 gas from CO_2-α-CD complex powders (CO) prepared by solid encapsulation under various conditions. (**a**) Different relative humidity levels and packaging, (**b**) Different ratios of CO_2-α-CD complex and water, and (**c**) Different ratios of CO_2-α-CD complex and oil. (Adapted with permission from Ho et al. 2016b)

3.3 Applications of Gas-CD Complexes in Agriculture and Food Production

Gas encapsulation into CD powder for agriculture and food production is still at an early stage of development. The first report on the application of gas-CD complexes was possibly C_2H_4-α-CD complex for ripening of tomatoes (Teijin Ltd. 1975). Almost three decades later, there have been other reports on the applications of C_2H_4-α-CD inclusion complex. Such a gas complex has successfully been tested for its applicability in horticulture to promote the ripening process of mangos and germination of mung bean sprouts. Only 100 g of this powder was able to trigger the ripening of 20 tons of mangoes in a container in-transit from a farm to a distant market (Ho et al. 2013, 2016a; Ho 2013). For germination of mung bean sprouts, soaking mung beans with this powder solution at gas concentration of 0.525–52.5 mg gas/100 mL solution significantly improved the rate of seed germination, hypocotyl thickness, and uniformity in length of mung bean sprouts (Khang et al. 2013).

Although encapsulation of CO_2 into α-CD was previously patented in Japan in 1987 (Trotta et al. 2011), this work did not lead to a commercial application probably due to the limited storage capacity achieved in the applied processes. Zeller and Kim (2008) investigated the use of CO_2-α-CD complexes in cappuccino frothing. Mixing of 1.5–3.0 g of complex powder (prepared at 3.4 MPa at 298.15 K for 7 days) to 11.5 g of instant cappuccino mix and then reconstituted with 150 mL hot water (361.15 K) increased frothing 2–3 times compared to a sample without added complexes. A similar result in froth height enhancement was observed by adding N_2O-α-CD complexes (encapsulated at 3.4 MPa, 298.15 K for 3 days) in hot instant cappuccino, cocoa, espresso, and coffee mixes. The flavor of products with N_2O-α-cyclodextrin complexes was much better than that with CO_2-α-CD complexes, which possibly induced carbonation and sensation flavor. The froth stability was mainly determined by foam stabilizers present in milk protein. In another study, mixing of approximately 3% of N_2O-α-CD complex into dough resulted in an increase of oven-baked pizza volume ingredient by 22% with a uniform internal foam structure.

For applications of CO_2-α-CD complexes in dairy products, mixing CO_2-α-CD complexes into cottage cheese curds helped control the growth of yeast and mould during storage in actual usage conditions, such as frequent opening of the container by the consumer, during which natural airborne mould is introduced to the product (Ho 2017; Ho et al. 2018). Due to presence of water in the cheese curds that ruptures the coating α-CD material, CO_2 gas from the complex powder was slowly released and solubilized and diffused throughout the sample, resulting in noticeable inhibition of mould and yeast growth. The study results indicated that when the samples were exposed to air, the control samples (without adding CO_2-α-CD complexes) were spoiled at day 5 and day 30 of storage at 25 and 7 °C, respectively. In contrast, after adding CO_2-α-CD complexes into the cheese curds at concentrations of 300, 600, and 900 ppm CO_2, there was no sign of mould and yeast growth until day 10 at 25 °C and up to 6 weeks at 7 °C. Figure 3.7 shows the cheese curd surface of

Fig. 3.7 Images showing the growth of mould and yeast in cottage cheese curds with or without added CO_2 complex powders and kept at 7 °C. Images were taken at every week during a 6-week storage period. (Adapted with permission from Ho et al. 2018)

samples mixed with CO_2-α-CD complexes at CO_2 concentrations of 300, 600, and 900 ppm and stored at 7 °C up to 6 weeks.

1-MPC-α-CD is a commercially available gas-α-CD complex powder. It is known as EthylBloc® (AgroFresh Inc.), which is primarily used for ornamental crops, and SmartFresh™ (AgroFresh Inc.), which is for edible crops (Kyaw 2019). Treating sweet cherry with 500–2000 ppb (in the headspace of an airtight container) released from 1-MPC-α-CD complex powder at normal temperature for 8 h and retarded the ripening process of sweet cherry during storage (Jinwei et al. 2019). After 12 days of storage at normal temperature, samples treated with 1-MPC-α-CD complex powder exhibited a decline in the browning process rate (1.6-fold reduction) and in titratable acidity and soluble protein content (1.25-fold reduction) when compared with control samples. In addition, inhibition of superoxide dismutase and ascorbate peroxidase activity was also observed in samples treated with 1-MPC-α-CD complex powder. Similar results in the delay of the ripening process of sweet cherry were also reported for 800–3200 ppb of 1-methyl-3-(2-methylcyclopropyl)-1-cyclopropene gas released from 1-methyl-3-(2-methylcyclopropyl)-1-cyclopropene/Cu-β-CD (Jinwei et al. 2019). In addition, the use of 1-MPC-α-CD complex powder to delay the climacteric process and to extend the shelf life of avocado was also reported (Amin et al. 2016). Avocados treated with 1-MPC gas (released from 1-MPC-α-CD complex powder) with headspace concentrations of 0.2–0.4 ppm and a solution with a gas-solubilized concentration of 0.3–0.6 ppm had a shelf life extension 2–2.5 times greater than control samples. No differences were observed between treatment with 1-MCP gas and immersion into a solution of 1-MCP. The effectiveness of 1-MPC-α-CD complex powder in delaying ripening and prolonging the shelf life of tomato has been investigated (Ariyanto et al. 2019).

3.4 Conclusion and Further Perspectives

Although there is great potential in the use of gas-CD complexes for food production, research on gas encapsulation using CD powders and their subsequent applications are still in an initial stage of research and development. This could be due to low encapsulation capacity. The amount of encapsulated gas in the gas-CD complexes is insufficient to replace conventional ways in the use of gas for certain applications. For example, the amount of CO_2 gas in CO_2-α-CD complex powder (e.g., 5% w/w) is not financially viable when compared with CO_2 gas obtained from high pressured cylinders for carbonated drinks. Therefore, further investigation should focus on improvement of gas encapsulation capacity. Manipulation of the crystal packing of gas complex powder into channel types, in which the cavities of the CD molecules are aligned like coins in a roll to produce very long channels in which the gas molecules are entrapped during encapsulation could be a direction for further studies. This type of crystal packing could increase gas encapsulation capacity. Another possibility to enhance encapsulation capacity is to modify the structure of CD molecules, especially β- and γ-CD, such that they can occupy more gas in their cavities.

References

Adisa OO, Cox BJ, Hill JM (2011) Encapsulation of methane molecules into carbon nanotubes. Physica B Condens Matter 406(1):88–93

Allan PK (2012) A study of metal-organic frameworks for the storage and release of medical gases. PhD thesis, University of St Andrews, UK

Amin I, Poerwanto R, Kartika J (2016) 1-MCP application to prolong avocado shelflife. Acta Hortic 1120:131–136

Anson A, Callejas M, Benito A et al (2004) Hydrogen adsorption studies on single wall carbon nanotubes. Carbon 42(7):1243–1248

Ariyanto HD, Yoshii H (2019) Effect of stepwise humidity change on the release rate constant of 1-methylcyclopropene (1-MCP) in a cyclodextrin inclusion complex powder. Food Packag Shelf Life 21:100322

Ariyanto H, Bercis M, Yoshii H (2019) Effect of 1-methylcyclopropene release from inclusion complexes powder on the quality of tomato by short-term treatment. IOP Conf Ser Earth Environ Sci 309:012065

Arvanitoyannis I (2012) Modified atmosphere and active packaging technologies. CRC Press, Boca Raton

Astray G, Gonzalez-Barreiro C, Mejuto JC et al (2009) A review on the use of cyclodextrins in foods. Food Hydrocoll 23(7):1631–1640

Avrami M (1940) Kinetics of phase change. II transformation-time relations for random distribution of nuclei. J Chem Phys 8(2):212–224
Bazzano M, Barolo C, Buscaino R et al (2016) Controlled atmosphere in food packaging using ethylene-α-cyclodextrin inclusion complexes dispersed in photocured acrylic films. Ind Eng Chem Res 55(3):579–585
Blankenship SM, Dole JM (2003) 1-Methylcyclopropene: a review. Postharvest Biol Technol 28(1):1–25
Caira MR (2001) On the isostructurality of cyclodextrin inclusion complexes and its practical utility. Rev Roum Chim 46(4):371–386
Cavalli R, Akhter AK, Bisazza A et al (2010) Nanosponge formulations as oxygen delivery systems. Int J Pharm 402(1–2):254–257
Chen M, Chen X, Yam K (2020) Encapsulation complex of chlorine dioxide in α-cyclodextrin: structure characterization and release property. Food Control 107:106783
Coleman AW, Nicolis I (1993) Inorganic salt modulation of the aqueous solubility of β-cyclodextrin. Supramol Chem 2(2–3):93–97
Cramer F, Henglein F (1956) Einschlussverbindungen der cyclodextrine mit gasen. Angew Chem 68(20):649–649
Dincer I (2002) Technical, environmental and exergetic aspects of hydrogen energy systems. Int J Hydrog Energy 27:265–285
Eddaoudi M, Coleman A, Junk P (1996) Solubilities of the cyclodextrins in the presence of transition metal salts. J Incl Phenom Macrocycl Chem 26(1–3):133–151
Guo T-X, Bedane AH, Pan Y et al (2017a) Characteristics of carbon dioxide gas adsorption on β-cyclodextrin derivative. Mater Lett 189:114–117
Guo T, Bedane AH, Pan Y et al (2017b) Characteristics of carbon dioxide gas on a solid acid derivative of β-cyclodextrin. Energy Fuel 31(4):4186–4192
Guzel-Seydim ZB, Greene AK, Seydim A (2004) Use of ozone in the food industry. LWT 37(4):453–460
Hall RC (1978) The nitrogen detector in gas chromatography. Crit Rev Anal Chem 7(7):323–380
Hanson WH, Smith R (1993) Sprayable dispensing system for viscous vegetable oils and apparatus therefore. US Patents: US5249747A
Hedges AR, Shieh WJ, Sikorski CT (1995) Use of cyclodextrins for encapsulation in the use and treatment of food products. In: Sara JR, Gary AR (eds) Encapsulation and controlled release of food ingredients, vol 590. ACS Publications, Washington, DC, pp 60–71
Ho BT (2013) Production of ethylene powder by encapsulation of ethylene gas into α-cyclodextrin and its application for the ripening of fruit. PhD thesis, University of Queensland, Australia
Ho MT (2017) Studies of the encapsulation and release of carbon dioxide from amorphous and crystalline alpha-cyclodextrin powders and its application in food systems. PhD thesis, University of Queensland, Australia
Ho BT, Bhandari BR (2016) Novel solid encapsulation of ethylene gas using amorphous α-cyclodextrin and the release characteristics. J Agric Food Chem 64(17):3318–3323
Ho BT, Joyce DC, Bhandari BR (2011a) Encapsulation of ethylene gas into α-cyclodextrin and characterisation of the inclusion complexes. Food Chem 127(2):572–580
Ho BT, Joyce DC, Bhandari BR (2011b) Release kinetics of ethylene gas from ethylene-α-cyclodextrin inclusion complexes. Food Chem 129(2):259–266
Ho BT, Bhandari B, Joyce D et al (2013) Ethylene gas from an ethylene-α-cyclodextrin inclusion complex powder ripens 'calypso' mango fruit. Acta Hortic 1012:331–336
Ho TM, Howes T, Bhandari BR (2014) Encapsulation of gases in powder solid matrices and their applications: a review. Powder Technol 259:87–108
Ho BT, Yuwono TD, Joyce DC et al (2015a) Controlled release of ethylene gas from the ethylene-α-cyclodextrin inclusion complex powder with deliquescent salts. J Incl Phenom Macrocycl Chem 83(3–4):281–288
Ho TM, Howes T, Bhandari BR (2015b) Characterization of crystalline and spray-dried amorphous α-cyclodextrin powders. Powder Technol 284:585–594

Ho TM, Howes T, Bhandari BR (2015c) Encapsulation of CO_2 into amorphous and crystalline α-cyclodextrin powders and the characterization of the complexes formed. Food Chem 187:407–415

Ho BT, Hofman PJ, Joyce DC et al (2016a) Uses of an innovative ethylene-α-cyclodextrin inclusion complex powder for ripening of mango fruit. Postharvest Biol Technol 113:77–86

Ho TM, Howes T, Bhandari BR (2016b) Dehydration of CO_2-α-cyclodextrin complex powder by desiccant adsorption method and its release properties. J Microencapsul 33(8):763–772

Ho TM, Howes T, Bhandari BR (2016c) Encapsulation of CO_2 into amorphous alpha-cyclodextrin powder at different moisture contents—part 1: encapsulation capacity and stability of inclusion complexes. Food Chem 203:348–355

Ho TM, Tuyen T, Howes T, Bhandari BR (2016d) Method of measurement of CO_2 adsorbed into α-cyclodextrin by infra-red CO_2 probe. Int J Food Prop 19(8):1696–1707

Ho TM, Howes T, Jack KS et al (2016e) Encapsulation of CO_2 into amorphous alpha-cyclodextrin powder at different moisture contents—part 2: characterization of complexed powders and determination of crystalline structure. Food Chem 206:92–101

Ho TM, Truong T, Bhandari BR (2017) Chapter 5: Spray-drying and non-equilibrium states/glass transition. In: Bhandari BR, Roos YH (eds) Non-equilibrium states and glass transitions in foods—processing effects and product-specific implications. Elsevier, Duxford, pp 111–136

Ho TM, Howes T, Bhandari BR (2018) An innovative approach to extend the shelf life of cottage cheese curds using food grade CO_2-α-cyclodextrin complex powder: a preliminary study. J Food Process Preserv 42(2):e13514

Hotchkiss JH, Werner BG, Lee EY (2006) Addition of carbon dioxide to dairy products to improve quality: a comprehensive review. Compr Rev Food Sci Food Saf 5(4):158–168

Irie T, Uekama K (1997) Pharmaceutical applications of cyclodextrins. III Toxicological issues and safety evaluation. J Pharm Sci 86(2):147–162

Jinwei Z, Jiangtao F, Wei Y (2019) Effects of inclusion compounds of 1-methylcyclopropene/α--cyclodextrin or 1-methyl-3-(2-methylcyclopropyl)-1-cyclopropene/Cu-β-cyclodextrin on the preservation of sweet cherry (Prunus avium L.). Curr Sci 117(10):1716

Khang N, Binh T, Bhandari B (2013) Application of ethylene powder for the germination of mung bean sprouts. In: Poster in 46th annual AIFST convention, Brisbane, Australia

Kostansek EC (2002) Delivery systems for cyclopropenes requiring less water. Taiwan Patents: TW589140B

Kyaw PN (2019) Regulation of ethylene production and postharvest fruit quality of stone fruit using different formulations of new ethylene antagonists. PhD thesis, Curtin University, Australia

Lipasek RA, Ortiz JC, Taylor LS et al (2012) Effects of anticaking agents and storage conditions on the moisture sorption, caking, and flowability of deliquescent ingredients. Food Res Int 45(1):369–380

Loftsson T, Duchêne D (2007) Cyclodextrins and their pharmaceutical applications. Int J Pharm 329(1–2):1–11

Masel RI (1996) Principles of adsorption and reaction on solid surfaces, vol 3. John Wiley & Sons, New York

McKinlay AC, Xiao B, Wragg DS et al (2008) Exceptional behavior over the whole adsorption- storage-delivery cycle for NO in porous metal organic frameworks. J Am Chem Soc 130(31):10440–10444

Neoh T-L, Yoshii H, Furuta T (2006) Encapsulation and release characteristics of carbon dioxide in α-cyclodextrin. J Incl Phenom Macrocycl Chem 56(1–2):125–133

Neoh TL, Yamauchi K, Yoshii H et al (2007) Kinetics of molecular encapsulation of 1-methylcyclopropene into α-cyclodextrin. J Agric Food Chem 55(26):11020–11026

Neoh TL, Yamauchi K, Yoshii H et al (2008) Kinetic study of thermally stimulated dissociation of inclusion complex of 1-methylcyclopropene with α-cyclodextrin by thermal analysis. J Phys Chem B 112(49):15914–15920

Pereva S, Himitliiska T, Spassov T et al (2015) Cyclodextrin-based solid–gas clathrates. J Agric Food Chem 63(29):6603–6613

Rudkevich DM (2004) Emerging supramolecular chemistry of gases. Angew Chem Int Ed 43(5):558–571
Saenger W (1985) Nature and size of included guest molecule determines architecture of crystalline cyclodextrin host matrix. Isr J Chem 25(1):43–50
Saenger W, Jacob J, Gessler K et al (1998) Structures of the common cyclodextrins and their larger analogues beyond the doughnut. Chem Rev 98(5):1787–1802
Saltveit ME (1999) Effect of ethylene on quality of fresh fruits and vegetables. Postharvest Biol Technol 15(3):279–292
Schneiderman E, Stalcup AM (2000) Cyclodextrins: a versatile tool in separation science. J Chromatogr B Biomed Sci Appl 745(1):83–102
Shomer R, Cogan U, Mannheim CH (1994) Thermal death parameters of orange juice and effect of minimal heat treatment and carbon dioxide on shelf-life. J Food Process Preserv 18(4):305–315
Singh M, Sharma R, Banerjee U (2002) Biotechnological applications of cyclodextrins. Biotechnol Adv 20(5–6):341–359
Szejtli J (1988) Cyclodextrin inclusion complexes. In: Cyclodextrin technology. Kluwer Academic Publishing, Dordrecht, pp 79–185
Szejtli J (1989) Downstream processing using cyclodextrins. Trends Biotechnol 7(7):170–174
Teijin Ltd. (1975) Ethylene alpha-cyclodextrin—useful as a sustained release plant growth regulator. Japan Patent: JP50058226-A
Trotta F, Cavalli R, Martina K et al (2011) Cyclodextrin nanosponges as effective gas carriers. J Incl Phenom Macrocycl Chem 71(1–2):189–194
Trotta F, Zanetti M, Cavalli R (2012) Cyclodextrin-based nanosponges as drug carriers. Beilstein J Org Chem 8(1):2091–2099
Wheatley PS, Butler AR, Crane MS et al (2006) NO-releasing zeolites and their antithrombotic properties. J Am Chem Soc 128(2):502–509
Wood JP, Ryan SP, Snyder EG et al (2010) Adsorption of chlorine dioxide gas on activated carbons. J Air Waste Manag Assoc 60(8):898–906
Yoon JH, Heo NH (1992) A study on hydrogen encapsulation in Cs2. 5-zeolite A. J Phys Chem 96(12):4997–5000
Zeller BL, Kim DA (2008) Gasified food products and methods of preparation thereof. WIPO (PCT) Patents: WO2008036744A1

Chapter 4
Encapsulation of Flavors

Thi Van Anh Nguyen and Hidefumi Yoshii

4.1 Introduction About Flavors

Generally, flavors are defined as a combination of odor and taste (Potdar et al. 2020). It is one of the most important properties determining consumers' preferences to food products. Flavors are organic compounds with their molecular weights are from 50 to 300 Da and boiling point are around 20–250 °C (Lopes et al. 2016). Flavors can be obtained from natural or artificial sources. Natural flavors exit in natural sources such as fruits, herbs, spices, and animal products or in fermentation processes and can be produced by distillation, extraction, hydrolysis, etc. while artificial (synthetic) flavors are produced by chemical synthesis through one or several steps (Trifković et al. 2016).

Even the use of flavors in food processing is more and more increasing (Estevinho and Rocha 2017; Lopes et al. 2016), some properties such as lability, low solubility, instability under processing conditions and environmental conditions have limited its applications. The high flavor intensity is unpleasant so it can have a negative impact on human health and affect customers' perception. Therefore, the amount of flavors in food products must be strictly controlled (Trifković et al. 2016). The encapsulation of flavors within a solid matrix can overcome the limited properties of flavors, help to control the release rate of flavor at appropriate time in food matrix, and mask undesired flavor of food products.

T. V. A. Nguyen (✉)
Department of Food Technology, Faculty of Engineering and Food Technology, Hue University of Agriculture and Forestry, Hue University, Phung Hung, Hue, Thua Thien Hue Province, Vietnam
e-mail: nguyenthivananh@huaf.edu.vn; nguyenthivananh@hueuni.edu.vn

H. Yoshii
Department of Food Science and Human Nutrition, Setsunan University, Osaka, Japan
e-mail: hidefumi.yoshii@setsunan.ac.jp

T. M. Ho et al. (eds.), *Functionality of Cyclodextrins in Encapsulation for Food Applications*, https://doi.org/10.1007/978-3-030-80056-7_4

Molecular encapsulation is one of chemical method in which cyclodextrins (CDs) have been used as encapsulants. Based on the unique structure of CDs, the guest molecule, which has appropriate size and geometry, could be encapsulated in CDs' cavity. The CD molecular encapsulation occurs in whole or partly the molecule of core material. The encapsulated form is known as inclusion complex. In flavor encapsulation, hydrophobic flavors can be easily encapsulated in hydrophobic cavity of CDs. The advantage of the encapsulation technique is that the solubility of flavor in water can be enhanced because the outside of CD molecule is hydrophilic. Other ones are that it can inhibit evaporation and degradation of flavors during processing or storage, and mask undesired odors (Hedges 1998; Rakmai and Cheirsilp 2016).

4.2 Preparation and Characteristics of Flavor-CD Complexes

4.2.1 Complex Formation of Flavors into CDs

The inclusion complexes of guests and CDs can be prepared by co-precipitation, kneading, slurry complexation, spray drying and freeze drying and so forth (Del Valle 2004). For flavor encapsulation, the main methods are the kneading and spray drying of the aqueous mixture of flavors and CD. Other methods such as the sublimation and the pressurization are occasionally applied. Spray drying method is quite simple as it involves the preparation and spray drying of flavor and CD mixture while kneading technique consists of the kneading of CD and flavor slurry, dehydration of the kneaded slurry and crushing of dried powder. When CD inclusion complexes of flavors are industrially prepared by spray-drying, the obtained powder is a mixture of the inclusion complex crystals. Figure 4.1 shows the manufacturing process of the inclusion complex powder of wasabi oil in CDs (Yoshii and Shiga 2013). Szente and Szejtli (2004) reported that the total amount of β-CD inclusion complex for various flavors was about 9–14% (w/w), as shown in Table 4.1 and suggested that formation of a mole guest in CD. Molecular weight of β-CD is 1134 g/mol and that of flavor compounds ranges from 112 to 174 g/mol. If flavor molar ratio of inclusion complex to CD was one, weight percentage of flavor to CD might be about 10% (w/w). Encapsulation percentage of the included flavor in CD is lower than that of the emulsified flavor of spray-dried powder. The guest flavor in spray-dried powder is encapsulated in the wall material such as proteins (whey proteins) and carbohydrates (maltodextrins, MD). On the other hand, molecular inclusion complex of flavors in CDs is included in CD cavity. The stability of the encapsulated oil with modified starch, arabic gum, β-CD and other wall materials was compared and the results indicated the shelf life of the included orange oil in β-CD was the most stable (Westing et al. 1988). The initial flavor contents of the model flavor containing 30 flavor components were higher in the order of γ-CD > β-CD > α-CD, while the flavor retentions were in the opposite order of

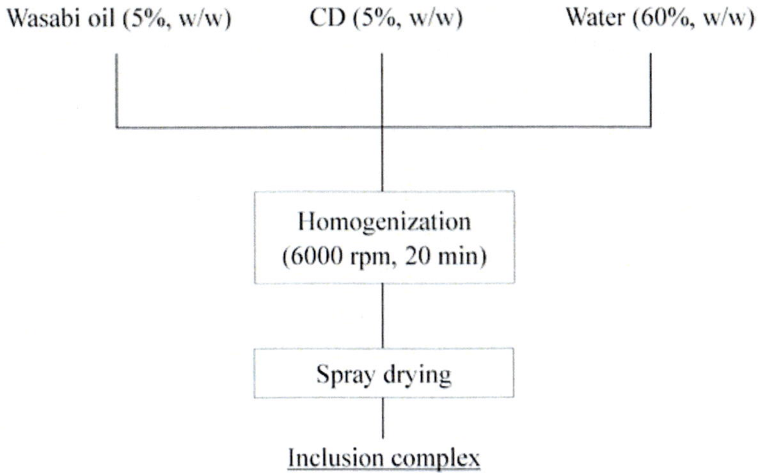

Fig. 4.1 Preparation of inclusion complex of wasabi oil in CDs

Table 4.1 Guest flavor content in inclusion complex powder (adapted with permission from Szente and Szejtli 2004)

Flavor	Flavor content in inclusion complex powder (%, w/w)
Citral	9.5
Citronellal	9
β-Ionone	13.3
Linalool	10.1
Linalyl acetate	9.8
l-Menthol	10.6
Sage oil	10.5
Cinnamon oil	10.4
Jasmine oil	11
Bergamot oil	9.8
Orange oil	8.9
Lemon oil	10.2
Lime oil	9.6
Onion oil	10
Garlic oil	9.4

α-CD > β-CD > γ-CD (Reineccius et al. 2002). The stability of inclusion complexes with 22 volatile nonelectrolytes (guests) was investigated and the stability was correlated to the Van der Waals volume of guest molecule(s) resident maximally in the CD cavity (Sanemasa et al. 1994). A twin-screw kneader was used to form the inclusion complex powder of flavor in CD and concluded that CD enhances flavor retention with a level of 70–100% (Kollengode and Hanna 1997). The inclusion complex powder of lemon oil in CDs was formed by using kneading method and a saturating solution method (Bhandari et al. 1998, 1999). They indicated flavor profiles of the included lemon oil in the β-CD was different from that of lemon oil. This difference of flavor profiles might be due to the inclusion ability (the stability constant). The inclusion competition reaction of multi-flavors like enzyme competitive reaction was carried out (Goubet et al. 2001; Shiga et al. 2002). Arora and Damodaran (2010) investigated competitive binding of off-flavor compounds with soy protein and β-CD in a ternary system, and β-CD can be used effectively for removing soy protein-bound off-flavor carbonyl compounds. The use of β-CD in food products have been summarized and the results pointed the useful technique of flavor inclusion in CDs (Hedges and McBride 1999). The headspace method to examine flavor release from flavor included CD solution was studied (Kant et al. 2004). Yang et al. (2020) studied off-flavor removal from thermal-treated watermelon juice by adsorbent treatment with β-CD.

Water is very important in the inclusion complex formation of guest compound to CDs and is usually added 20–60% (w/w) to total amount of CD slurry. Yoshii et al. (1998) investigated formation of inclusion complex between β-CD and *d*-limonene under high shear stresses of kneading and proposed a mathematical simulation model. The affinity of flavor compounds to CDs is affected by various factors such as the guest and CD concentrations, and kneading or homogenization conditions (rotation speed and pressure and the temperature). We investigated the effect of stirring time of the mixed solution of CD and flavor on the crystal structure of the inclusion complex of *d*-limonene and CDs. Figure 4.2 shows powder X-ray diffractograms patterns of inclusion complexes after stirring time 0, 1 and 24 h. These X-ray diffractograms suggest the formation time of inclusion complex might be very short within 2 h. Inclusion is thought to occur in a very short time (Fig. 4.2). Food flavors contain multi components. These multi-component flavors occurred

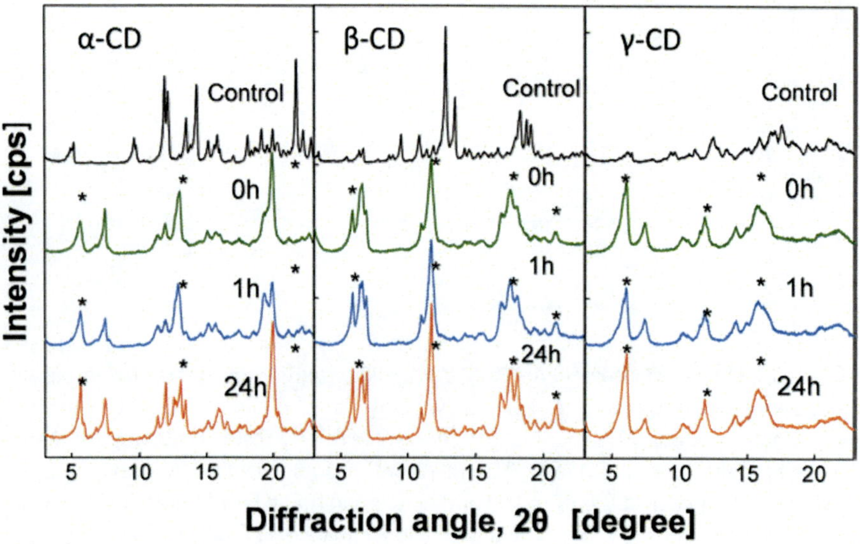

Fig. 4.2 X-ray diffractogram of *d*-limonene inclusion complex powder for various mixing time

the competitive inclusion reactions like the competitive enzyme reactions, Shiga et al. (2002) shows the competitive inclusion of *d*-limonene and methyl caproate into CDs. The excessive addition of CD amount to flavor might affect the included composition in CDs and impair flavor quality. Commercial flavor liquids contain organic solvent and/or surfactants. These compounds also might affect the included composition in CDs. The production of slow-release flavors by formation inclusion complex with CDs and their derivatives was reviewed (Zhu et al. 2019). The result indicates that inclusion complex method is an efficient way for encapsulation of flavor compounds.

4.2.2 Flavor Inclusion Reaction as a Substitution Reaction with Crystal Water in CD Cavity

Yoshii et al. (1994) investigated minimum number of required water molecules to encapsulate *d*-limonene (the guest molecule) in α-, β-, and γ-CD by means of a micro-aqueous method, which is used to study enzymatic reactions in organic solvents (Yamane et al. 1988). This micro-aqueous method requires that the inclusion process can proceed in a liquid guest such as *d*-limonene, in which the CD is suspended. A given amount of water was added to the resulting cloudy suspension of CD to obtain an inclusion complex. With this method it is possible to prevent the formation of large agglomerates of CD paste in which the water cannot be distributed uniformly.

Flavor inclusion process in an aqueous solution could be described the following reaction model (Eq. 4.1).

$$\mathrm{CD}\cdot \mathrm{nH_2O} + \mathrm{Guest\ flavor} \rightarrow [\mathrm{CD}\cdot \mathrm{Guest}] + \mathrm{nH_2O} \tag{4.1}$$

where [CD·Guest] denotes the inclusion complex, and nH_2O on the right are the water molecules released from the CD cavity during the inclusion process. Equation (4.1) was applied to analyze the inclusion process in the micro-aqueous system (Yoshii et al. 1994). The release of this included water would then become the driving force for the inclusion of *d*-limonene into the CD cavity, according to Eq. (4.1). The reaction scheme for Eq. (4.1) implies that the inclusion process is autocatalyzed by water. By this autocatalytic inclusion process with water, the molar ratio of included *d*-limonene to CD (the inclusion fraction) could be expressed the following equation:

$$R = \frac{R_{max}}{1 + a\exp(-bw)} \tag{4.2}$$

where R and $R_{\max}$ are the inclusion fraction and the maximum inclusion fraction. w is the water content. R_0 is the inclusion fraction at w = 0. a and b are the parameters, which mean a = $(R_{\max} - R_0)/R_0$ and b = $kR_{\max}$. k is a proportional constant of the autocatalytic inclusion reaction. By using the intercept of the tangent at the inflection point of R at 0.5, the minimum water content required for inclusion, w_{min} can be calculated as follows.

$$w_{min} = \frac{\ln a - 2}{b} \tag{4.3}$$

w_{min} could be determined with the parameters a and b in Eq. (4.2). Inclusion complexes between *d*-limonene and CDs were prepared with various proportions of water by the micro aqueous method. Figure 4.3 shows the formation of inclusion complexes between *d*-limonene and CDs with various water contents.

The inclusion complexes were barely formed at zero water content with all the CDs. The inclusion ratio increased gradually with the increase of water content, but over a specific moisture content, which is roughly 2, 4, and 10 for α-, β-, and γ-CD, respectively. The inclusion ratio increased exponentially and reached a maximal plateau. The maximum inclusion ratios, R_{max} were 0.4 for α-CD, 0.7 for β-CD, and 1.68 for γ-CD. These inclusion ratios suggested that the inclusion stoichiometries between CDs and *d*-limonene were roughly 2:1 ($R_{max} = 0.5$), 1:1 (1.0), and 2:3 (1.5) for α-, β-, and γ-CD, respectively. *R* fit well with Eq. (4.2). Minimum water content, w_{min}, of each CD was calculated with Eq. (4.3) and is listed in the second column of Table 4.2 as the number of water molecules released from the CD cavity on the inclusion of *d*-limonene.

In Table 4.2, first and third columns indicate the numbers of included water molecules determined by crystal analysis and the numbers of released water molecules for the inclusion calculated using the stoichiometric inclusion ratios of *d*-limonene to for α-, β-, and γ-CD. w_{min} calculated with Eq. (4.3) were comparable with those released to include *d*-limonene in the CD cavity in the crystal state, as shown in the third column of Table 4.2. These results suggested that the inclusion process could be simulated with the auto-catalytic reaction model of water.

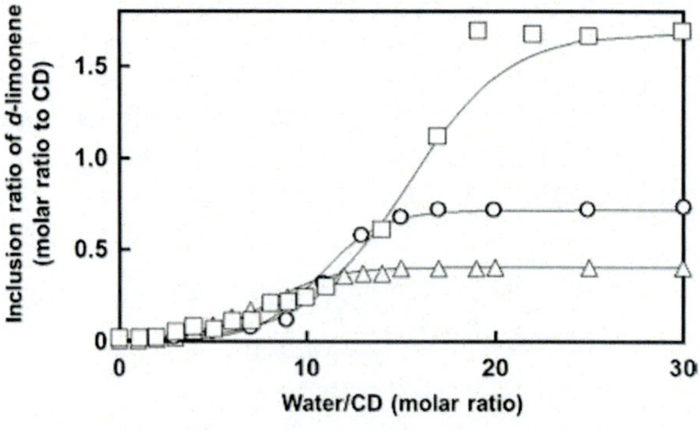

Fig. 4.3 The formation of inclusion complexes between *d*-limonene and CDs with various water contents. △, α-CD; ○, β-CD; □, γ-CD. Solid lines were calculated with Eq. (4.2)

Table 4.2 Minimum numbers of water molecules needed for the formation of inclusion complexes between α-, β-, and γ-CD, and *d*-limonene (adapted with permission from Yoshii et al. 1994)

CD type	Inclusion water molecules/cavity (molar ratio)	w_{min} by Eq. (4.3)	Released water molecular/*d*-limonene (molar ratio)
α-CD	2	4.45	4
β-CD	7	7.43	7
γ-CD	12	9.93	8

Formation of flavor inclusion complexes of CD was also investigated with ethanol under anhydrous conditions (Yoshii et al. 1998). The inclusion ratios of *d*-limonene, acetophenone, or phenyl ethanol in CDs were investigated with ethanol and methanol. This inclusion process could be resembled an enzyme reaction of Michaelis-Menten type with substrate inhibition as following.

$$G + AOH + CD \overset{K_m}{\leftrightarrow} G/AOH/CD \overset{a}{\rightarrow} Complex \tag{4.4}$$

$$CD + AOH \overset{K_i}{\leftrightarrow} CD \bullet AOH \tag{4.5}$$

$$CD \bullet AOH + AOH \overset{K_i}{\leftrightarrow} CD \bullet (AOH)_2 \tag{4.6}$$

$$CD \bullet (AOH)_2 + AOH \overset{K_i}{\leftrightarrow} CD \bullet (AOH)_3 \tag{4.7}$$

$$CD \bullet (AOH)_{n-1} + AOH \overset{K_i}{\leftrightarrow} CD \bullet (AOH)_n \tag{4.8}$$

where G is the guest flavor and AOH is ethanol or methanol. Complex is the inclusion complex. K_m and K_i are the stability constant and inhibition constant, respectively and a is the inclusion reaction constant. n is the number of alcohol molecules adsorbed on CD is assumed the adsorbed alcohol number to free OH in CD, 18 for α-CD, 21 for β-CD, and 24 for γ-CD. By using above model, the inclusion ratio, R could be described in the analogy of enzyme kinetics with substrate inhibition.

$$R = \frac{a}{1 + \frac{K_m}{S} + \frac{S/K_i\left(1 - (S/K_i)^n\right)}{(1 - S/K_i)}} \tag{4.9}$$

where S is the molar ratio of alcohol to CD.

This equation could be well fitted to the inclusion ratio of flavor in alcohol (methanol or ethanol). The adsorption model of alcohol on CD, analogous to the substrate inhibition model of enzyme kinetics could be correlated the inclusion ratio with the amount of alcohol added to CD.

4.2.3 *Flavor Encapsulation by the Blended Encapsulant of CD by Spray Drying*

Encapsulation of hydrophobic flavor compounds, an emulsified flavor solution is commonly dried by spray drying. Spray drying is the good choice of dry flavor manufactures because of its ease and low cost process (Ré 1998). The formation of flavor powder by spray drying, using the blend of β-CD and gum arabic (GA) and MD as encapsulants and the effect of the applied pressure of homogenizer on inclusion of flavors into β-CD was investigated (Shiga et al. 2001).

Two kinds of the inclusion complex slurries of 10% β-CD concentration were prepared with Polytron homogenizer and homogenizing by both Polytron homogenizer and Microfluidizer (82.8 MPa). After spray drying, the inclusion flavor, ethyl *n*-hexanoate or *d*-limonene was measured by FID gas chromatography. Table 4.3 shows the molar ratios of ethyl *n*-hexanoate and *d*-limonene to β-CD encapsulated in the spray-dried powders for the two emulsifying processes. Molar ratios of the included flavor to β-CD increased using the microfluidizer. The inclusion complex was effectively formed at a high pressure applied in Microfluidizer to the feed liquid of spray dryer.

Figure 4.4 shows flavor retentions of ethyl *n*-hexanoate and *d*-limonene during spray drying against total solid concentration in the feed liquid. Flavor retention of *d*-limonene in spray-dried powder was higher than that of ethyl *n*-hexanoate. Spray-dried powder emulsified by GA had higher flavor retention than the powder using β-CD. Figure 4.5 shows SEM picture of the spray-dried powder containing ethyl *n*-hexanoate. Spray-dried powder of β-CD was aggregated crystal of β-CD and the included and emulsified powder of β-CD/GA and β-CD/GA/MD were covered by GA or the blend mixture of GA and MD. The powder of emulsified ethyl *n*-hexanoate had big wrinkles on the surface of the powder.

The formation of rice flavor powder with α-CD by spray drying using the binary mixtures of α-CD and highly branched cyclic dextrin (HBCD) as wall material was investigated (Kawakami et al. 2009). The composition of α-CD and HBCD influenced the flavor retention and the amount of surface oil on the resultant spray-dried powders. The optimum α-CD content was 67%, at which a spray-dried powder of the highest flavor retention and the least amount of surface oil was obtained.

Table 4.3 Effect of microfluidizer on flavor encapsulation yield in CD (adapted with permission from Shiga et al. 2001)

	Homogenizer		Microfluidizer	
Flavor/β-CD molar ratio	Ethyl *n*-hexanoate	*d*-Limonene	Ethyl *n*-hexanoate	*d*-Limonene
1	0.60	0.59	0.77	0.78
2	0.77	0.73	0.84	0.93

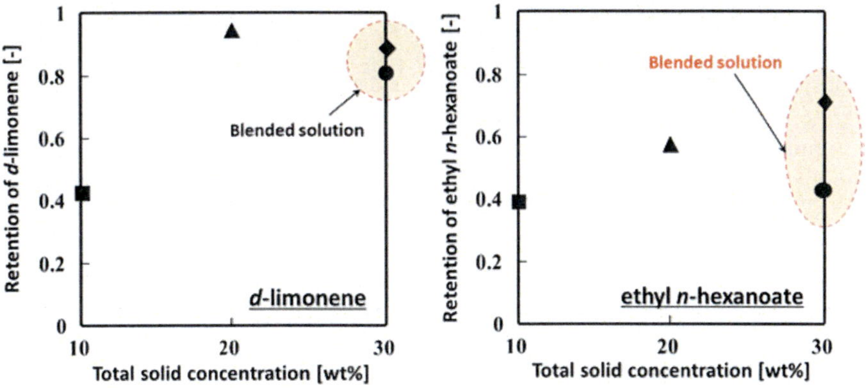

Fig. 4.4 Flavor retentions of ethyl *n*-hexanoate and *d*-limonene during spray drying. ■, β-CD 10% (w/w); ▲, GA 10% (w/w)/MD 10% (w/w); ●, β-CD 10% (w/w)/GA 20% (w/w); ◆, β-CD 10% (w/w)/GA 10% (w/w)/MD 10% (w/w). (Adapted with permission from Shiga et al. 2001)

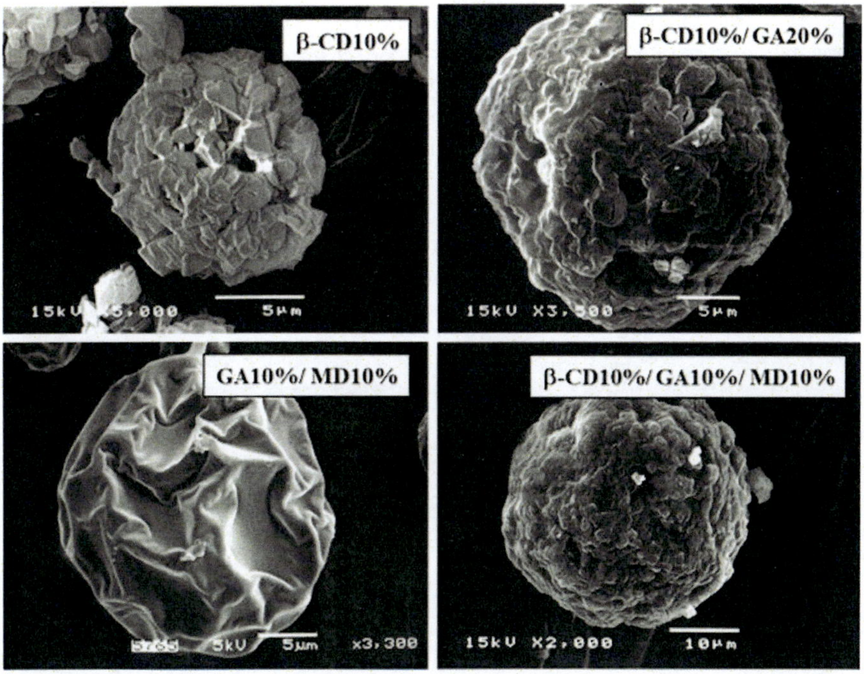

Fig. 4.5 SEM of spray-dried powder containing ethyl *n*-hexanoate

4.2.4 Release Property of the Complex

Inclusion complexes of CDs are considered as the effective system for encapsulating the volatile flavors. Release of flavor from an inclusion complex is important factor imply the success of molecular encapsulation system because it presents the stability of encapsulated flavor during the storage or processing conditions. The release behavior of encapsulated flavor could be controlled based on the desired storage conditions or applications.

The release rate of flavors depends on types of flavors as well as type of CDs. Besides, the temperature, water activity and relative humidity (RH) of storage environment or processing conditions are common factors, which affect the release rate constant of flavors from their inclusion complexes.

Allyl isothiocyanate (AITC), hinokitiol and *d*-limonene were prepared inclusion complex powder by kneading method. Shiga et al. (2000) investigated the flavor release behaviors at various temperature and humidity conditions. Figure 4.6 shows the release behaviors of AITC, *d*-limonene, and hinokitiol included in β-CD at 50 °C and 75% or 90% RH. Release rate was higher in the order of hinokitiol, *d*-limonene, and AITC. Release behavior significantly depended on the type of flavor and CD.

Hinokitiol exhibited an extended release and a good controlled released property even in a higher relative humidity. AITC shows a rapid release, and *d*-limonene has a moderate release rate. Hinokitiol included in β-CD was very stable and was hardly released slowly at high humid condition. This high stability of hinokitiol might be due to the tight inclusion of hinokitiol molecule into CD cavity. Figure 4.7 shows release of *d*-limonene included in α-, β- and γ-CD at 70 °C and 75% RH. *d*-limonene included in α-CD behaves a good controlled released property compare to β- and γ-CD.

Effects of temperature and RH on the release rate of the flavor included in CD were investigated. Figure 4.8 shows effect of RH at 70 °C and temperatures at 75% RH on the release of *d*-limonene included in β-CD. As RH was increased, the

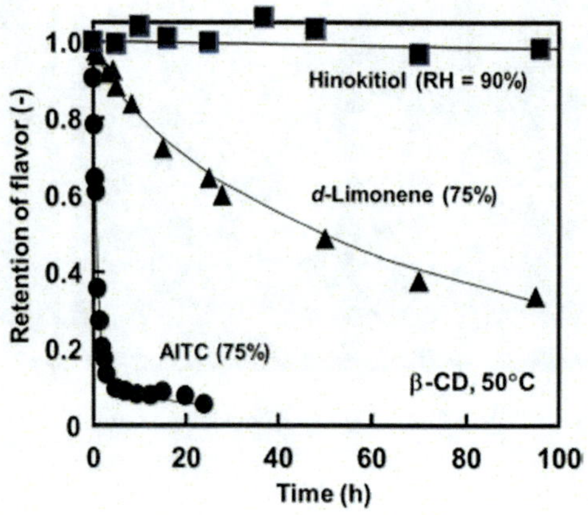

Fig. 4.6 Release of different flavors included in β-CD at 50 °C and 75% RH or 90% RH. ●, AITC (75% RH); ▲, *d*-limonene (75% RH); ■, hinokitiol (99% RH)

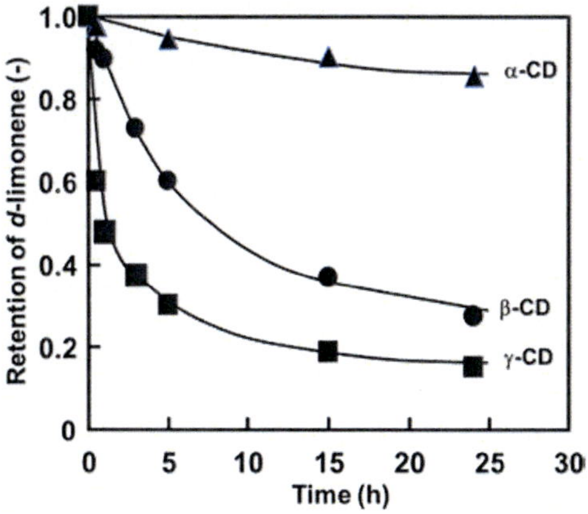

Fig. 4.7 Release of *d*-limonene included in α-, β- and γ-CD at 70 °C and 75% RH

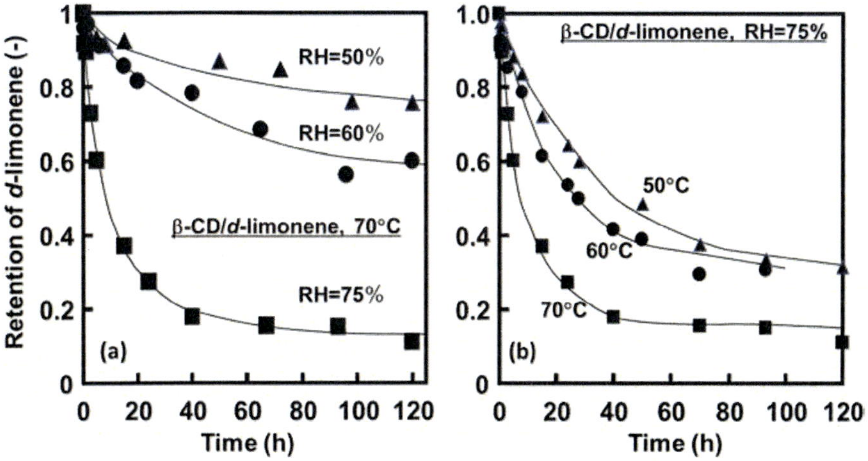

Fig. 4.8 Effect of RH (**a**) and temperature (**b**) on the release of *d*-limonene included in β-CD. (**a**) ▲, 50% RH; ●, 60% RH; ■, 75% RH at 70 °C. (**b**) ▲, 50 °C; ●, 60 °C; ■, 70 °C at 75% RH

release of *d*-limonene was greatly accelerated, particularly between 60% and 75% RH. The effect of RH on the release was more pronounced than temperature. This suggested that the release of *d*-limonene was closely related to the presence of water molecules surrounding the powder.

4.2.4.1 Weibull Model

Weibull model or Avrami equation is an empirical equation which first described by Weibull (1951) for adapting to the dissolution or release process. This equation was applied successfully to drug dissolution or release of pharmaceutical dosage form and the Weibull equation was expressed by accumulation of drug (Costa and Lobo 2001).

$$m = 1 - \exp\left(\frac{-(t - T_i)^b}{a}\right) \tag{4.10}$$

where m is the released fraction, a is a scale parameter, T_i is location parameter (represent the lag time before the onset of the dissolution or release process, and equal zero in most case), b characterizes the curve. If $b = 1$ the curve will be exponential, $b > 1$ the curve will be sigmoid, S-shaped, or $b < 1$ the curve will be parabolic.

The Weibull model was applied with modification for the present of solid-gas and solid-liquid reaction scheme. It was used to describe the release of encapsulated ethyl butyrate (Yoshii et al. 2001), *d*-limonene (Soottitantawat et al. 2010), ethyl *n*-hexanoate and *d*-limonene (Shiga et al. 2001). The Weibull model or Avrami equation was applied to the flavor release behavior in the encapsulated powders as follows:

$$R = \exp\left(-(kt)^n\right) \tag{4.11}$$

where R is the retention of encapsulated flavor, t is storage time, k is release rate constant, and n is the parameter that represents the release mechanism of encapsulated flavor. The parameter $n = 1$ represent the first-order kinetic.

If taking double logarithm of both side of equation, the following equation obtain:

$$\ln\left(-\ln R\right) = n \ln k + n \ln t \tag{4.12}$$

From this equation, the parameter n could be obtained as a slope of $\ln(-\ln R)$ vs $\ln t$ and the release rate constant k can be got from the intercept at $\ln t = 0$.

In the inclusion process, the exchange of guest compound and water molecules in cavity might be essential as described in Sect. 4.2. The release of guest flavor included in CD powder might be the exchange reaction of flavor and moisture water in the environmental condition. The retention of flavor included in CD, R could be calculated with Avrami-Erofeev equation as Eqs. (4.11) and (4.12).

The plots of $\ln(-\ln R)$ against $\ln t$ were shown in Fig. 4.9a for *d*-limonene release from α-, β-, and γ-CD powder at 60 °C and 75% RH. These plots could be obtained the straight lines and release rate constants from the intercept and the mechanism numbers, n from the slope. These mechanism numbers, n was plotted to water vapor concentration for α-, β-, and γ-CD powder as shown in Fig. 4.9b. n = 0.67 for β-CD powder indicated the diffusion release mechanism of *d*-limonene and n = 0.32 and 0.31 for α-CD and γ-CD powder indicated the fast release at the initial storage time and rapidly decreased the release rate to the slow release from CD powder.

Figure 4.10 shows the log-log plot of the release rate constants for the release of AITC (a) and *d*-limonene (b) included in α-, β-, and γ-CD powder against water vapor concentration. In the release rate constant of AITC, release rate constants of

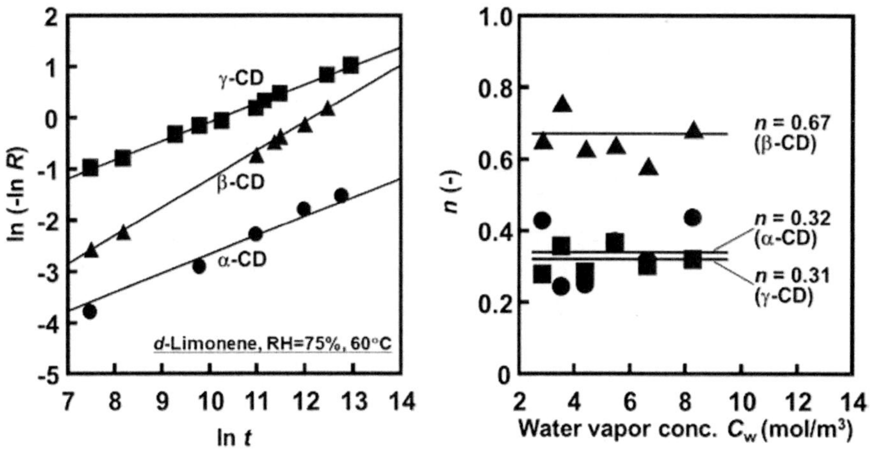

Fig. 4.9 Avrami-Erofeev plots of the *d*-limonene release from α-, β-, and γ-CD powder at 60 °C (**a**) and 75% RH and the relationship between the mechanism number n and water vapor concentration (**b**). ■, γ-CD; ▲, β-CD; ●, α-CD

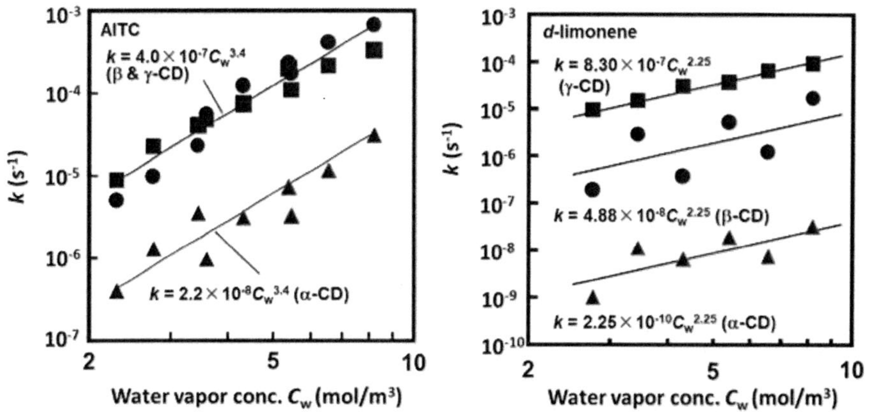

Fig. 4.10 The log-log plot of the release rate constants for the release of AITC (**a**) and *d*-limonene (**b**) included in α-, β-, and γ-CD powder against water vapor concentration. ■, γ-CD; ▲, β-CD; ●, α-CD

β-, and γ-CD powder correlated with same straight lines and were larger than those of α-CD powder. In the release of *d*-limonene, the release rate constants for α- and β-CD were a little scattered and those for γ-CD has a good straight line. Exponential dependences of water vapor concentration for AITC and *d*-limonene were 3.4 and 2.25, respectively. This value for AITC, 3.4 is larger than that for *d*-limonene. This might be the fitness of the guest flavor in CD cavity. Straight molecule of AITC might be easy release from CD cavity.

Although Weibull model was applied for correlation of release behavior of many flavors from CDs, it could not correlate successfully for flavor release with two-step mechanism.

4.2.4.2 Statistical Perceptive Model

Kahlrausch-William-Watt (KWW) equation has also been used for estimating the release behavior of encapsulated materials. The KWW equation was first applied to express the relaxation phenomena in polymer. The release of flavor from spray-dried powders has been considered as the release of hundred oil-droplet with different size distributed in the spray-dried particles. Therefore, with the statistical perspective, the distribution of activation energy of release rate constant was considered and the Gaussian distribution as a function of release rate constant was applied for the flavor release from spray-dried powder (Furuta et al. 2011).

The flavor release from spray-dried powder during storage can be considered as the relaxation of the amorphous particles containing different size of oil droplets. Under the statistical perceptive view, the distribution of different sized oil droplets may follow the distribution of activation energy for release rate and the following equation was suggested for the correlation of the complicated time-dependent phenomena (Furuta et al. 2011).

The flavor release in spray-dried might be correlated with the first-order model with Gaussian distribution of activation energy for release rate constant. This statistical model was the first developed by Kawamura et al. (1981) for express the inactivation kinetics of α-chymotrypsin and glucoamylase covalently bond to water insoluble support in an aqueous system. This equation is also applied successfully to understand the oxidation kinetics of ethyl eicosapentaenoate (EPA) included in CDs and polysaccharide/CD mixtures (Yoshii et al. 2003) and linoleic acid powder (Ishido et al. 2002).

In brief, the oxidation reaction kinetics can be assumed as first-order kinetic:

$$\phi = \exp\left(-k_1 t\right) \tag{4.13}$$

where ϕ is the retention of the core oil in the powder, t is the storage time. The rate constant k_1 is related to the free energy of activation and Eyring equation can be used for reaction rate constant as follow:

$$k_1 = \left(\frac{\kappa_B T}{h}\right)\exp\left(-\frac{\Delta G^{\neq}}{RT}\right) \tag{4.14}$$

where $\Delta G^{\neq}$ is the free energy of activation, κ, h and R are the Boltzmann's constant, Plank's constant, and gas constant, respectively, T is absolute temperature. The free energy of activation $\Delta G^{\neq}$ was assumed to follow the Gaussian with the standard deviation of σ and mean value $\Delta \overline{G}^{\neq}$:

$$f\left(\Delta G^{\neq}\right)=\frac{1}{\sqrt{2\pi}\sigma}\exp\left(-\frac{\left(\Delta G^{\neq}-\Delta\bar{G}^{\neq}\right)^2}{2\sigma^2}\right)=\frac{1}{\sqrt{2\pi}\sigma}\exp\left(-\frac{R^2T^2(lnk_1-ln\bar{k}_1)^2}{2\sigma^2}\right) \quad (4.15)$$

where $\bar{k}_1$ is the average value of k_1 and σ is the standard deviation (J/mol). From these equations the unoxidized fraction of EPA at time *t* could be calculated by the following equation:

$$\phi=\frac{C}{C_0}=\frac{\int_{-\infty}^{+\infty}f(\Delta G^{\neq})\exp(-k_1t)\,d\Delta G^{\neq}}{\int_{-\infty}^{+\infty}f(\Delta G^{\neq})\,d\Delta G^{\neq}}$$
$$=\frac{1}{\sqrt{2\pi}\sigma}\int_{-\infty}^{+\infty}\exp\left(-\frac{R^2T^2\left(lnk_1-ln\bar{k}_1\right)^2}{2\sigma^2}\right)\exp\left(-k_1t\right)\,d\left(lnk_1\right) \quad (4.16)$$

The initial rate of the oxidation reaction can be expressed as follows:

$$k_{10}=-\left.\frac{d\left(\frac{C}{C_0}\right)}{dt}\right|_{t=0}=\bar{k}_1\exp\left(\frac{\sigma^2}{2R^2T^2}\right) \quad (4.17)$$

where k_{10} is an initial rate constant of reaction. Therefore, the unoxidized fraction of EPA can be calculated as follows:

$$\phi=\frac{RT}{\sqrt{2\pi}\sigma}=\exp\left(\frac{-R^2T^2\left(lnk_1-lnk_{10}+\frac{\sigma^2}{2R^2T^2}\right)^2}{2\sigma^2}\right)\exp(-k_1t)\;d\left(lnk_1\right) \quad (4.18)$$

This equation could apply for the simulation of flavor release from CD inclusion complex. The release of allyl sulfide from three types of CD inclusion complexes correlated well with the first-order release rate equation with the Gaussian distribution of ΔG of the release rate constant (Figs. 4.11 and 4.12) (Nguyen and Yoshii 2018). The release rate constants of allyl sulfide from CDs inclusion complexes and standard deviation of activation energies at 100 °C were shown in Table 4.4.

The release time-course of allyl sulfide from the α-, β-, and γ-CD inclusion complex spray-dried powders correlated well using Eq. (4.18) with Gaussian distribution of ΔG of the release rate constant. The data from Table 4.4 showed that the release rate constant k_{10} of the allyl sulfide/α-CD inclusion complex was 5.95×10^{-2} (day^{-1}) and smaller than that of β- and γ-CD (7.51×10^{-1} (day^{-1})). This finding

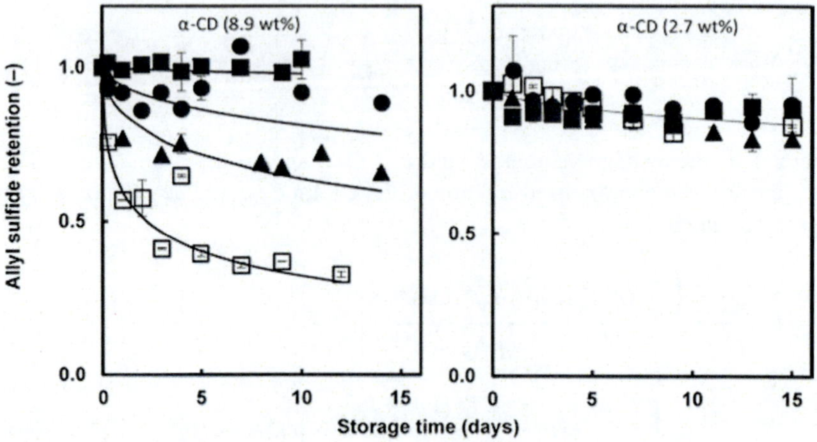

Fig. 4.11 Release behavior of allyl sulfide in α-CD powders with initial moisture content of 8.9 and 2.7 (w/w), respectively, at different RH and 50 °C (■, ●, ▲, □: α-CD with 6%, 40%, 54%, 73% RH, respectively). The solid lines were calculated by Eq. (4.18). (Adapted with permission from Nguyen and Yoshii 2018)

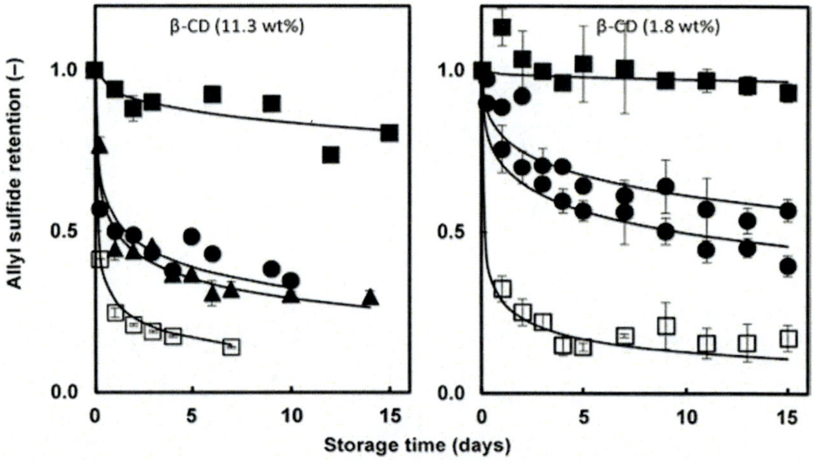

Fig. 4.12 Release behavior of allyl sulfide in β-CD powders with initial moisture content of 11.3 and 1.8 (w/w), respectively, at different RH and 50 °C (■, ●, ▲, □: β-CD with 6%, 40%, 54%, 73% RH, respectively). The solid lines were calculated by Eq. (4.18). (Adapted with permission from Nguyen and Yoshii 2018)

implies that the allyl sulfide/α-CD inclusion complex was much more stable than that of β- and γ-CD when stored in hot conditions.

Under the different RH (6%, 40%, 54%, and 73%) and at 50 °C, the release rate of allyl sulfide from α-, β-, and γ-CD inclusion complex had two steps, an initial rapid release rate followed by a slow release rate. The obtained data (Figs. 4.11 and 4.12) was fit with two-steps release mechanism that could describe by using a

Table 4.4 Value of release rate constant k_{10} and standard deviation of ΔG's distribution σ of allyl sulfide release from inclusion complex powders at 100 °C (adapted with permission from Nguyen and Yoshii 2018)

Types of CD	Alpha	Beta	Gamma
k_{10} (day^{-1})	5.95×10^{-2}	7.51×10^{-1}	7.51×10^{-1}
σ	0.8	0.99	0.99

first-order release rate equation with Gaussian distribution of activation energy of release rate constant.

The effect of initial moisture content of CD-inclusion complexes on the allyl sulfide release rate was investigated using α-, β-, and γ-CD powders with two moisture contents (2.7% and 8.9% (w/w) for α-CD; 1.8% and 11.3% (w/w) for β-CD; and 2.7% and 8.8% (w/w) for γ-CD) at 50 °C and 6%, 40%, 54%, and 73% RH (Figs. 4.11 and 4.12). In all inclusion complexes, the release of allyl sulfide followed the two-steps mechanism. The standard deviation of ΔG's distributions σ was 3, 3.9, and 2.3 for α-, β-, and γ-CD, respectively. The release rate of allyl sulfide from β-CD inclusion complex was significantly affected by the increase in RH at the two investigated initial moisture content of the inclusion complexes (Fig. 4.12). Similarity, the rate of allyl sulfide from γ-CD inclusion complex had the same trend (data not shown). In the case of α-CD, when the initial moisture content 2.7% (w/w), RH of environment did not affect the release rate (Fig. 4.11). The different of allyl sulfide release behavior from the CD-inclusion complexes at the lower initial moisture content might be involved in the adsorption of water molecule at hydroxyl group located on the surface or intermolecular space of cavity of CDs.

Generally, flavor release from inclusion complex depends on environment conditions as well as types of CDs and flavors. The release rate could be correlated by empirical or statistic perceptive model.

4.3 Applications of the Flavor-CD Complexes in Food Production

Flavor-CD-inclusion complexes have found an important place in the food industry due to their characteristics to protect, stabilize, solubilize, mask, and control flavor release. CDs could pass through human digestion system (Astray et al. 2020) therefore many industrial flavor/CDs complexes-based products are produced and sold on market. They are powdered flavors (e.g., citrus fruits, vanilla, and apple), spices (ginger, garlic, mustard, horseradish), and other herbs (peppermint or basil, and green tea). These products play important roles in bakery, beverages, instant food production (Trifković et al. 2016).

Table 4.5 Some recent researches of application of flavor-CD complexes in food industry

Applications	Role of CD	References
Yoghurt	Masks goaty flavor	Young et al. (2012)
Soy milk	Improve the flavor quality	Shi et al. (2017)
Soy milk	Remove soy protein-bound off-flavor carbonyl compounds	Arora and Damodaran (2010)
Soy milk	Remove soy protein isolated bound precursors and the volatile compounds	Damodaran and Arora (2013)
Non-alcoholic lemon juice model	Aroma improvement	do Carmo et al. (2017)
Fish oil	Improvement sensory profile	Serfert et al. (2010)
Baking application	Enhance the heat stability of cinnamon flavor	Gong et al. (2020)

There are many examples of flavor-CD inclusion complexes in food industry. Cyroma-line is a product of flavored sugar for baking produced in Hungary. In this product, CDs play a role as a flavor preservation during heating. In Japan, Flavono—a chewing gum produced with CDs in order to stabilize flavor or Gymet—a Japanese commercial dietary fiber drink in which unpleasant taste is masked by CD complexation. Another product is a Japanese instant tea drink (Stick Lemon) in which CDs act as flavor preservation. FlavorAktiv standard kit, in which CDs are also used, to preserve beer flavor standards produced in Great Britain. In Germany, a complex of curcumin with γ-CD was used as food supplement with the trade name of CAVAMAX® W8 Curcumin. In bakery production, Vivid® Baking Flavor Powder is designed for maintaining flavor under high temperature (Trifković et al. 2016). Besides, some recent applications of flavor-CD inclusion complexes (in the last 10 years) are presented in Table 4.5.

Recently, applications of flavor encapsulation in CD in food industry have been developed. Most of them are in confectionery and instant drink with purpose of prolonging the desired flavor, enhancing the stability or masking off-flavor.

4.4 Conclusion and Future Perspectives

Flavor inclusions with CD are also widely used in non-food products. In the case of food, CD is limited to the use of α-, β-, and γ-CD and their applications are also limited. Recently, in some countries, the Aging population with low ability of olfactory ability has been increasing. Therefore, developments of foods with a slow release of flavors during chewing are carried. The use of flavors in foods has also increased since it became clear that flavors affect the taste of foods, especially in salt sensitivity. By the increasing number of vegans need, the research on flavoring foods and flavor stabilization using cyclodextrin are essential.

References

Arora A, Damodaran S (2010) Competitive binding of off-flavor compounds with soy protein and β-cyclodextrin in a ternary system: a model study. J Am Oil Chem Soc 87(6):673–679

Astray G, Mejuto J, Simal-Gandara J (2020) Latest developments in the application of cyclodextrin host-guest complexes in beverage technology processes. Food Hydrocoll 106:105–882

Bhandari BR, D'Arc BR, Thi Bich LL (1998) Lemon oil to β-cyclodextrin ratio effect on the inclusion efficiency of β-cyclodextrin and the retention of oil volatiles in the complex. J Agric Food Chem 46(4):1494–1499

Bhandari BR, D'Arc BR, Padukka I (1999) Encapsulation of lemon oil by paste method using β-cyclodextrin: encapsulation efficiency and profile of oil volatiles. J Agric Food Chem 47(12):5194–5197

Costa P, Lobo JMS (2001) Modeling and comparison of dissolution profiles. Eur J Pharm Sci 13(2):123–133

Damodaran S, Arora A (2013) Off-flavor precursors in soy protein isolate and novel strategies for their removal. Annu Rev Food Sci Technol 4:327–346

Del Valle EM (2004) Cyclodextrins and their uses: a review. Process Biochem 39(9):1033–1046

do Carmo CS, Pais R, Simplício AL et al (2017) Improvement of aroma and shelf-life of non-alcoholic beverages through cyclodextrins-limonene inclusion complexes. Food Bioprocess Technol 10(7):1297–1309

Estevinho BN, Rocha F (2017) Chapter 1: A key for the future of the flavors in food industry: nanoencapsulation and microencapsulation. In: Grumezescu A, Oprea AE (eds) Nanotechnology applications in food: flavor, stability, nutrition, and safety. Academic Press, Elsevier, London, pp 1–19

Furuta T, Soottitantawat A, Neoh TL et al (2011) Chapter 1: Effect of microencapsulation on food flavors and their releases. In: Dehavastin S (ed) Physicochemical aspects of food engineering and processing. CRC Press, Boca Raton, FL, pp 1–40

Gong C, Lee MC, Godec M et al (2020) Ultrasonic encapsulation of cinnamon flavor to impart heat stability for baking applications. Food Hydrocoll 99:105–316

Goubet I, Dahout C, Sémon E et al (2001) Competitive binding of aroma compounds by β-cyclodextrin. J Agric Food Chem 49(12):5916–5922

Hedges AR (1998) Industrial applications of cyclodextrins. Chem Rev 98(5):2035–2044

Hedges A, McBride C (1999) Utilization of β-cyclodextrin in food. Cereal Foods World 44(10):700–704

Ishido E, Hakamata K, Minemoto Y et al (2002) Oxidation process of linoleic acid encapsulated with a polysaccharide by spray-drying. Food Sci Technol Res 8(1):85–88

Kant A, Linforth RS, Hort J et al (2004) Effect of β-cyclodextrin on aroma release and flavor perception. J Agric Food Chem 52(7):2028–2035

Kawakami K, Fujita A, Mikami T et al (2009) Formation of rice flavor powder with α-cyclodextrin by spray drying. Eur Food Res Technol 229(2):239–245

Kawamura Y, Nakanishi K, Matsuno R et al (1981) Stability of immobilized α-chymotrypsin. Biotechnol Bioeng 23(6):1219–1236

Kollengode AN, Hanna MA (1997) Cyclodextrin complexed flavors retention in extruded starches. J Food Sci 62(5):1057–1060

Lopes DB, Speranza P, Macedo GA (2016) Chapter 18: A new approach for flavor and aroma encapsulation. In: Grumezescu AM (ed) Novel approaches of nanotechnology in food. Academic Press, Elsevier, London, pp 623–661

Nguyen TVA, Yoshii H (2018) Release behavior of allyl sulfide from cyclodextrin inclusion complex of allyl sulfide under different storage conditions. Biosci Biotechnol Biochem 82(5):848–855

Potdar SB, Landge VK, Barkade SS et al (2020) Chapter 16: Flavor encapsulation and release studies in food. In: Sonawane SH, Bhanvase BA, Sivakumar M (eds) Encapsulation of active molecules and their delivery system. Elsevier, New Delhi, pp 293–321

Rakmai J, Cheirsilp B (2016) Continuous production of β-cyclodextrin by cyclodextrin glycosyltransferase immobilized in mixed gel beads: comparative study in continuous stirred tank reactor and packed bed reactor. Biochem Eng J 105:107–113

Ré MI (1998) Microencapsulation by spray drying. Dry Technol 16(6):1195–1236

Reineccius T, Reineccius GA, Peppard T (2002) Encapsulation of flavors using cyclodextrins: comparison of flavor retention in alpha, beta, and gamma types. J Food Sci 67(9):3271–3279

Sanemasa I, Wu Y, Koide Y et al (1994) Stability on drying of cyclodextrin precipitates of volatile nonelectrolytes. Bull Chem Soc Jpn 67(10):2744–2750

Serfert Y, Drusch S, Schwarz K (2010) Sensory odour profiling and lipid oxidation status of fish oil and microencapsulated fish oil. Food Chem 123(4):968–975

Shi X, Lü Y, Guo S (2017) Effects of heat treatment and β-cyclodextrin addition on soymilk flavor. Trans Chin Soc Agric Eng 33(8):293–300

Shiga H, Furuta T, Yoshii H et al (2000) Release characteristics of allyl isothiocyanate encapsulated in cyclodextrins. In: Proceedings of the 10th international symposium on cyclodextrins, Ann Arbor, MI, pp 553–558

Shiga H, Yoshii H, Nishiyama T et al (2001) Flavor encapsulation and release characteristics of spray-dried powder by the blended encapsulant of cyclodextrin and gum arabic. Dry Technol 19(7):1385–1395

Shiga H, Nishimura H, Yoshii H et al (2002) Competitive inclusion of binary flavor mixtures in α-, β-and γ-cyclodextrin. J Chem Eng Jpn 35(5):468–473

Soottitantawat A, Yoshii H, Furuta T (2010) Chapter 42: Effect of water activity on the release characteristics of encapsulated flavor. In: Reid DS, Sajjaanantakul T, Lillford PJ, Charoenrein S (eds) Water properties in food, health, pharmaceutical and biological systems: ISOPOW 10. Wiley-Blackwell, Singapore, pp 499–505

Szente L, Szejtli J (2004) Cyclodextrins as food ingredients. Trends Food Sci Technol 15(3–4):137–142

Trifković K, Đorđević V, Balanč B et al (2016) Chapter 9: Novel approaches in nanoencapsulation of aromas and flavors. In: Grumezescu AM (ed) Encapsulations: nanotechnology in the agri-food industry, vol 2. Academic Press, Elsevier, Amsterdam, pp 363–419

Weibull W (1951) A statistical distribution function of wide applicability. J Appl Mech 18:293–297

Westing LL, Reineccius GA, Caporaso F (1988) Chapter 12: Shelf life of orange oil: effects of encapsulation by spray-drying, extrusion, and molecular inclusion. In: Risch SJ, Reineccius GA (eds) Flavor encapsulation. ACS Publications, Washington, DC, pp 110–121

Yamane T, Kojima Y, Ichiryu T et al (1988) Biocatalysis in a microaqueous organic solvent. Ann N Y Acad Sci 542:282–293

Yang X, Yang F, Liu Y et al (2020) Off-flavor removal from thermal-treated watermelon juice by adsorbent treatment with β-cyclodextrin, xanthan gum, carboxymethyl cellulose sodium, and sugar/acid. LWT 131:109–775

Yoshii H, Shiga H (2013) Flavor release from the inclusion complex of flavor in cyclodextrins and its application (Japanese). In: Terao K, Ikeda T (eds) Science and technology of cyclodextrin. CMC Publishing Co. Ltd., Tokyo, pp 233–242

Yoshii H, Furuta T, Yasunishi A et al (1994) Minimum number of water molecules required for inclusion of *d*-limonene in the cyclodextrin cavity. J Biochem 115(6):1035–1037

Yoshii H, Kometani T, Furuta T et al (1998) Formation of inclusion complexes of cycldextrin with ethanol under anhydrous conditions. Biosci Biotechnol Biochem 62(11):2166–2170

Yoshii H, Soottitantawat A, Liu XD et al (2001) Flavor release from spray-dried maltodextrin/gum arabic or soy matrices as a function of storage relative humidity. Innov Food Sci Emerg Technol 2(1):55–61

Yoshii H, Furuta T, Fujiwara M et al (2003) Oxidation stability of powdery ethyl eicosapentaenoate included in cyclodextrins and polysaccharide/cyclodextrin mixtures. Jpn J Food Eng (Japan) 4(1):25–31

Young O, Gupta R, Sadooghy-Saraby S (2012) Effects of cyclodextrins on the flavor of goat milk and its yogurt. J Food Sci 77(2):S122–S127

Zhu G, Zhu G, Xiao Z (2019) A review of the production of slow-release flavor by formation inclusion complex with cyclodextrins and their derivatives. J Incl Phenom Macrocycl Chem 95:17–33

Chapter 5
Encapsulation of Colors and Pigments

Afroza Sultana and Hidefumi Yoshii

5.1 Introduction

A wide range of colors are used in food products. Color compounds are characterized by their chemical structures. These compounds have various physiological functions in addition to exhibiting those colors. We can visually perceive electromagnetic wavelengths from 380 to 780 nm, called visible light. The chromophores of biological substances (organic compounds) are absorbed in the visible light region. Color is an important characteristic that determines the values and acceptability of fruits and vegetables. The basic carbon skeletons of carotenoids consist of repeating isoprene units. A single bond between double bonds in a conjugated double bond (Fig. 5.1) shows physicochemical properties as a double bond. Chemical compounds with a conjugated double bond can also be represented by a structural formula in which π electrons are spread throughout the molecule (for example, benzene). As a result, these compounds absorb light from the ultraviolet to the visible regions and exhibit a color characteristic for each substance and lipophilic property.

Natural pigments can be classified according to their chemical structure, as shown in Fig. 5.2. Pigment categories include porphyrin, carotenoid, flavonoid, quinoid and other color compounds.

Carotenoids are tetraterpene pigments that exhibit the yellow, orange, red and purple colors of flowers, fruits, vegetables and plants. These pigments are an

A. Sultana
Department of Food Processing and Engineering, Faculty of Food Science and Technology, Chattogram Veterinary and Animal Sciences University, Chattogram, Bangladesh
e-mail: afroza-sultana@cvasu.ac.bd

H. Yoshii (✉)
Department of Food Science and Human Nutrition, Setsunan University, Osaka, Japan
e-mail: hidefumi.yoshii@setsunan.ac.jp

T. M. Ho et al. (eds.), *Functionality of Cyclodextrins in Encapsulation for Food Applications*, https://doi.org/10.1007/978-3-030-80056-7_5

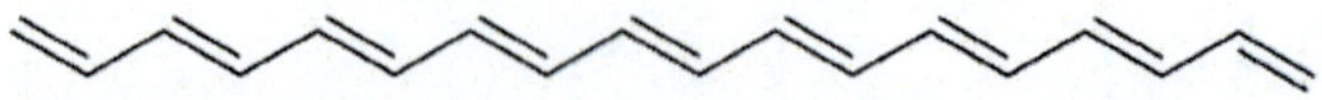

Fig. 5.1 Conjugated double bond

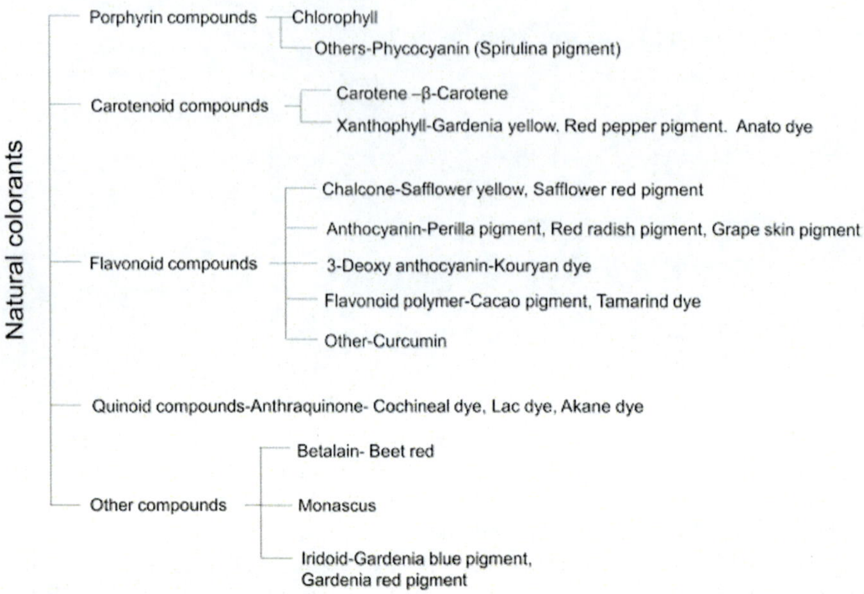

Fig. 5.2 Categories of natural colorants

important group of lipid-soluble pigments that are widely distributed in nature. Carotenoids are present in photosynthetic bacteria, some species of archaea and fungi, algae, plants, and animals. Most carotenoids consist of eight isoprene units with a 40-carbon skeleton. Their general structures commonly consist of a polyene chain with nine conjugated double bonds and an end group at both ends of the polyene chain, as shown in Fig. 5.3. Mortensen (2006) summarized carotenoids and other pigments as natural colorants. Tanaka et al. (2012) reviewed cancer chemoprevention by carotenoids and indicated the beneficial effects of carotenoid-rich vegetables and fruits in relation to cancer risk. These carotenoids have antioxidant action, deactivating free radicals and quenching reactive oxygen species. The beneficial effects of carotenoids have powerful appeal for their application in food products.

Flavonoids are polyphenol compounds and widely distributed in the plant kingdom as pigments for flowers and leaves. The basic skeletons of flavonoids have 15 carbon atoms and a 1,3-diphenylpropanoid skeleton ((E)-1,3-diphenylprop-2-en-1-one), in which two benzene rings are linked by three carbon atoms (Fig. 5.4). Flavonoid is a general term for substances with this skeleton. Many compounds are flavonoids. Among them, chalcone, flavone, flavonol, and anthocyanidinine skeletons contain compounds that exist as pigments.

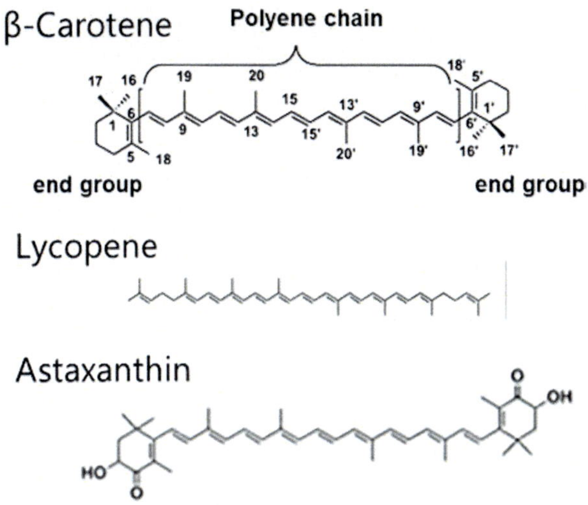

Fig. 5.3 Chemical structures of β-carotene, lycopene and astaxanthin

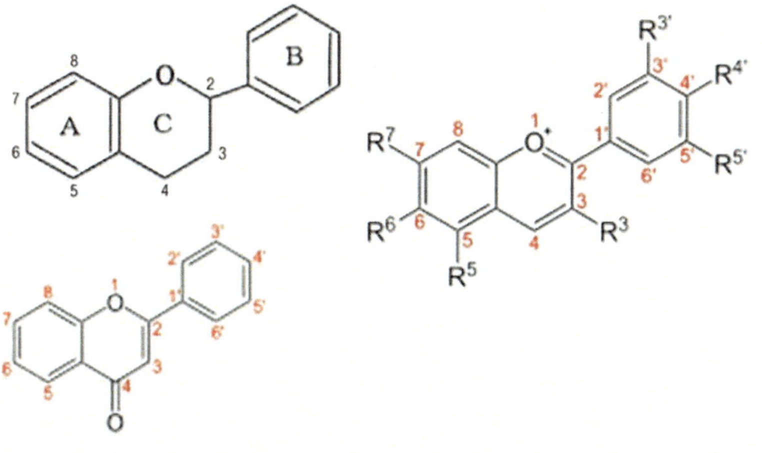

Fig. 5.4 Basic flavonoid skeleton (flavonoids can be subdivided into different subgroups depending on the carbon of the C ring on which the B ring is attached and the degree of unsaturation and oxidation of the C ring)

Anthocyanins are saccharides linked to anthocyanidins (glycosides). Cocoa pigments are condensed tannins with flavonoids. Since the structure is not clear, it is classified as a flavonoid system.

Curcumin is the turmeric pigment and a hydrophobic yellow-orange polyphenol derived from the rhizome of the herb *Curcuma longa*. Curcumin is also a very interesting pharmacological compound due to its many pharmacological applications (Mohan et al. 2012; Singh et al. 2010; Yallapu et al. 2010).

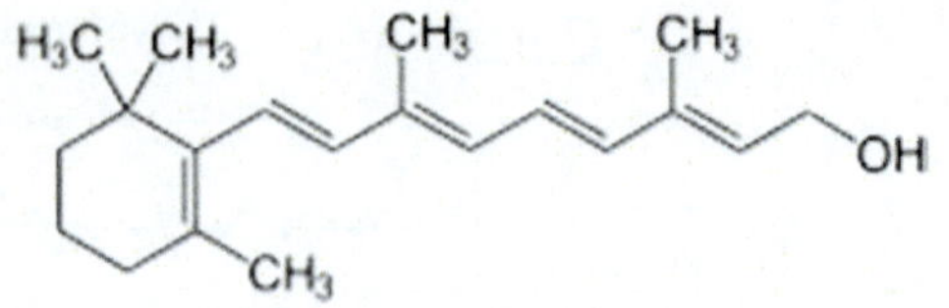

Fig. 5.5 Chemical structure of retinyl palmitate (vitamin A)

Vitamin A is a lipid-soluble vitamin that is required for vision, reproduction, and maintenance of epithelial barriers and immunity. Vitamin A does not exist in plants. However, carotenoid pigments in plants can be converted into vitamin A, although the vitamin A equivalency of carotene is very low (Van Loo-Bouwman et al. 2014). The most common forms of vitamin A precursors present in food are retinol and retinyl esters, which are stored as esters of fatty acids, predominantly as retinyl palmitate. The absorption spectrum of β-carotene (a carotenoid pigment) includes violet and blue-green light, as is indicated by its peaks at around 450 and 475 nm. The chemical structure of retinyl palmitate is shown in Fig. 5.5.

5.2 Preparation and Characteristics of Color- and Pigment-Cyclodextrin Complexes

Yoshii (2004) reviewed the application of cyclodextrins (CDs) in food processing. There are several types of formation methods for inclusion complexes between pigment-based compounds and CDs, such as (1) spray drying, (2) freeze drying, (3) kneading, (4) crystal conversion, (5) chemical inclusion, and (6) sublimation methods using CDs. The physical inclusion method is a vitrification method using carbohydrates or proteins as wall material containing CDs. The steps of the inclusion method consist of (1) the formation of inclusion complexes using saturated solutions of CDs, slurry solutions, or pastes containing CDs and saccharides, (2) powdering with drying and (3) pulverization. Powdering using crystal conversion consists of two processes: a cocrystallization method and a solidification method with pulverization. Figure 5.6 shows the powder structures obtained by various formation methods of inclusion complexes (Fang and Bhandari 2010). They showed the technologies of encapsulation of polyphenols, including spray drying, coacervation, liposome entrapment, inclusion complexation, cocrystallization, nanoencapsulation, freeze drying, yeast encapsulation and emulsion.

In the kneading method, CD syrups are used as by-products to form pure CDs, which contain a mixture of CDs and low molecular weight dextrins. By mixing and kneading the functional ingredients and CD syrups, functional ingredients such as color pigments are included in CD cavities. After kneading or mixing, the kneaded solution or paste is dried by spray drying, vacuum drying or freeze drying. In this section, the characteristics of kneading are described.

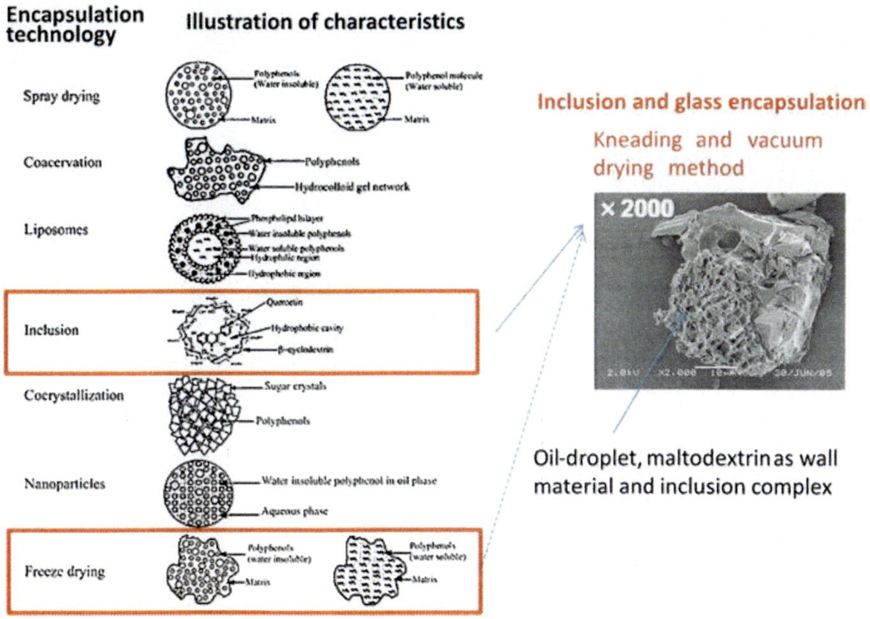

Fig. 5.6 Illustration of powder structures with various formation methods. (Adapted with permission from Fang and Bhandari 2010)

5.2.1 Techniques for the Complex Preparation of Vitamin A

CD syrup is produced as a by-product in the CD manufacturing process. The characteristics of CD syrup, with the commercial name SL-20 (Nippon Shokuhin Kako), are indicated. This syrup contains 19 wt% equal amounts of a mixture of α-, β-, and γ-CD and low molecular weight maltodextrins of approximately 20 DE. The water content of the syrup is approximately 25%. In the formation of inclusion complexes such as vitamin A, this high-concentration syrup and vitamin A are mixed, and the mixture paste is kneaded at a low water content. High shear force enables efficient formation of inclusion complexes of CD and functional compounds with low water content. Kneading has the advantage of decreasing the drying energy of the kneaded paste. This kneading method is suitable for small-scale production.

The effect of kneading time on the encapsulated content of vitamin A (retinyl palmitate, RETP) is shown in Fig. 5.7. Formation of the inclusion complex in the kneading method needs kneading time to increase contact opportunities of vitamin A and CD. The encapsulant agents used were maltodextrin (MD) (DE = 11 or 28), CD syrup, a mixture of 7.5% or 15% MD (DE = 11) and CD syrup and a mixture of 15 wt% γ-CD to CD syrup and a mixture of γ-CD and MD (DE = 28). Table 5.1 shows the mixing compositions of encapsulant reagents and RETP (Koeda et al. 2014). A high-concentration syrup of this mixture (approximately 440 g) was added to the kneader (PNV-1H, Irie Shokai Co., Ltd., Tokyo) at 70 °C

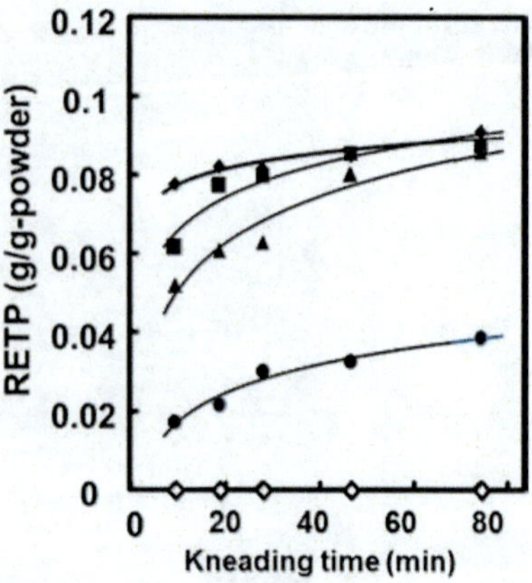

Fig. 5.7 Effect of kneading time on the encapsulation content of RETP (◇, γ-CD 15 wt% + MD; ●, CD syrup; ▲, CD syrup + MD 7.5 wt%; ■, CD syrup + MD 15 wt%; ◆, CD syrup + γ-CD 15 wt%). (Adapted with permission from Koeda et al. 2014)

Table 5.1 Composition of RETP encapsulant

Encapsulant	CD syrup	CD syrup + MD (DE = 11) 7.5 wt%	CD syrup + MD (DE = 11) 15 wt%	CD syrup + γ-CD 15 wt%	MD (DE = 28) + γ-CD 15 wt%	Spray drying
CD syrup	88	80.5	73	73		60
MD (DE = 11)		7.5	15		58	11
MD (DE = 28)					15	
γ-CD				15	15	
Capsul						13
RETP	12	12	12	12	12	16

for 20 min. Then, the kneader was cooled to 50 °C. Approximately 60 g of RETP was added at 60 °C, and the syrup was kneaded with RETP at 65 rpm for a predetermined time. After the predetermined time, the kneading paste was sampled and vacuum dried at room temperature. The dried powder was crushed into small particles. The RETP content depended more strongly on the mixture excipients used than the kneading time.

The viscosity of the kneading paste, the RETP content of the dry powder, and the reconstituted average oil droplet diameter are shown in Table 5.2 (Koeda et al. 2014). In the case of the encapsulant reagent with MD (DE = 28) containing 15% γ-CD, RETP could not be included at all. The reconstituted oil droplet diameter for γ-CD and MD (DE = 28) as the encapsulant was approximately 37 μm. Conversely, the reconstituted oil droplet diameter for CD syrup and mixtures with CD syrup was approximately 1–2 μm. These phenomena suggest that α-CD and β-CD can form RETP inclusions and have emulsifying ability.

Table 5.2 Physical properties of the encapsulant powder

Encapsulant	Paste viscosity (mPa s)	RETP content in powder (g-RETP/g-powder)	Reconstituted oil-droplet size (μm)
CD syrup	112	0.092	1.64
CD syrup + MD (DE = 11) 7.5 wt%	131	0.09	1.45
CD syrup + MD (DE = 11) 15 wt%	252	0.089	1.01
CD syrup + γ-CD 15 wt%	143	0.04	1.08
MD (DE = 28) + γ-CD 15 wt%	147	0	37

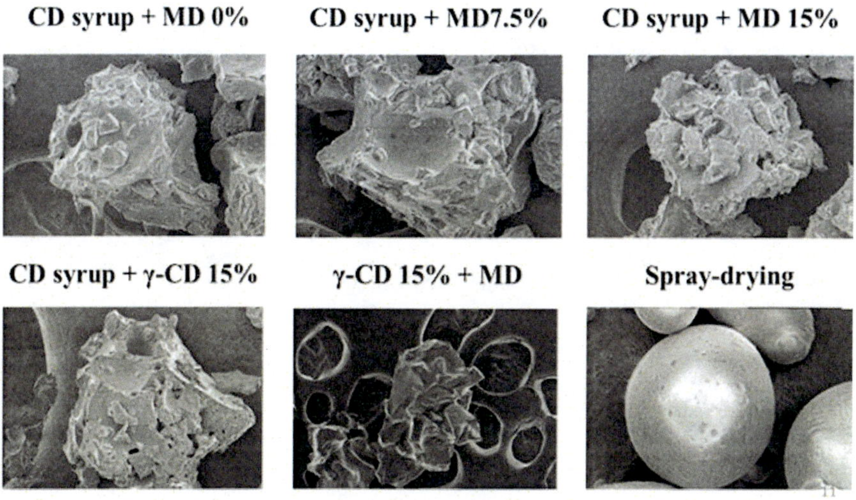

Fig. 5.8 Scanning electron micrographs using various encapsulants containing CD syrup

5.2.2 RETP Stability in the Encapsulant Using CD Syrup

Figure 5.8 shows scanning electron micrographs of RETP powder prepared with various encapsulants and surface photographs of spray-dried powder using a modified starch Capsul®. In the kneading method, kneading pastes were vacuum-dried. CD syrups can entrain air bubbles and create foam in the paste. The crushed product had large bubbles and fine pores of oil droplets.

The RETP stabilities of those powders were investigated at 60 °C. Figure 5.9 shows the RETP stability in various encapsulants. The solid lines in the figure are correlated with the Avrami equation as follows:

$$R = \exp\left[-(kt)^{n}\right] \tag{5.1}$$

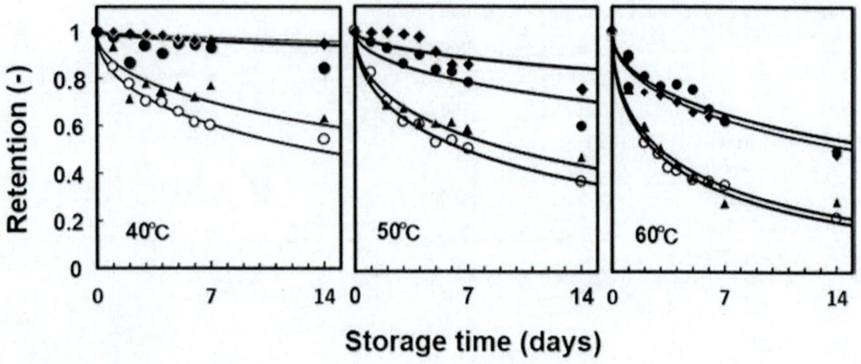

Fig. 5.9 RETP stability in various encapsulants at 40, 50 and 60 °C (○ CD syrup; ▲ CD syrup and MD 7.5%; ● CD syrup and MD 15%; △ CD syrup and γ-CD 15%; ◆ spray drying). (Adapted with permission from Koeda et al. 2014)

Fig. 5.10 Arrhenius plot for the activation energies of the degradation rate constants of RETP in the powders (○ CD syrup; □ CD syrup and MD 7.5%; ● CD syrup and MD 15%; △ CD syrup and γ-CD 15%; □ spray drying). (Adapted with permission from Koeda et al. 2014)

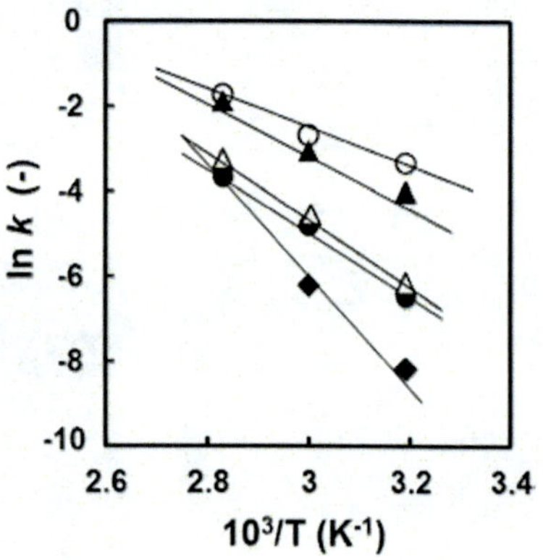

Here, R is the RETP retention, k is the RETP decomposition rate constant in the powder, n is a parameter of the RETP decomposition mechanism, and t is time. When n is a value near 0.54, it is considered that RETP is decomposed by the diffusion mechanism.

The RETP stability in the spray-dried powder using the modified starch Capsul was the most stable. RETP oil droplets were encapsulated in spray-dried powder. The RETP stability behaviors using CD syrup + 15 wt% MD (DE = 28) was almost the same as those of the spray-dried powder. RETP oil was exposed to crushed powder after vacuum drying and pulverization. These data suggest that the included RETP in CDs might be stable. The decomposition rate constants of RETP are shown in Fig. 5.10.

The activation energies of the decomposition rate constants are 35.7 kJ/mol for CD syrup only, 47.9 kJ/mol for CD syrup and 7.5% MD (DE = 28), and 62.6 kJ/mol for CD syrup and MD 15%. These activation energies depended on the MD content and had a linear relationship with the MD content, as shown in Fig. 5.11. This

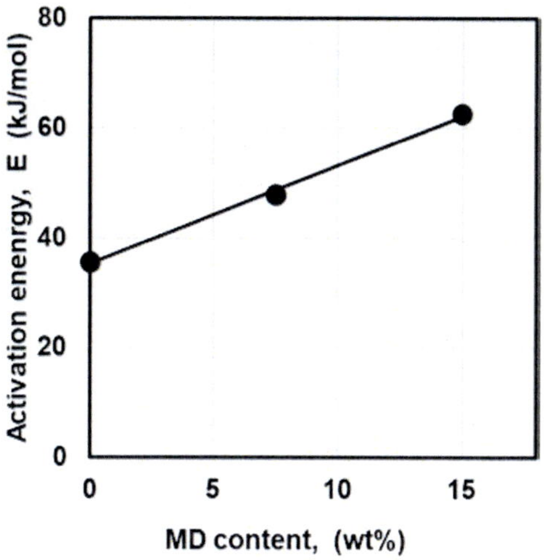

Fig. 5.11 Activation energy for MD (DE = 11) content in the powders. (Adapted with permission from Koeda et al. 2014)

finding suggests that the decomposition rate constant might be dependent on the oxygen diffusion coefficient in the powder.

The natural logarithms of these frequency factors, which were obtained with these Arrhenius plots, were correlated with the activation energies by the following equation:

$$\ln k_0 = 0.314E - 1.084 \tag{5.2}$$

The good correlation suggests that the relationship between the RETP decomposition rate constant in the encapsulant powder of RETP using CD syrup and the activation energy had a chemical compensation effect. This relationship suggests that oxygen diffusion in the encapsulant powder might be very important.

5.3 Applications of CD Complex in Colors and Pigments

5.3.1 Lightly Pickled Eggplant

In the storage of lightly pickled eggplants, color transfer of the eggplant was observed from the color of the eggplant pickles to the pickle liquid during shop storage. The pickle solution became bluish purple during the storage period. Furuta et al. (2004) minimized the purple coloring of eggplant pickles by the addition of high-molecular-weight sugars. Yoshii (2004) indicated β-CD application to lightly pickled eggplant in a review of applications for the food processing of CD. Figure 5.12 shows the effect of β-CD on the color change of lightly pickled eggplant during storage.

This anti-coloring might be due to the binding of the pigment of eggplant pulp, nasunine (delphinidin-3-(*p*-coumary-L-rhamnosyl-glucoside)-5-glucoside), and

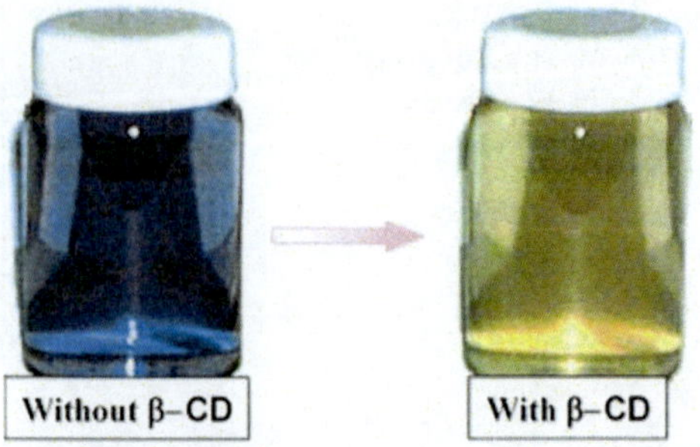

Fig. 5.12 Inhibition of color change in the lightly pickled eggplant solution with β-CD

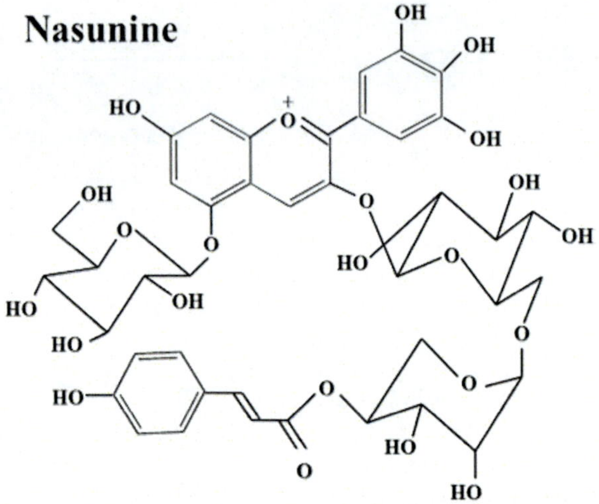

Fig. 5.13 Chemical structure of nasunine (delphinidin-3-(*p*-coumaryl-L-rhamnosyl-glucoside)-5-glucoside)

alum in the pickle solution to form a bluish-purple pigment. The chemical structure of nasunine is shown in Fig. 5.13. β-CD was used to inhibit the bluish-purple coloration of lightly pickled eggplant in this solution by diffused color pigment complexation with CD. Inhibitions of color transfer with β-CD and various saccharide compounds have been investigated, and this mechanism was considered by Yoshii (2004).

To investigate the color change of the lightly pickled eggplant, 21 g of eggplant skins was immersed in 6.3 g of 9% NaCl and 70 mL of 50 mM potassium phosphate buffer solution, pH 6.0, containing 32 mM L-ascorbic acid and 2 mM aluminium potassium sulfate (Yoshii 2004). The effects of additives such as β-CD, glucose, or maltodextrin with several average molecular weights, 960, 1700, and 680, were investigated on the color of the immersing solution (model pickle solution), with and without eggplant skin, at 4 °C storage temperature. After storage of the lightly pickled eggplant, the absorbance at 580 nm of the 20-fold diluted solution of the immersing solution were measured with a 50 mM potassium phosphate buffer solution of pH 6.0 by a spectrophotometer. Nasunine has its peak absorbance at pH 6.0.

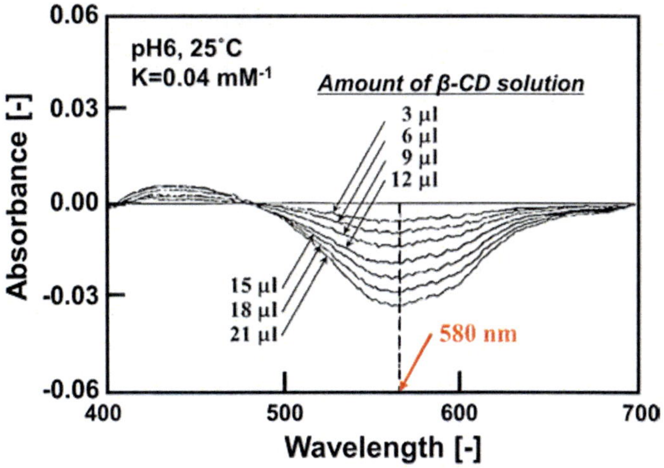

Fig. 5.14 Different spectral changes with β-CD titration to the lightly pickled eggplant solution at pH 6.0

Figure 5.14 shows the different spectral changes with β-CD titration to the lightly pickled eggplant solution at pH 6.0. The titration of β-CD caused the color change at 580 nm. This might be due to the inclusion of some parts in nasunine to CD or the inhibition of color stacking of nasunine. The absorbance at 580 nm also decreased with the addition of glucose or other MDs to the lightly pickled eggplant solution. This result suggests that the effect of β-CD on the inhibition of color change during storage might be due to the stacking inhibition of the color pigment nasunine in the lightly pickled eggplant solution. Figure 5.15 indicates the absorbance change as the difference values from the original absorbance after 7 days of storage at 0% and 9% sodium chloride concentrations. CD and CD polymer could inhibit the color formation of nasunin and the amounts of intermolecular forces such as β-CD and β-CD conjugate were in direct proportion to the anti-coloring.

The slope values of the absorbance difference were plotted to the saccharide molecular mass as shown in Fig. 5.16. The slope values were linearly correlated with the molecular mass of the added saccharides. This might be due to the stacking inhibition of color change, which depends on the size of the added saccharides.

β-CD was found suitable additive with a high molecular mass to inhibit the stacking inhibition of the color pigment nasunine. The effect of saccharide addition with 9% sodium chloride in pickle solution was lower than that without sodium chloride since the color leakage from eggplant skin might depend on the osmotic pressure of the pickling solution. The mechanism of stacking inhibition by various saccharides is shown in Fig. 5.17. β-CD might include the benzene ring of a chemical part of the color pigment side structure. The slope value of β-CD might be slightly higher than the linear line due to this inclusion effect. With the addition of saccharides to the lightly pickled eggplant solution, the saccharide might enter the

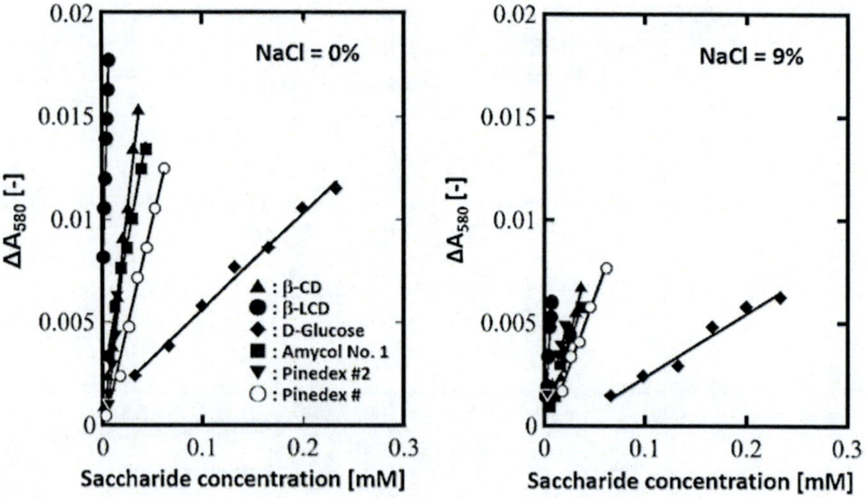

Fig. 5.15 Absorbance changes at 580 nm of the titration of various saccharides after 7 days of storage

Fig. 5.16 Slope values of absorbance differences with the addition of various saccharides

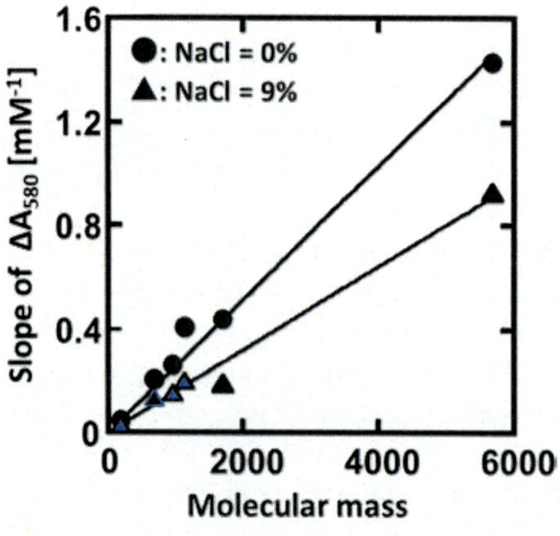

gap of the copigmentation of the intermolecular plane structure and inhibit the stacking of the color pigments.

5.3.2 *Curcumin*

An inclusion complex of β-CD and curcumin was formed to investigate the stability of the colorant (Mangolim et al. 2014). Curcumin, a natural yellow-orange colorant, is a hydrophobic polyphenol that is widely used in food applications. Coprecipitation could be a better option to complex curcumin with β-CD than freeze drying and

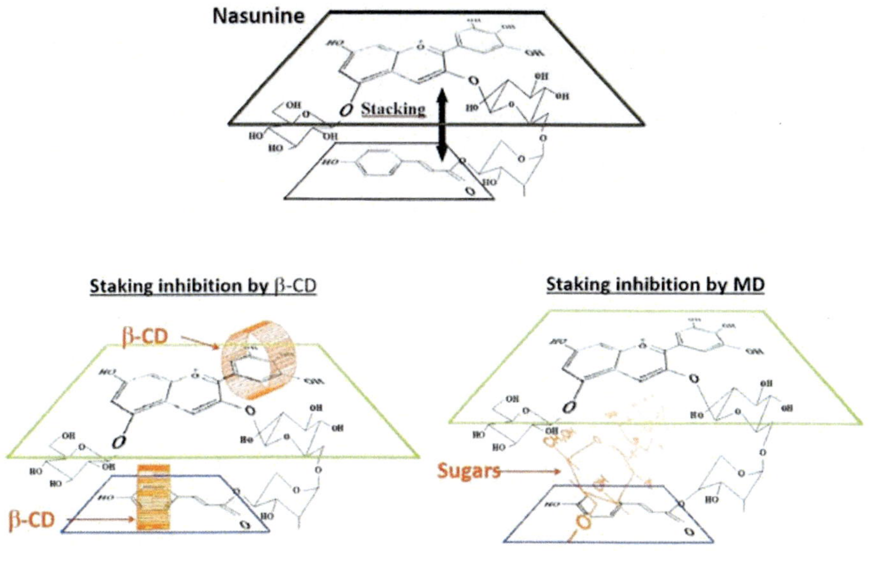

Fig. 5.17 The mechanism of stacking inhibition by various saccharides. (Adapted with permission from Yoshii 2004)

solvent evaporation methods. CD-complexed curcumin had better stability than pure curcumin against photodegradation. It was reported that in addition to the physical barrier, the complexation effect played a role in improving the stability of the complexed colorant. Some controversial results may also be obtained (no/less stability against light) (Paramera et al. 2011) due to the intermolecular hydrogen bond formation between CD molecules and curcumin (Tonnesen et al. 2002). However, the complexation method (coprecipitation) might improve the stability against photodegradation (Mangolim et al. 2014). Additionally, CD-complexed curcumin was 2.7-fold more stable than the pure colorant at pH 1–7. β-CD has little effect on curcumin stability in strong alkali media (Tang et al. 2002). However, curcumin absorbance is high in an acidic medium.

5.3.3 Fresh Pear Juice

Polyphenol oxidase is one of the major enzymes that causes browning reactions and changes the color of fresh juices. The addition of CD to pear juices could reduce the browning reaction and increase the color stability (López-Nicolas and García-Carmona 2007). The authors reported that the addition of 90 mM α-CD could retain the original greenish color after the first hour of juicing. However, juices significantly turned from green to yellow without the addition of CD. Compared to β-CD, α-CD could inhibit the enzymatic browning reaction of fresh juices. Due to the poor solubility of β-CD in aqueous medium, a small amount of this CD may be applied in juices, which may be the reason for the low or lack of inhibition properties of enzymatic browning. In addition, the inner diameter of the hydrophobic cavity of α-CD leads to a better interaction between CD and phenols present in the pear juice. However, modified β-CD, such as maltosysl-β-CD, could positively respond to increase the inhibition effect of the browning reaction of pear juices (López-Nicolas and García-Carmona 2007). According to the authors, maltosysl-β-CD might increase the protective barrier of ascorbic acid oxidation and act as a secondary antioxidant to reduce pear juice browning.

5.3.4 γ-CD.Gingerol in Yoghurt

Gingerols, phenolic ketones with a range of unbranched alkyl chain lengths, are the main active ingredients of ginger. Due to the health benefits and antioxidation properties of ginger, its juice and dry forms have recently been used in yogurt and yogurt butter (Felfoul et al. 2017; Arslan et al. 2009). Better homogenization of γ-CD. gingerol was found compared to gingerol alone fortified in yogurt (Pais et al. 2020). In terms of the color change of fortified yogurt, γ-CD.gingerol yogurt showed undetectable color changes to the average consumer compared to plain yogurt. However, a detectable color change was found in gingerol yogurt.

5.3.5 Yellow Bell Pepper Pigments in Isotonic Drinks

Yellow bell pepper is rich in flavonoids, carotenoids, phenolic acids and sterols. Due to the higher concentrations of pigments and aroma, it is used commercially in the food industry (Oboh et al. 2011). Lutein, zeaxanthin, α-cryptoxanthin, α-carotene and β-carotene are the main pigments found abundantly in yellow bell pepper (Lobo et al. 2018). Several studies have been carried out to investigate the effect of CD complexes on the color stability of food products (Marcolino et al. 2011; López-Nicolas and García-Carmona 2007). However, the complexing method and CD types affect the color stability of active compounds. Coprecipitation and molecular inclusion methods were applied to complex the CD with yellow bell pepper pigments (Lobo et al. 2018). The color stability of isotonic drinks was analysed in the presence (1400 lux) and absence of light at 25 and 31 °C for 21 days. The color stability of isotonic drinks with CD complexes obtained by the molecular inclusion method was higher for up to 21 days under both storage conditions due to the coprecipitation method.

5.4 Conclusion and Further Perspectives

The structure of CD has a molecular cavity with hydrophobic properties composed of hydrogen and glycoside oxygen. Various guest substances are included in this molecular cavity and are encapsulated. By using this inclusion property, hydrophobic substances such as aromatic substances, lipids and antibacterial substances can be included in a powder to inhibit physical and chemical decomposition and denaturation by heat, light, and oxygen. Molecular inclusion of hydrophobic food ingredients with CDs has significantly improved the storage stability of food characteristics such as flavors, vitamins, color and unsaturated fatty acid-containing oils. In the food industry, CDs are useful food additives to improve the physical and chemical properties of food.

In this chapter, the encapsulation of RETP with CD syrup was summarized with the inclusion of RETP to CD by the kneading method. This inclusion of functional compounds in CD can be applied to produce supplemental powders of functional food compounds. β-CD can be applied as a food additive to inhibit the stacking of the color pigment nasunine. CD can also be used as a food additive as a high-mass saccharide.

CD usage methods depend on the type of the inclusion complexes of color and pigments since color and pigments need clear color, colorless products, or moderate color products. Further studies should go on to investigate the regulation method of the interaction between CD and color pigment compounds.

References

Arslan D, Sert D, Ayar A et al (2009) Shelf life determination of Yayik butter fortified with spice extracts. Int J Dairy Technol 62:189–194

Fang Z, Bhandari B (2010) Encapsulation of polyphenols—a review. Trends Food Sci Technol 21(10):510–523

Felfoul I, Borchan M, Samet-Bali O et al (2017) Effect of ginger (*Zingiber ocinalis*) addition on fermented bovine milk: rheological properties, sensory attributes and antioxidant potential. J New Sci 44:2400–2409

Furuta T, Miyawakil E, Yoshii H, et al (2004) Minimizing the purple coloring of eggplant pickles by the addition of high molecular sugars. In: IFT annual meeting, Jul 12–16, 17D-7

Koeda T, Wada Y, Neoh TL et al (2014) Encapsulation of retinyl palmitate with a mixture of cyclodextrins and maltodextrins by the kneading method. Food Sci Technol Res 20(3):529–535

Lobo FAT, Silva V, Domingues J et al (2018) Inclusion complexes of yellow bell pepper pigments with β-cyclodextrin: preparation, characterisation and application as food natural colorant. J Sci Food Agric 98(7):2665–2671

López-Nicolas JM, García-Carmona F (2007) Use of cyclodextrins as secondary antioxidants to improve the color of fresh pear juice. J Agric Food Chem 55(15):6330–6338

Mangolim CS, Moriwaki C, Nogueira AC et al (2014) Curcumin-β-cyclodextrin inclusion complex: stability, solubility, characterisation by FT-IR, FT-Raman, X-ray diffraction and photoacoustic spectroscopy, and food application. Food Chem 153:361–370

Marcolino VA, Zanin GM, Durrant LR et al (2011) Interaction of curcumin and bixin with beta-cyclodextrin: complexation methods, stability, and applications in food. J Agric Food Chem 59:3348–3357

Mohan PRK, Sreelakshmi G, Muraleedharan CV et al (2012) Water soluble complexes of curcumin with cyclodextrins: characterization by FT-Raman spectroscopy. Vib Spectrosc 62:77–84

Mortensen A (2006) Carotenoids and other pigments as natural colorants. Pure Appl Chem 78(8):1477–1491

Oboh G, Ademiluyi AO, Faloye YM (2011) Effect of combination on the antioxidant and inhibitory properties of tropical pepper varieties against α-amylase and α-glucosidase activities in vitro. J Med Food 14:1152–1158

Pais JM, Pereira B, Paz FAA et al (2020) Solid γ-cyclodextrin inclusion compound with gingerols, a multi-component guest: preparation, properties and application in yogurt. Biomolecules 10(2):344

Paramera EI, Konteles SJ, Karathanos VT (2011) Stability and release properties of curcumin encapsulated in *Saccharomyces cerevisiae*, β-cyclodextrin and modified starch. Food Chem 125:913–922

Singh G, Kapoor IPS, Singh P et al (2010) Comparative study of chemical composition and antioxidant activity of fresh and dry rhizomes of turmeric (*Curcuma longa* Linn.). Food Chem Toxicol 48:1026–1031

Tanaka T, Shnimizu M, Moriwaki H (2012) Cancer chemoprevention by carotenoids. Molecules 17(3):3202–3242

Tang B, Ma L, Wang H et al (2002) Study on the supramolecular interaction of curcumin and β-cyclodextrin by spectrophotometry and its analytical application. J Agric Food Chem 50:1355–1361

Tonnesen HH, Másson M, Loftsson T (2002) Studies of curcumin and curcuminoids. XXVII. Cyclodextrin complexation: Solubility, chemical and photochemical stability. Int J Pharm 244:127–135

Van Loo-Bouwman CA, Naber THJ, Schaafsma G (2014) A review of vitamin A equivalency of β-carotene in various food matrices for human consumption. Br J Nutr 111(12):2153–2166

Yallapu MM, Jaggi M, Chauhan SC (2010) β-Cyclodextrin-curcumin self-assembly enhances curcumin delivery in prostate cancer cells. Colloids Surf B Biointerfaces 79:113–125

Yoshii H (2004) Application to food processing of cyclodextrin. Nippon Shokuhin Kagaku Kogaku Kaishi 51(12):647–655

Chapter 6
Encapsulation of Polyphenols, Plant Bioactive Compounds

Diana Alves and Eva Pinho

6.1 Polyphenols

Plants have long been the subject of great interest, as they provide unlimited opportunities for the identification of biologically active compounds with great potential to be applied in food, cosmetic and pharmaceutical industries (Soto et al. 2015; Ballesteros et al. 2017; Atanasov et al. 2021). Included in the plant bioactive compounds, phenolic compounds or polyphenols stand out in current research, with more than 8000 phenolic structures identified (Durazzo et al. 2019). Polyphenols are secondary metabolites synthesized by many plants, playing essential roles for their growth and development, as they confer protection against microorganisms, insects, herbivores and UV light. They are, also, responsible for plant odors, pigmentation and astringency as well as plants' flavors (Durazzo et al. 2019). These compounds are embedded in human and animal diets due to polyhenols well-established health-promoting effects, including antioxidant, antibacterial, antiviral and anti-inflammatory properties (Cory et al. 2018). Indeed, the consumption of a polyphenol-rich diet has been associated with a lower risk of chronic human diseases, mainly those associated with oxidative stress, such as cardiovascular diseases, diabetes, cancer, aging-related diseases, hypertension and neurodegenerative diseases (Fraga et al. 2019). In the food industry, polyphenols have been exploited mainly due to their

D. Alves
CEB—Centre of Biological Engineering, LIBRO—Laboratório de Investigação em Biofilmes Rosário Oliveira, University of Minho, Braga, Portugal
e-mail: dianalves@ceb.uminho.pt

E. Pinho (✉)
National Institute for Agrarian and Veterinarian Research (INIAV), Vila do Conde, Portugal
e-mail: eva.pinho@iniav.pt

T. M. Ho et al. (eds.), *Functionality of Cyclodextrins in Encapsulation for Food Applications*, https://doi.org/10.1007/978-3-030-80056-7_6

antioxidant and antimicrobial properties. These activities make them suitable as preservatives, to increase the shelf-life of food products (Bouarab Chibane et al. 2019).

In terms of chemical features, polyphenols have at least one aromatic ring, bonded to one or more hydroxyl group (Abbas et al. 2017). This class of natural compounds can vary from simple, single aromatic ring, low molecular-mass compounds to complex and large structures (Munin and Edwards-Lévy 2011). Such diversity results from variations on the basic chemical skeleton, like degrees of oxidation, hydroxylation, methylation, glycosylation and conjugation with further molecules, mainly lipids, proteins and other phenolics or biomolecular metabolites (Conte et al. 2016). The chemistry of polyphenols, more specifically, the presence of hydroxyl groups, is responsible for their antioxidant and anti-inflammatory properties. Oxidative stress can be defined as a disparity between the production of reactive oxygen species (ROS) in the cells and tissues and their elimination by protective mechanisms. The excess of free radicals causes oxidative damage to proteins, lipids and DNA, which can lead to many chronic diseases (Schieber and Chandel 2014). An antioxidant can be defined as a substance able to transfer electron and/or hydrogen atom during free radical neutralization (Brewer 2011). The antioxidant activity of phenolic compounds depends on the number of hydroxyl groups and their arrangements. It has been considered that the higher the number of hydroxyl groups, the better their antioxidant features (Olszowy 2019). Furthermore, depending on the arrangement of hydroxyl groups, these compounds are also able to chelate metals with redox-activity (Santos et al. 2017) and activate antioxidative enzymes (Procházková et al. 2011). In line with polyphenols antioxidant activity, recent studies have also shown their modulatory effects on cellular biomarkers related to inflammation. In particular, polyphenols' ability to inhibit pro-inflammatory cytokines and directly attenuate the nuclear factor kappa B (NFkB)-mediated or oxidative stress-induced inflammatory signaling pathways (Zhang and Tsao 2016). In addition to antioxidant and anti-inflammatory features, polyphenols have also raised the interest of scientific community for their antimicrobial effects (Daglia 2012). There is a great number of polyphenols whose antimicrobial activity has been investigated against a wide range of microorganisms, including Gram-positive and Gram-negative bacteria (Bouarab-Chibane et al. 2019) as well as fungi (Salazar-Aranda et al. 2015). Although the mechanisms of antimicrobial action are not yet fully elucidated, several approaches have been reported, including membrane disruption, inhibition of biofilm formation and of cell envelope, and nuclei acid synthesis (Górniak et al. 2019). Furthermore, there is growing evidence that some polyphenols act synergistically with traditional antibiotics against multidrug-resistant microorganisms (Hossain et al. 2020).

Polyphenolic compounds have been classified in two major groups: flavonoids and non-flavonoids which can be further sub-divided into phenolic acids, quinones, stilbenes, tannins, coumarins and lignans.

6.1.1 *Flavonoids*

Flavonoids comprise the class of polyphenols with the largest number of isolated, identified and diversified structures present in the plant kingdom, including vegetables, fruits, cereals and dry legumes (Mutha et al. 2021).

The basic structure of flavonoids is composed by two benzene rings (A and B) joined together by a pyrane ring (C) (Fig. 6.1). Depending on the presence or absence of a carbonyl at position 4, a double bond between positions 2 and 3, or a hydroxyl group at position 3, they are usually further classified into flavan-3-ols, flavones, flavonols, flavanones, isoflavones and anthocyanins (Table 6.1) (Thilakarathna and Vasantha Rupasinghe 2013).

Flavonoids are recognized as the pigments responsible for the colors of flowers, fruits and seeds. Other biological functions played by these compounds include protection against UV radiation, pathogens and oxidative cell injury, auxin transport, and male fertility (Falcone Ferreyra et al. 2012). Furthermore, they exhibit a wide range of biological activities, including anti-inflammatory, antioxidant, antitumor, hepatoprotective, antimicrobial, antithrombotic and antiviral properties (Zhao et al. 2019). The great number of flavonoids groups is due to the many substitutional modifications possible on the basic structure backbone. Sugars are very frequent, as evidenced by most flavonoids existing as glycosides. Both sugars and hydroxyl groups increase the water solubility of these compounds whereas others substituents, such as methyl groups and isopentyl units, increase flavonoids lipophilic features (Kumar and Pandey 2013; Durazzo et al. 2019).

6.1.2 *Non-flavonoids*

Non-flavonoids are comprised by all the other polyphenolics, including phenolic acids, quinones, stilbenes, tannins, coumarins and lignans (Table 6.2).

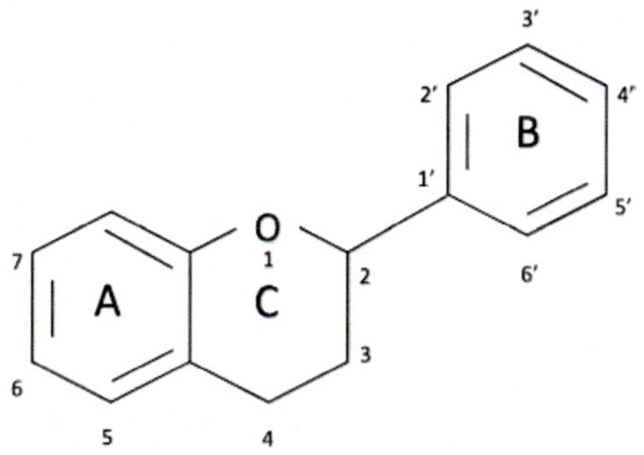

Fig. 6.1 General chemical structure of flavonoids

Table 6.1 Classification, general chemical structure (Chem3D Pro 12.0 software) and main sources of plant-derived flavonoids

Class	Structure backbone	Examples	Main sources
Flavan-3-ols	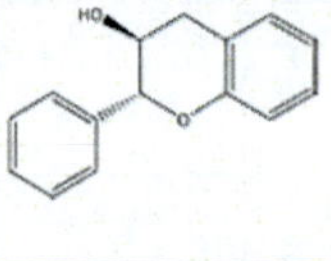	Catechin Epicatechin gallate Epigallocatechin Epigallocatechin gallate	Berries, cacao, grapes, green tea
Flavone	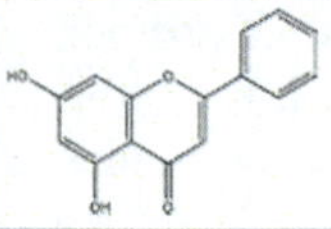	Apigenin Tangeretin Baicalein Luteolin	Fruits, vegetables, citrus fruits, *Scutellaria Baicalensis* Georg, broccoli, celery, green peppers
Flavonol	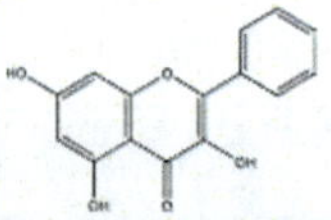	Kaempferol Myricetin Quercetin	Cabbage, cauliflower, propolis, spinach, berries, herbs, vegetables, apples, berries, onion, pears
Flavanone	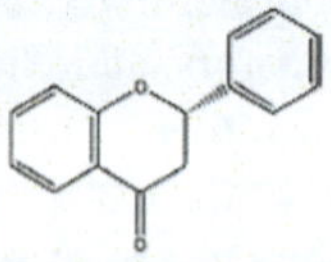	Hesperidin Naringin Taxifolin	Citrus fruits, grapefruit, bark and seeds
Isoflavone	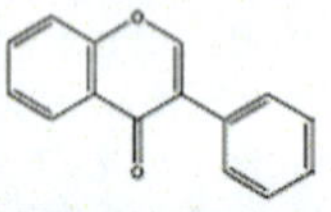	Genistin Genistein Daidzin Daidzein	Soybeans, soy-products
Anthocyanin	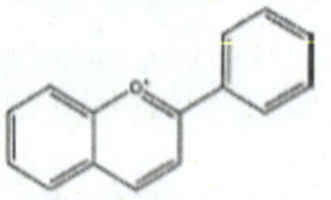	Apigenidin Cyanidin Malvidin	Berries, grapes, cacao

Phenolic acids represent the simplest class of polyphenols and their basic structure is characterized by a carboxyl group linked to one aromatic ring. They can be divided in two groups, hydroxybenzoic acids (C_1–C_6) and hydroxycinnamic acids (C_3–C_6), depending on whether they are derived from benzoic or cinnamic acids, respectively. Examples of hydroxybenzoic acids include *p*-hydroxybenzoic, gallic and ellagic acid whereas caffeic, ferulic, sinapic and *p*-coumaric acids belong to the group of hydroxycinnamic acids (Durazzo et al. 2019; Kumar and Goel 2019). This group of molecules rarely occur in their free form but in the form of esters, glycosides or amides in fruits and cereal grains. Hydroxybenzoic acids are less common than hydroxycinnamic acids and they are frequently found in the form of glucosides in tea and red fruit. Phenolic acids are sensitive to temperature, elevated pH, oxidation and light. In terms of water solubility, hydroxybenzoic acids are water-soluble while hydroxycinnamic acids are poorly soluble in water (Munin and Edwards-Lévy 2011). Current interest on phenolic acids is attributed to their biological activities, mainly as antimicrobial (Sánchez-Maldonado et al. 2011), antioxidant and antitumor. Phenolic acids' antimicrobial activity is determined by their chemical structure, especially by saturated chain length, position and number of substitutions in the benzene ring (Kumar and Goel 2019).

Table 6.2 Chemical structure and some examples of non-flavonoid compounds (Chem3D Pro 12.0 software)

Class	Examples	
Phenolic acids	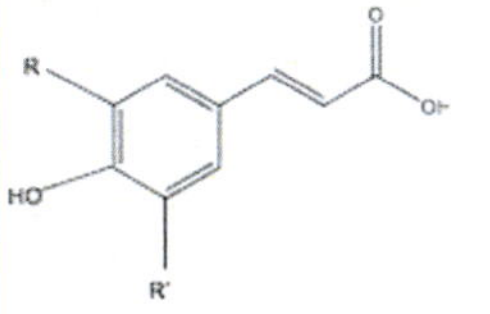 Benzoic acid derivatives	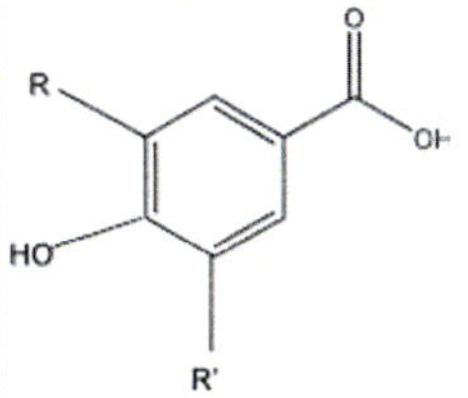 Cinnamic acid derivatives
Quinones	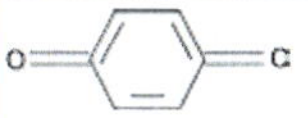 Quinone	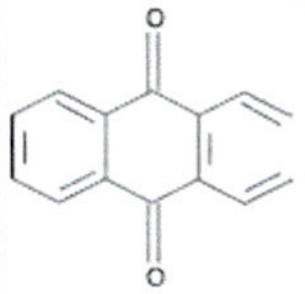 Anthraquinone
Stilbenes	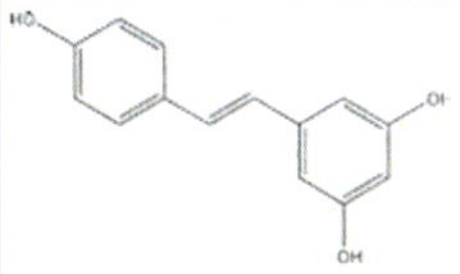 Resveratro	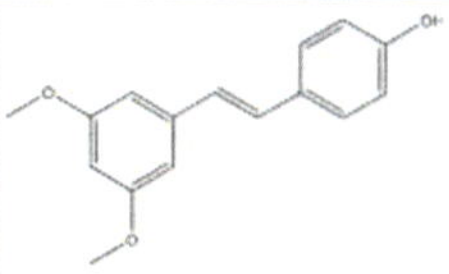 Pterostilbene
Tannins	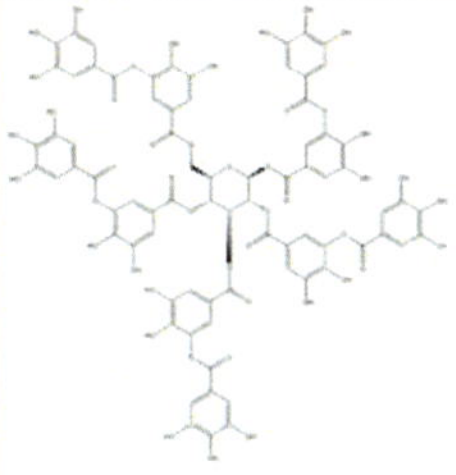 Gallotannin	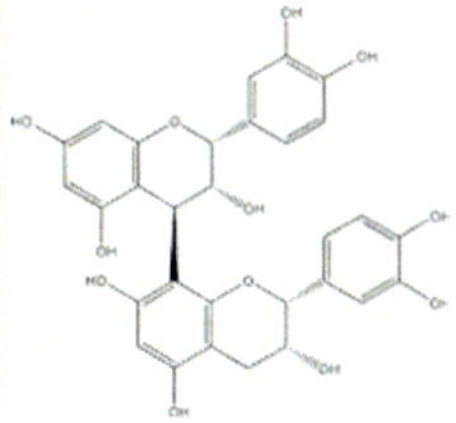 Procyanidin B2
Lignans	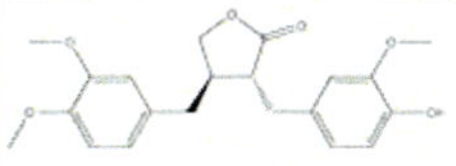 Arctigenin	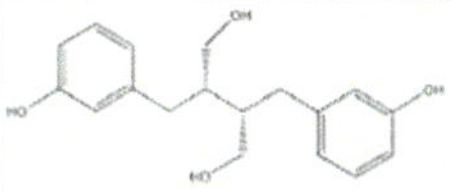 Enterodiol

Quinones are highly reactive, oxidised phenolic compounds that contain two adjacent or separated carbonyl groups in a six-member unsaturated ring structure (Son et al. 2016). Some quinones exert antioxidant, antimicrobial, anti-inflammatory and anticancer activities. These biological effects are attributed to two important properties: first, quinones can undergo reversible oxidation-reduction reactions and, second, many of them can undergo nucleophilic attack due to their electrophilic characters (El-Najjar et al. 2011).

Stilbenes are characterized by the chemical structure C_6–C_2–C_6, with two benzene rings connected by a double bond (Albuquerque et al. 2021). These class of compounds exists as two diastereoisomeric forms, *E*-1,2-diphenylethylene (trans-configuration) and *Z*-1,2-diphenylethylene (cis-configuration) (Pecyna et al. 2020).

Resveratrol is probably the most common stilbene and has been widely studied for its anti-ageing and antioxidant properties (Pannu and Bhatnagar 2019).

Tannins are molecules with relatively large size and their chemical structure consists of two or three phenolic hydroxyl groups on a phenyl ring. These compounds are responsible to confer astringent and bitter taste to some fruits (Albuquerque et al. 2021). Tannins are usually classified into two groups: hydrolysable and non-hydrolysable tannins, also called condensed tannins or proanthocyanidins (Durazzo et al. 2019). Hydrolysable tannins comprise a mixture of simple phenols (e.g., ellagic and gallic acids) and carbohydrates, which can be separated by hydrolysis when subjected to acid or basic conditions. Condensed tannins are polymers or oligomers formed by two or more monomers of flavan-3-ols units (Okuda and Ito 2011). These particular compounds have recently attracted attention due to their biological effects such as antioxidant, anticancer, antidiabetic, neuroprotective, and antimicrobial (Rauf et al. 2019).

The basic chemical structure of coumarins is characterized by a 1,2-benzopyrone ring system, and their substitution pattern results in different biological activities, including anti-thrombotic, anti-inflammatory, antimicrobial, antiviral, anticoagulant and antioxidant (Stefanachi et al. 2018).

Lignans are composed by the combination of phenylpropanoid dimers (C_6–C_3) liked by the central carbons of the side chains. The major biological activities attributed to this class of polyphenols are antioxidant, antitumor, anti-inflammatory and antiviral (Teponno et al. 2016).

6.2 Polyphenol-Cyclodextrin Inclusion Complexes

As mentioned above, polyphenols gather several pharmacological activities, such as antioxidant, antimicrobial, cardioprotective among others, which allow their use as nutraceuticals and preservatives in food industry (Cory et al. 2018). Despite the growing interest on these molecules, polyphenols application has been challenged due to limited water solubility, poor bioavailability, and degradation by environmental factors such as temperature, pH and light (Pinho et al. 2014). Cyclodextrins (CD), molecules capable of forming inclusion complexes with a wide range of bioactive compounds, have been used as an encapsulation strategy to safeguard polyphenols stability, bioactivity, and bioavailability (Tian et al. 2020). CDs' inclusion complex formation induces conformational alterations on the guest physical-chemical properties, for example, it enhances the solubility, stabilization against derivatizing agents (like oxygen, visible or ultra-violet light, and heat), control of the volatility and sublimation properties, physical isolation of incompatible compounds (via chromatographic separation), taste modification by masking potentially adverse flavors, control odors and the sustained release of such encapsulated compounds (Pinho et al. 2014).

Most of the publications in this area of expertise concern the encapsulation of polyphenols with β-CDs and its derivatives to improve the polyphenols solubility and stability (Table 6.3).

Table 6.3 Published works regarding the inclusion of polyphenolics by CDs and derivatives

Polyphenols	CD	Complexation method	Encapsulation capacity	Release properties	References
Curcumin	βCD HPβCD	Solubility	ND	6 h	de Castro et al. (2018)
	βCD	Kneading Coprecipitation Solubility	1:2	ND	Marcolino et al. (2011)
	βCD	Oil-in-water emulsion	ND	180 h	Wu et al. (2018)
	βCD	Coprecipitation	EE = 74%	ND	Mangolim et al. (2014)
Ferulic acid	βCD	Freeze-drying	1:1	ND	Simsek et al. (2019)
	HPβCD	Freeze-drying	1:1 EE = 87.74%	ND	Kim (2020)
	HPβCD	Spray-drying	EE = 68.06%	ND	Olga et al. (2015)
	HPβCD	Freeze-drying	EE = 100%	10 min	Sharif et al. (2018)
Gallic acid	HPβCD	Spray-drying	EE = 89.22%	ND	Olga et al. (2015)
	βCD	Freeze-drying	EE = 80.00%	ND	da Rosa et al. (2013)
	βCD	Freeze-drying	ND	ND.	Rezaee et al. (2018)
Ferulic acid Gallic acid	HPβCD	Spray-drying	EE = 68.92/77.34%	ND	Olga et al. (2015)
p-Coumaric acid	βCD	Freeze-drying	1:1	ND	Kim (2020)
	HPβCD	Freeze-drying	1:1 EE = 81.71%	ND	
Caffeic acid	βCD	Freeze-drying	1:1	ND	Simsek et al. (2019)
	αCD βCD γCD	Solubility	EE = 0.69–3.52%	ND	Kfoury et al. (2019)
Trans-cinnamaldehyde	βCD	Coprecipitation	EE = 76–91%	ND	Canales et al. (2019)
	βCD	Coprecipitation	ND	192 h	Chen et al. (2019)
Piperine	βCD	Solubility	EE = 42.6%	ND	Garrido et al. (2019)
	βCD	Ultrasound assisted complexation followed by freeze drying	EE = 18.61–73.50%	ND	Quilaqueo et al. (2019)

(continued)

Table 6.3 (continued)

Polyphenols	CD	Complexation method	Encapsulation capacity	Release properties	References
Cyanidin-3-*O*-glucoside	βCD	Solubility Freeze drying	ND	ND	Fernandes et al. (2018)
	βCD	Solubility	ND	ND	Fernandes et al. (2013)
Epigallocatechin gallate	SBEβCD	Solubility˙	1:1 EE = 93%	ND	Liu et al. (2016)
Neohesperidin dihydrochalcone	GβCD	Ultrasonic-assisted technique in the ethanol aqueous solution	1:2 EE = 74.6%	ND	Dong et al. (2017)
Fisetin	βCD γCD	Solubility	1:1	ND	Zhang et al. (2015)
p-Hydroxybenzoic acid	HPβCD	Freeze-drying	1:1 EE = 91.34%	ND	Kim (2020)
Vanillic acid	HPβCD	Freeze-drying	1:1 binding stoichiometry EE = 89.31%	ND	Kim (2020)
Chlorogenic acid	βCD HPβCD	Freeze-drying	1:1	ND	Shao et al. (2014)
Hydroxycinnamic acids	βCD	Solubility	1:1	ND	Budryn et al. 2015)
α-Methylstilbene	αCD βCD HPβCD MβCD γCD	ND	1:1	ND	Matencio et al. (2017)
Quercetin	SBEβCD	Solubility	EE = 0.002–0.107%	ND	Azzi et al. (2018)
p-Cymene	HPβCD HPγCD	Solubility	EE = 44.0–54.5%	ND	Celebioglu et al. (2018)
Daidzein	βCD	Complexation in water and acetone and freeze drying	EE = 15.248%	ND	Fateminasab et al. (2018)
Trans-anethole	βCD	Coprecipitation	ND	ND.	Ye et al. (2017)
Hexahydro-β-acids	MβCD	Grinding method	EE = 95%	ND	Lu et al. (2019)

αCD alpha-cyclodextrin, *βCD* beta-cyclodextrin, *γCD* gamma-cyclodextrin, *HPβCD* hydroxypropyl-beta-cyclodextrin, *HPγCD* hydroxypropyl-gamma-cyclodextrin, *MβCD* methyl-beta-cyclodextrin, *GβCD* glucosyl-beta-cyclodextrin, *SBEβCD* sulfobutylether-beta-cyclodextrin, *ND* not determined, *EE* encapsulation efficiency

6.2.1 Inclusion Complex: Assembly and Characterization

In nature is possible to find CDs comprising six (α), seven (β), and eight (γ) glucopyranose units with a molecular weight range from 1000 to 2000 Da. All three molecules share the same cavity height but have a distinguished volume and diameter (Fig. 6.2) (Pinho et al. 2014). The inclusion complex formation, acceptance of

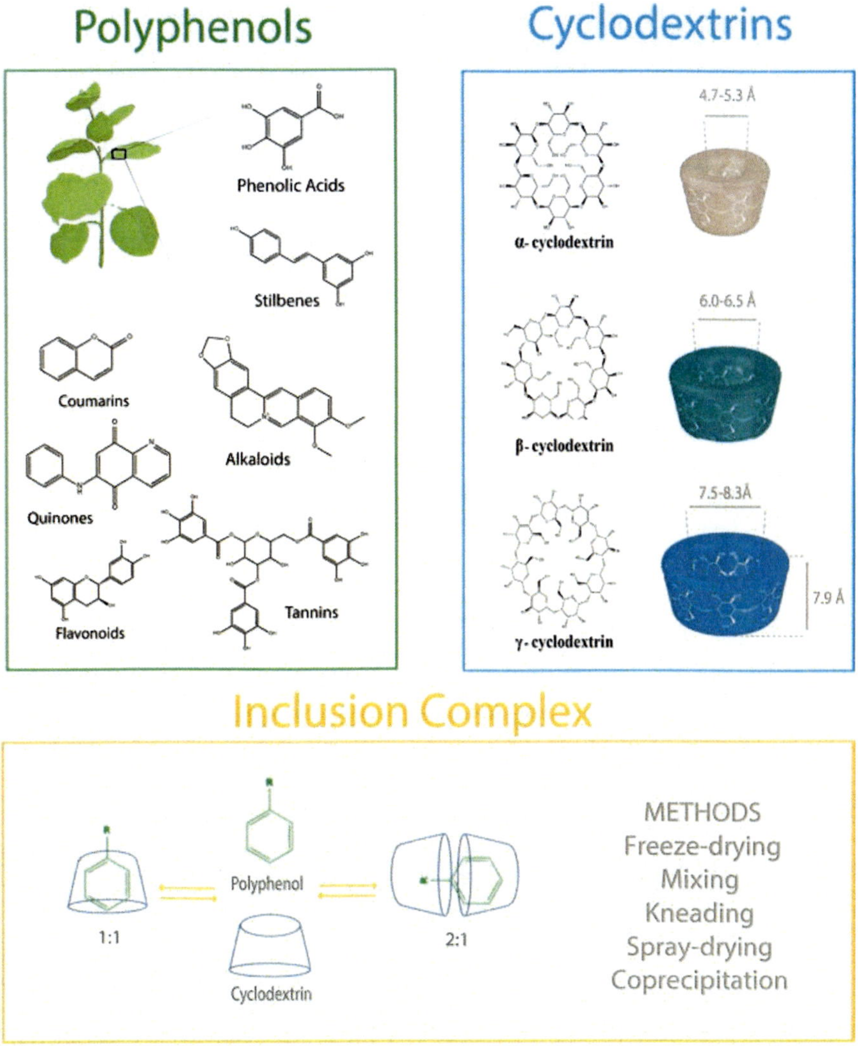

Fig. 6.2 Polyphenol-CD inclusion complex outcomes from a dimensional fit between the polyphenol (or part of it) and CDs' inner cavity. Inclusion complex stability is set by the polyphenol and CD concentrations, polyphenol solubility, the binding constant, the size of the polyphenol towards the cavity of the CD and the technique used to prepare the complex

the polyphenol (or part of it) into CDs' inner cavity, results from a dimensional fit between both players. The process is stabilized by non-covalent bonds established or broken and by the equilibrium between the molecules in complex and in solution (Pinho et al. 2015a). Once inside the cavity, the polyphenol suffers conformational adjustments to maximize the contribution of weak van der Wall's forces. Other molecular interactions, such as electrostatic, hydrogen bond or charge-transfer, are also involved on this process (Sharma and Baldi 2016). Accordingly, the type, number and position of functional substitutions on the CDs affect their physicochemical characteristics, including CDs' capacity to form inclusion complexes with polyphenols (Pinho et al. 2015a, b). The βCD has been the most used CD for the encapsulation of polyphenols for food applications (Marcolino et al. 2011; da Rosa et al. 2013; Tao et al. 2014; Shao et al. 2014; Budryn et al. 2015; Santos et al. 2015; Gong et al. 2016; Matencio et al. 2017; Saldanha do Carmo et al. 2017a, b; Wu et al. 2018; de Castro et al. 2018; Simsek et al. 2019; Jiang et al. 2019) since it has the lowest cost production, higher availability and enhanced inclusion complex-forming properties (Liu et al. 2020). βCD derivatives, specially HPβCD (hydroxypropyl-β-cyclodextrin) (Shao et al. 2014; Olga et al. 2015; de Castro et al. 2018; Sharif et al. 2018; Yildiz et al. 2018; Kim 2020), have also been widely applied in food industry to enhance polyphenols properties (Crini et al. 2018).

Despite the diversity of the polyphenol's molecules, most of the works report inclusion complexes characterized by a 1:1 ratio, in which a single polyphenol accommodates itself inside one CD cavity (da Rosa et al. 2013; Tao et al. 2014; Budryn et al. 2015; Santos et al. 2015; Liu et al. 2016; Matencio et al. 2017; Yildiz et al. 2018; de Castro et al. 2018; Sharif et al. 2018; Jiang et al. 2019; Simsek et al. 2019; Kim 2020). Some works go further on the complexes characterization and identify the orientation of the polyphenol molecule inside the CD. For instance, the chlorogenic acid had the aromatic ring and the ethylene side chain deeply included inside CDs' cavity, being the HPβCD complexes more stable than βCD complexes (Shao et al. 2014). A similar orientation was described for fisetin (Zhang et al. 2015) and hydroxycinnamic acid (Budryn et al. 2014) when encapsulated by βCD. Otherwise, two βCDs molecules were required to form a inclusion complex with curcumin, and the reported results suggested that each aromatic ring was contained within one βCD' cavity (Marcolino et al. 2011). A similar complex was described to another flavonoid. In this case, two molecules of the GβCD (glucosyl-β-cyclodextrin) were needed to encapsulate neohesperidin dihydrochalcone, a bicyclic flavonoid with sweetener properties. The proposed inclusion complex implies that the aromatic rings of the flavonoid were, also, deeply inserted in the cavity of GβCD from the wide side (Dong et al. 2017).

Moreover, the inclusion complex between polyphenols and CDs have been, frequently, characterized by the encapsulation efficiency (da Rosa et al. 2013; Tao et al. 2014; Olga et al. 2015; Santos et al. 2015; Dong et al. 2017; Saldanha do Carmo et al. 2017a, b; Sharif et al. 2018; Kim 2020). This quantitative parameter is obtained by experimental data and reflects the liaison between the initial concentration of polyphenol added to the system and the amount encapsulated by the CDs (Nasirullah et al. 2011). It is influenced by a wide range of factors, including polyphenol and CD concentrations, polyphenol solubility, the binding constant, the size of the polyphenol towards the cavity of the CD and the technique used to prepare the complex (Fig. 6.2) (Olga et al. 2015). The presence of two possible polyphenols can, also, affect the encapsulation efficiency. However, the characterization of inclusion complexes in an antagonistic environment, i.e., with two or more possible guest molecules have been poorly studied. Olga and co-workers described the co-encapsulation of ferulic acid and gallic acid by HPβCD. They found that the presence of ferulic acid had an antagonistic effect on the gallic acid encapsulation efficiency. The encapsulation efficiency was 89.22% for gallic acid and 68.06% for ferulic acid when the inclusion complexes were assembled in separate, but the gallic acid encapsulation efficiency decreased to 68.92% for the co-encapsulation (Olga et al. 2015).

There are several methods available for inclusion complex assembly between CDs and polyphenols. The method selection is, usually, based on the properties of the guest molecule, the facilities available and the cost involved. The method used to prepare the inclusion complexes should be chosen with caution, as it is probably the factor with the greatest impact on encapsulation efficiency (Pinho et al. 2014). Ferulic acid encapsulation by HPβCD was prepared by freeze-drying (Sharif et al. 2018) and spray-drying (Olga et al. 2015). Appling the first method, the encapsulation efficiency obtained was near 88% while for the second technique was 68%. The freeze-drying is the most used method to prepare inclusion complex between CDs and polyphenols for food application (da Rosa et al. 2013; Shao et al. 2014; Tao et al. 2014; Santos et al. 2015; Sharif et al. 2018; Simsek et al. 2019; Kim 2020). Solubility (Marcolino et al. 2011; Budryn et al. 2015; Gong et al. 2016; Liu et al. 2016; Matencio et al. 2017; Yildiz et al. 2018), kneading (Marcolino et al. 2011; Tao et al. 2014; Santos et al. 2015), spray-drying (Olga et al. 2015; Saldanha do Carmo et al. 2017a, b), coprecipitation (Marcolino et al. 2011; Jiang et al. 2019) and ultrasonic assisted technique (Dong et al. 2017) have also been used to prepare inclusion complexes of polyphenols and CDs to be used by in food products. Details regarding the inclusion complexes formations methods can be found in the review by dos Santos Lima et al. (2019).

6.2.2 Inclusion Complexes: Polyphenols Sustained Release

Polyphenols bioactive properties have been enhanced by CDs encapsulation. In food industry, this strategy has been used to improve polyphenols stability against environmental factors enhancing their solubility and consequently boosting polyphenols' antimicrobial and antioxidant properties and masking unpleasant odors or flavors (Hu et al. 2021). Such improvements are set by the inclusion complex stability and ability to release the polyphenol (Pinho et al. 2014, 2015a, b). Usually, the release profiles of CDs inclusion complexes and polyphenolics are characterized by two phases: a burst release on the first minutes flowed by a slower release (Astray et al. 2020; Liu et al. 2020). The inclusion complex stability must be tailored to the end-use of the inclusion complex. For the enhancement of antimicrobial or antioxidant properties, a compromise between the stability of the complex and the polyphenol release must be found. The CD must be available to sustain the polyphenols release over time. Otherwise, for concealing odors or flavors the inclusion complex polyphenols/CD must be permanent, since its dissolution may result on the loss of the desired effect (Salústio et al. 2011, 2013).

Although the appropriate assessment of the polyphenol-CD inclusion complex release profile represents a central question for the validation of such complexes application on food industry, most of the published works regarding this bioactive molecules by CDs for food applications lack the release profile assessment (Marcolino et al. 2011; Budryn et al. 2014; Shao et al. 2014; Zhang et al. 2015; Olga et al. 2015; Santos et al. 2015; Gong et al. 2016; Liu et al. 2016; Dong et al. 2017; Matencio et al. 2017; Saldanha do Carmo et al. 2017a, b; Yildiz et al. 2018; Simsek et al. 2019; Kim 2020). One exception reported a catechin/βCD inclusion complex with a temperature-release behavior. The release profile was characterized by a burst release for the first 30 min, regardless the temperature. Higher temperatures values resulted on the completely release of catechin from the CD cavity (Jiang et al. 2019).

The assessment of the release profile becomes more relevant when the polyphenols inclusion complexes are integrated on materials for the development of bioactive food packaging. For instance, the release of ferulic acid from gliadin electrospun fibers was superior when encapsulated by HPβCD, mainly due to the improvement of this polyphenol solubility by the inclusion complex formation (Sharif et al. 2018).

6.3 Polyphenol-CD Inclusion Complexes in Food Products

Food industry has long recognized the value of polyphenols due to their aforementioned health benefits to humans, and it has identified their encapsulation on CDs as a promising approach to enhance polyphenols benefits and overcome important drawbacks associated to their use as free compounds (Fig. 6.3) (Tian et al. 2020). Polyphenol-CD inclusion complexes have been applied in this field with several purposes: to improve the functionality of polyphenolic compounds (Kfoury et al. 2019), to control the release of flavors and mask unpleasant odors (Aree and Jongrungruangchok 2018; Pellicer et al. 2018), to develop intelligent and edible food packaging materials (Rodríguez-López et al. 2020) and to assist in the green extraction of polyphenols from plants (Diamanti et al. 2017).

In the last years, a great number of studies have been reported regarding the encapsulation of natural polyphenolic compounds by CDs to be further applied in

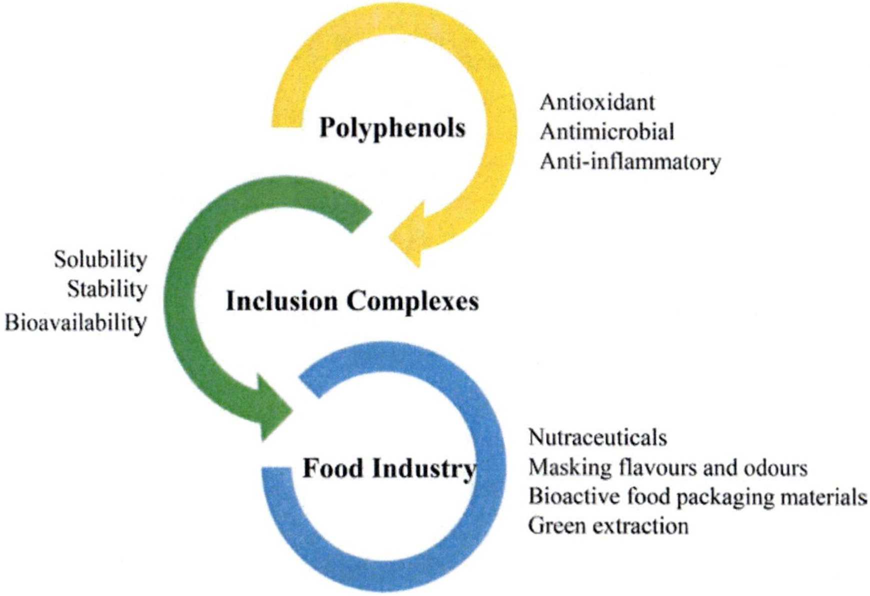

Fig. 6.3 Schematic of application of polyphenolic-CD inclusion complexes on food industry

food industry (Hu et al. 2021). An overview of these applications is summarized in Table 6.4 and some of them are analysed with more detail in the following sub-chapters.

6.3.1 Improving the Functional Properties of Polyphenols

Bioactive compounds such as polyphenols often have undesirable characteristics that limit their use as nutraceuticals, namely substances that are consumed as food or as part of a food and provide health benefits (Piccolella et al. 2019). The low solubility of polyphenols hinders their absorption by physiological tissues. Moreover, their poor stability against physiological, processing and storage conditions (temperature, light, oxygen, and pH) as well as interactions with food matrixes, results in considerable limited bioavailability (Munin and Edwards-Lévy 2011). Several studies have reported the improvement of the stability, water solubility and bioavailability of phenolic nutraceuticals through their complexation with CDs (Table 6.4).

Caffeic acid, for instance, is a non-flavonoid with several health benefits, mainly attributed to its high antioxidant activity. Its potential to prevent cancer, diabetes and cardiovascular diseases has increased the interest on its use as a nutraceutical food ingredient. However, it has poor aqueous solubility. Its complexation with α-, β-, and γ-CDs in water and using ethanol as cosolvent resulted in increased solubility

Table 6.4 Applications of polyphenol-CD inclusion complexes in the food industry

Applications	Polyphenols	CDs	Main findings	References
Nutraceuticals	Catechin	βCD	Improved stability against alkaline pH and high humidity. Higher antioxidant retention in protein solutions	Ho et al. (2017)
		βCD	Protection from the degradation caused by interaction with proteins	Ho et al. (2019a)
		βCD	Increased stability in a solid matrix and bitterness masking	Ho et al. (2019b)
		βCD	Enhanced stability	Jiang et al. (2019)
	Ferulic acid *ρ*-Coumaric acid *ρ*-Hydroxybenzoic acid Vanillic acid	HPβCD	Improved solubility, durability and stability. Reduced volatility and toxicity. Stabilization against light, UV, thermal decomposition and oxidation. Masking of unpleasant odors and bitter tastes	Kim (2020)
	Ferulic acid Gallic acid	HPβCD	Oxidation inhibition in food systems	Olga et al. (2015)
	Caffeic acid	αCD βCD γCD	Increased solubility and antioxidant activity	Kfoury et al. (2019)
	α-Tocopherol	βCD	Prolong the shelf-life of beverages due to its antioxidant properties and good dispersibility and solubility	Saldanha do Carmo et al. (2017a, b)
	Quercetin	SBEβCD	Increased solubility	Azzi et al. (2018)
	Piperine	βCD	Increased antioxidant activity and bioaccessibility but decreased effect on masking the pungent flavor	Quilaqueo et al. (2019)
	Daidzein	βCD	Conservation of antioxidant activity and enhancement of bioavailability, solubility, and stability	Fateminasab et al. (2018)
	Fisetin	βCD γCD	Enhanced solubility	Zhang et al. (2015)
	α-Methylstilbene	αCD βCD HPβCD MβCD γCD	Enhanced solubility	Matencio et al. (2017)
	Curcumin	βCD	Color stability against sunlight, pH and temperature. Intensified the color of vanilla ice creams with good sensorial acceptance	Mangolim et al. (2014)
	Cyanidin-3-*O*-glucoside	βCD	Despite the color diminish, molecular inclusion with CDs may lead to the anthocyanin stabilization, which could promote its use as food colorant	Fernandes et al. (2013)

Active packaging	Hexahydro-β-acids	MβCD	Improved water solubility and antimicrobial activity against *Listeria mocytogenes*	Lu et al. (2019)
	Gallic acid	βCD	Controlled release in a short duration, enabling high antioxidant activity	Rezaee et al. (2018)
	Curcumin	βC HPβCD	Antimicrobial activity against *Bacillus subtilis*	de Castro et al. (2018)
	Curcumin	βCD	Enhanced solubility and preservation effect on the freshness of apple juice	Wu et al. (2018)
	Ferulic acid	HPβCD	Enhanced solubility and stability	Sharif et al. (2018)
Food systems	Cyanidin-3-*O*-glucoside	βCD	Increased thermal stability	Fernandes et al. (2018)
	ρ-Cymene	HPβCD HPγCD	Enhanced solubility and thermal stability	Celebioglu et al. (2018)
	Piperine	βCD	Protection of bioactive properties and thermal stability	Garrido et al. (2019)
	Epigallocatechin gallate	SBEβCD	Oxidation inhibition	Liu et al. (2016)
	Neohesperidin dihydrochalcone	GβCD	Masking the bitterness	Dong et al. (2017)
	Caffeic acid Trans-Ferulic acid *ρ*-Coumaric acid	βCD	Improved solubility, durability, and stability and reduced bitter taste	Simsek et al. (2019)
	Chlorogenic acid	βCD HPβCD	Enhanced antioxidant activity and copigments	Shao et al. (2014)
	Hydroxycinnamic acids	βCD	Reducing unfavorable protein–polyphenol interactions during processing of foods enriched with phenolic acids	Budryn et al. (2015)
	Gallic acid	βCD	Antioxidant capacities	da Rosa et al. (2013)
	Curcumin	βCD	Intensification of color and increased solubility	Marcolino et al. (2011)

αCD alpha-cyclodextrin, *βCD* beta-cyclodextrin, *γCD* gamma-cyclodextrin, *HPβCD* hydroxypropyl-beta-cyclodextrin, *HPγCD* hydroxypropyl-gamma-cyclodextrin, *MβCD* methyl-beta-cyclodextrin, *GβCD* glucosyl-beta-cyclodextrin, *SBEβCD* sulfobutylether-beta-cyclodextrin and enhanced antioxidant activity (Kfoury et al. 2019). Flavonoids such as fisetin, have also been considered for nutritional supplements but its application is limited, also, by its low solubility and bioavailability. Fisetin complexation with β- and γ-CDs, using a freeze-drying method, proved to enhance its water solubility (Zhang et al. 2015). Fernandes et al. (2018) used β-CDs to encapsulate cyanidin-3-*O*-glucosid, an anthocyanin from strawberries, which resulted in its enhanced thermal stability. Other polyphenolic compounds with potential to be used as nutraceuticals and successfully encapsulated on CDs include catechin (Jiang et al. 2019), ferulic acid, gallic acid, vanillic acid, *p*-coumaric acid (Olga et al. 2015; Kim 2020) and α-methylstilbene (Matencio et al. 2017).

Besides the improvement of the stability of polyphenols to withstand harsh conditions during storage, processing and gastrointestinal digestion of food, complexation of these bioactive compounds with CDs can also reduce undesirable interactions that might take place in complex food matrixes. For instance, it has been reported that the free flavonoid catechin was significantly degraded in a protein rich matrix, as compared to a sugar or oil-based matrix (Ho et al. 2019a). This phenomenon was attributed to hydrogen bound interaction between the compound and the protein amide groups. However, this degradation was significantly avoided after catechin encapsulation on β-CDs.

Polyphenol-CD inclusion complexes have also been applied to induce organoleptic modification, such as masking undesirable odours, reducing color loss, controlling the browning of food and the release of flavors (Hu et al. 2021). Carmo and co-workers, for instance, reported the development of a micro-scale delivery system of α-tocopherol using CDs and zein to improve the shelf-life and color stability of fruit beverages. Encapsulation of this polyphenolic compound in β-CDs, did not only increased its solubility in fruit juice beverages but also resulted in reduced color losses, which was attributed to suppressing the degradation of anthocyanins (Saldanha do Carmo et al. 2017a, b). Another example of the use of polyphenol-CD inclusion complexes to mask undesirable tastes comprises the encapsulation of catechin in β-CDs that proved to mask its bitterness (Ho et al. 2019b). Furthermore, CDs can also be used to improve flavor retention by encapsulation of volatile compounds (Ciobanu et al. 2013).

6.3.2 Active Packaging and Preservatives

The consumers' growing health-conscious has led the food industry to produce and provide safe food products with added functionality and extended shelf-lives. The development of active packaging has played an essential role to achieve these goals. In this approach, compounds are added to the packages to prolong the shelf-life of food products (Yildirim et al. 2018). The antimicrobial and antioxidant properties of polyphenols make them suitable to be applied in this specific field, and their encapsulation on CDs allows their compatibility with most packaging materials for food matrixes.

Polyphenols with antimicrobial properties to prevent microbial contamination associated to foodborne diseases as well as food spoilage have been widely investigated, as well as, their encapsulation by CDs for the development of bioactive food packaging. For instance, the antimicrobial activity of hexahydro-β-acids/HPβCD inclusion complex against *Listeria monocytogenes* was higher when compared with the free compound (Lu et al. 2019).

In addition to microbial contamination, oxidative damage also plays a crucial role in food spoilage. Rezaee et al. developed chitosan-gelatin based films to which they added gallic acid encapsulated in β-CDs. The controlled release of this bioactive compound in a short duration enabled high antioxidant activity (Rezaee et al. 2018).

6.3.3 Extraction of Polyphenols

Extraction of polyphenols usually involves the use of organic solvents such as methanol, ethanol, acetone, ethyl acetate or their mixtures. Consumers' growing concerns about environmental pollution and toxicity associated to these solvents underpinned the development of green and less expensive techniques to extract natural polyphenols. CDs have been used to extract compounds such as phenolic acids, flavonoids and stilbenes from different natural sources. Their application is possible due to the host-guest molecular recognition capability between CDs and polyphenolic compounds (Cai et al. 2018).

Zhu et al. (2016) used MβCD to extract several polyphenols (gallic acid, protocatechuic acid, catechin, chlorogenic acid, caffeic acid, sinapic acid, epicatechin and rutin) from jujube fruit. The extraction yield was higher than the one achieved when methanol was used as the organic solvent. It has also been shown that both βCD and HPβCD were more efficient in the extraction of anthocyanins from mulberry when compared to water or ethanol as solvents. Results also showed that the use of CDs increased the thermal stability of polyphenolic compounds during extraction (Xie et al. 2019).

6.4 Conclusion and Future Perspectives

CDs, as naturally derived functional molecules, have attracted great attention in food industry. Host–guest interactions between CDs and polyphenols dominantly contribute to the enhanced stability and bioactivity of polyphenols and, at the same time, guarantee their compatibility with food or packaging matrixes. A great number of studies on the preparation and characterization of inclusion complexes between polyphenols and CDs have been published. The results demonstrated the superior potential of CDs in enhancing the polyphenols' bioactive properties. However, because polyphenols comprise a group of molecules quite diverse, it is not possible to generalize the molecular mechanisms of interaction between CDs and polyphenols. Moreover, the efficiency of the encapsulation depends, essentially, on the size of the CD cavity and the method used to prepare the complexes. The use of substituted CDs has been found to improve the physicochemical properties of these bioactive molecules.

However, most of the developments highlighted in this chapter are still in the embryonic stage of product development, and the methods used to characterize the inclusion complexes lacks normalization, hampering the comparison among works. Most of them lack the evaluation of polyphenols release from the inclusion complex and research polyphenol/CD inclusion complexes' safety and toxicity, which limits their practical application in the food industry. The applications of novel polyphenol/CD inclusion complexes in the food industry must be comprehensively evaluated with a focus on their safety in the future research. The food industry would also benefit from a standardization of the methods used to characterize the inclusion complexes.

Acknowledgments Authors acknowledge the support of the Portuguese Foundation for Science and Technology (FCT) under the scope of the strategic funding of UIDB/04469/2020 unit.

References

Abbas M, Saeed F, Anjum FM et al (2017) Natural polyphenols: an overview. Int J Food Prop 20:1689–1699. https://doi.org/10.1080/10942912.2016.1220393

Albuquerque BR, Heleno SA, Oliveira MBP et al (2021) Phenolic compounds: current industrial applications, limitations and future challenges. Food Funct 12:14–29

Aree T, Jongrungruangchok S (2018) Structure–antioxidant activity relationship of β-cyclodextrin inclusion complexes with olive tyrosol, hydroxytyrosol and oleuropein: deep insights from X-ray analysis, DFT calculation and DPPH assay. Carbohydr Polym 199:661–669. https://doi.org/10.1016/j.carbpol.2018.07.019

Astray G, Mejuto JC, Simal-Gandara J (2020) Latest developments in the application of cyclodextrin host-guest complexes in beverage technology processes. Food Hydrocoll 106:105882. https://doi.org/10.1016/j.foodhyd.2020.105882

Atanasov AG, Zotchev SB, Dirsch VM et al (2021) Natural products in drug discovery: advances and opportunities. Nat Rev Drug Discov 20:200–216. https://doi.org/10.1038/s41573-020-00114-z

Azzi J, Jraij A, Auezova L et al (2018) Novel findings for quercetin encapsulation and preservation with cyclodextrins, liposomes, and drug-in-cyclodextrin-in-liposomes. Food Hydrocoll 81:328–340. https://doi.org/10.1016/j.foodhyd.2018.03.006

Ballesteros LF, Ramirez MJ, Orrego CE et al (2017) Encapsulation of antioxidant phenolic compounds extracted from spent coffee grounds by freeze-drying and spray-drying using different coating materials. Food Chem 237:623–631. https://doi.org/10.1016/j.foodchem.2017.05.142

Bouarab Chibane L, Degraeve P, Ferhout H et al (2019) Plant antimicrobial polyphenols as potential natural food preservatives. J Sci Food Agric 99:1457–1474. https://doi.org/10.1002/jsfa.9357

Bouarab-Chibane L, Forquet V, Lantéri P et al (2019) Antibacterial properties of polyphenols: characterization and QSAR (quantitative structure-activity relationship) models. Front Microbiol 10:1–23. https://doi.org/10.3389/fmicb.2019.00829

Brewer MS (2011) Natural antioxidants: sources, compounds, mechanisms of action, and potential applications. Compr Rev Food Sci Food Saf 10:221–247. https://doi.org/10.1111/j.1541-4337.2011.00156.x

Budryn G, Nebesny E, Pałecz B et al (2014) Inclusion complexes of β-cyclodextrin with chlorogenic acids (CHAs) from crude and purified aqueous extracts of green Robusta coffee beans (Coffea canephora L.). Food Res Int 61:202–213. https://doi.org/10.1016/j.foodres.2013.10.013

Budryn G, Zaczyńska D, Rachwał-Rosiak D, Oracz J (2015) Changes in properties of food proteins after interaction with free and β-cyclodextrin encapsulated hydroxycinnamic acids. Eur Food Res Technol 240:1157–1166. https://doi.org/10.1007/s00217-015-2419-9

Cai R, Yuan Y, Cui L et al (2018) Cyclodextrin-assisted extraction of phenolic compounds: current research and future prospects. Trends Food Sci Technol 79:19–27. https://doi.org/10.1016/j.tifs.2018.06.015

Canales D, Montoille L, Rivas LM et al (2019) Fungicides films of low-density polyethylene (LDPE)/inclusion complexes (carvacrol and cinnamaldehyde) against botrytis cinerea. Coatings 9:1–17

Celebioglu A, Yildiz ZI, Uyar T (2018) Electrospun nanofibers from cyclodextrin inclusion complexes with cineole and p-cymene: enhanced water solubility and thermal stability. Int J Food Sci Technol 53:112–120. https://doi.org/10.1111/ijfs.13564

Chen H, Li L, Ma Y et al (2019) Development of active packaging film containing bioactive components encapsulated in β-cyclodextrin and its application. Food Hydrocoll 90:360–366. https://doi.org/10.1016/j.foodhyd.2018.12.043

Ciobanu A, Landy D, Fourmentin S (2013) Complexation efficiency of cyclodextrins for volatile flavor compounds. Food Res Int 53:110–114. https://doi.org/10.1016/j.foodres.2013.03.048

Conte R, Calarco A, Napoletano A et al (2016) Polyphenols nanoencapsulation for therapeutic applications. J Biomol Res Ther 5. https://doi.org/10.4172/2167-7956.1000139

Cory H, Passarelli S, Szeto J et al (2018) The role of polyphenols in human health and food systems: a mini-review. Front Nutr 5:1–9. https://doi.org/10.3389/fnut.2018.00087

Crini G, Fourmentinn S, Lichtfouse E (2018) Cyclodextrin fundamentals, reactivity and analysis, 1st edn. Springer International Publishing, Cham

da Rosa CG, Borges CD, Zambiazi RC et al (2013) Microencapsulation of gallic acid in chitosan, β-cyclodextrin and xanthan. Ind Crop Prod 46:138–146. https://doi.org/10.1016/j.indcrop.2012.12.053

Daglia M (2012) Polyphenols as antimicrobial agents. Curr Opin Biotechnol 23:174–181. https://doi.org/10.1016/j.copbio.2011.08.007

de Castro DO, Tabary N, Martel B et al (2018) Controlled release of carvacrol and curcumin: bio-based food packaging by synergism action of TEMPO-oxidized cellulose nanocrystals and cyclodextrin. Cellulose 25:1249–1263. https://doi.org/10.1007/s10570-017-1646-6

Diamanti AC, Igoumenidis PE, Mourtzinos I et al (2017) Green extraction of polyphenols from whole pomegranate fruit using cyclodextrins. Food Chem 214:61–66. https://doi.org/10.1016/j.foodchem.2016.07.072

Dong Q, Wang Y, Wen J et al (2017) Inclusion complex of neohesperidin dihydrochalcone and glucosyl-β-cyclodextrin: synthesis, characterization, and bitter masking properties in aqueous solutions. J Mol Liq 241:926–933. https://doi.org/10.1016/j.molliq.2017.05.090

dos Santos Lima B, Shanmugam S, de Souza Siqueira Quintans J et al (2019) Inclusion complex with cyclodextrins enhances the bioavailability of flavonoid compounds: a systematic review. Phytochem Rev 18:1337–1359. https://doi.org/10.1007/s11101-019-09650-y

Durazzo A, Lucarini M, Souto EB et al (2019) Polyphenols: a concise overview on the chemistry, occurrence, and human health. Phyther Res 33:2221–2243. https://doi.org/10.1002/ptr.6419

El-Najjar N, Gali-Muhtasib H, Ketola RA et al (2011) The chemical and biological activities of quinones: overview and implications in analytical detection. Phytochem Rev 10:353–370. https://doi.org/10.1007/s11101-011-9209-1

Falcone Ferreyra ML, Rius SP, Casati P (2012) Flavonoids: biosynthesis, biological functions, and biotechnological applications. Front Plant Sci 3:1–15. https://doi.org/10.3389/fpls.2012.00222

Fateminasab F, Bordbar AK, Shityakov S, Gholami S (2018) Diadzein complexation with unmodified cyclodextrins: a detailed experimental and theoretical study. J Mol Liq 271:80–95

Fernandes A, Sousa A, Azevedo J et al (2013) Effect of cyclodextrins on the thermodynamic and kinetic properties of cyanidin-3-O-glucoside. Food Res Int 51:748–755. https://doi.org/10.1016/j.foodres.2013.01.037

Fernandes A, Rocha MAA, Santos LMNBF et al (2018) Blackberry anthocyanins: β-cyclodextrin fortification for thermal and gastrointestinal stabilization. Food Chem 245:426–431. https://doi.org/10.1016/j.foodchem.2017.10.109

Fraga CG, Croft KD, Kennedy DO, Tomás-Barberán FA (2019) The effects of polyphenols and other bioactives on human health. Food Funct 10:514–528. https://doi.org/10.1039/c8fo01997e

Garrido B, González S, Hermosilla J et al (2019) Carbonate-β-cyclodextrin-based nanosponge as a nanoencapsulation system for piperine: physicochemical characterization. J Soil Sci Plant Nutr 19:620–630. https://doi.org/10.1007/s42729-019-00062-7

Gong L, Li T, Chen F et al (2016) An inclusion complex of eugenol into β-cyclodextrin: preparation, and physicochemical and antifungal characterization. Food Chem 196:324–330. https://doi.org/10.1016/j.foodchem.2015.09.052

Górniak I, Bartoszewski R, Króliczewski J (2019) Comprehensive review of antimicrobial activities of plant flavonoids. Phytochem Rev 18:241–272. https://doi.org/10.1007/s11101-018-9591-z

Ho S, Thoo YY, Young DJ, Siow LF (2017) Cyclodextrin encapsulated catechin: effect of pH, relative humidity and various food models on antioxidant stability. LWT Food Sci Technol 85:232–239. https://doi.org/10.1016/j.lwt.2017.07.028

Ho S, Thoo YY, Young DJ, Siow LF (2019a) Probing the interaction of catechin and its β-CD inclusion complex with different food models. Lwt 100:368–373. https://doi.org/10.1016/j.lwt.2018.10.069

Ho S, Thoo YY, Young DJ, Siow LF (2019b) Stability and recovery of cyclodextrin encapsulated catechin in various food matrices. Food Chem 275:594–599. https://doi.org/10.1016/j.foodchem.2018.09.117

Hossain MA, Park HC, Park SW et al (2020) Synergism of the combination of traditional antibiotics and novel phenolic compounds against escherichia coli. Pathogens 9:1–21. https://doi.org/10.3390/pathogens9100811

Hu Y, Qiu C, Qin Y et al (2021) Cyclodextrin–phytochemical inclusion complexes: promising food materials with targeted nutrition and functionality. Trends Food Sci Technol 109:398–412. https://doi.org/10.1016/j.tifs.2020.12.023

Jiang L, Yang J, Wang Q et al (2019) Physicochemical properties of catechin/β-cyclodextrin inclusion complex obtained via co-precipitation. CYTA J Food 17:544–551. https://doi.org/10.1080/19476337.2019.1612948

Kfoury M, Geagea C, Ruellan S et al (2019) Effect of cyclodextrin and cosolvent on the solubility and antioxidant activity of caffeic acid. Food Chem 278:163–169. https://doi.org/10.1016/j.foodchem.2018.11.055

Kim J-S (2020) Synthesis and characterization of phenolic acid/hydroxypropyl-β-cyclodextrin inclusion complexes. Prev Nutr Food Sci 25:440–448. https://doi.org/10.3746/pnf.2020.25.4.440

Kumar N, Goel N (2019) Phenolic acids: natural versatile molecules with promising therapeutic applications. Biotechnol Rep 24:e00370. https://doi.org/10.1016/j.btre.2019.e00370

Kumar S, Pandey AK (2013) Chemistry and biological activities of flavonoids: an overview. Sci World J. https://doi.org/10.1155/2013/162750

Liu F, Antoniou J, Li Y et al (2016) Chitosan/sulfobutylether-β-cyclodextrin nanoparticles as a potential approach for tea polyphenol encapsulation. Food Hydrocoll 57:291–300. https://doi.org/10.1016/j.foodhyd.2016.01.024

Liu Y, Chen Y, Gao X et al (2020) Application of cyclodextrin in food industry. Crit Rev Food Sci Nutr:1–15. https://doi.org/10.1080/10408398.2020.1856035

Lu N, Xu H, Liu Y (2019) Characterization and antimicrobial activity of a 2-O-methyl-b-cyclodextrin inclusion complex containing hexahydro-b-acids. J Mater Sci 54:4287–4296

Mangolim CS, Moriwaki C, Nogueira AC et al (2014) Curcumin-β-cyclodextrin inclusion complex: stability, solubility, characterisation by FT-IR, FT-Raman, X-ray diffraction and photoacoustic spectroscopy, and food application. Food Chem 153:361–370. https://doi.org/10.1016/j.foodchem.2013.12.067

Marcolino VA, Zanin GM, Durrant LR et al (2011) Interaction of curcumin and bixin with β-cyclodextrin: complexation methods, stability, and applications in food. J Agric Food Chem 59:3348–3357. https://doi.org/10.1021/jf104223k

Matencio A, Hernández-García S, García-Carmona F, López-Nicolás JM (2017) An integral study of cyclodextrins as solubility enhancers of α-methylstilbene, a resveratrol analogue. Food Funct 8:270–277. https://doi.org/10.1039/c6fo01677d

Munin A, Edwards-Lévy F (2011) Encapsulation of natural polyphenolic compounds; a review. Pharmaceutics 3(4):793–829

Mutha RE, Tatiya AU, Surana SJ (2021) Flavonoids as natural phenolic compounds and their role in therapeutics: an overview. Fut J Pharm Sci 7:1–13. https://doi.org/10.1186/s43094-020-00161-8

Nasirullah, Kumar P, Shariff R (2011) Development of nutraceutical carriers for functional food applications. Nutr Food Sci 41:34–43. https://doi.org/10.1108/00346651111102874

Okuda T, Ito H (2011) Tannins of constant structure in medicinal and food plants-hydrolyzable tannins and polyphenols related to tannins. Molecules 16:2191–2217. https://doi.org/10.3390/molecules16032191

Olga G, Styliani C, Ioannis RG (2015) Coencapsulation of ferulic and gallic acid in hp-b-cyclodextrin. Food Chem 185:33–40. https://doi.org/10.1016/j.foodchem.2015.03.058

Olszowy M (2019) What is responsible for antioxidant properties of polyphenolic compounds from plants? Plant Physiol Biochem 144:135–143. https://doi.org/10.1016/j.plaphy.2019.09.039

Pannu N, Bhatnagar A (2019) Biomedicine & pharmacotherapy resveratrol: from enhanced biosynthesis and bioavailability to multitargeting chronic diseases. Biomed Pharmacother 109:2237–2251. https://doi.org/10.1016/j.biopha.2018.11.075

Pecyna P, Wargula J, Murias M (2020) More than resveratrol: new insights into stilbene-based compounds. Biomolecules 10:1–40

Pellicer JA, Fortea MI, Trabal J et al (2018) Optimization of the microencapsulation of synthetic strawberry flavour with different blends of encapsulating agents using spray drying. Powder Technol 338:591–598. https://doi.org/10.1016/j.powtec.2018.07.080

Piccolella S, Crescente G, Candela L, Pacifico S (2019) Nutraceutical polyphenols: new analytical challenges and opportunities. J Pharm Biomed Anal 175:112774. https://doi.org/10.1016/j.jpba.2019.07.022

Pinho E, Grootveld M, Soares G, Henriques M (2014) Cyclodextrins as encapsulation agents for plant bioactive compounds. Carbohydr Polym 101:121–135. https://doi.org/10.1016/j.carbpol.2013.08.078

Pinho E, Soares G, Henriques M (2015a) Evaluation of antibacterial activity of caffeic acid encapsulated by β-cyclodextrins. J Microencapsul 32:804–810. https://doi.org/10.3109/02652048.2015.1094531

Pinho E, Soares G, Henriques M (2015b) Cyclodextrin modulation of gallic acid in vitro antibacterial activity. J Incl Phenom Macrocycl Chem 81:205–214. https://doi.org/10.1007/s10847-014-0449-8

Procházková D, Boušová I, Wilhelmová N (2011) Antioxidant and prooxidant properties of flavonoids. Fitoterapia 82:513–523. https://doi.org/10.1016/j.fitote.2011.01.018

Quilaqueo M, Millao S, Luzardo-Ocampo I et al (2019) Inclusion of piperine in β-cyclodextrin complexes improves their bioaccessibility and in vitro antioxidant capacity. Food Hydrocoll 91:143–152. https://doi.org/10.1016/j.foodhyd.2019.01.011

Rauf A, Imran M, Abu-Izneid T et al (2019) Proanthocyanidins: a comprehensive review. Biomed Pharmacother 116:1–6. https://doi.org/10.1016/j.biopha.2019.108999

Rezaee M, Askari G, EmamDjomeh Z, Salami M (2018) Effect of organic additives on physiochemical properties and anti-oxidant release from chitosan-gelatin composite films to fatty food simulant. Int J Biol Macromol 114:844–850. https://doi.org/10.1016/j.ijbiomac.2018.03.122

Rodríguez-López MI, Mercader-Ros MT, Pellicer JA et al (2020) Evaluation of monoterpene-cyclodextrin complexes as bacterial growth effective hurdles. Food Control 108:106814. https://doi.org/10.1016/j.foodcont.2019.106814

Salazar-Aranda R, Granados-Guzmán G, Pérez-Meseguer J et al (2015) Activity of polyphenolic compounds against candida glabrata. Molecules 20:17903–17912. https://doi.org/10.3390/molecules201017903

Saldanha do Carmo C, Maia C, Poejo J et al (2017a) Microencapsulation of α-tocopherol with zein and β-cyclodextrin using spray drying for colour stability and shelf-life improvement of fruit beverages. RSC Adv 7:32065–32075. https://doi.org/10.1039/c6ra25946d

Saldanha do Carmo C, Pais R, Simplício AL et al (2017b) Improvement of aroma and shelf-life of non-alcoholic beverages through cyclodextrins-limonene inclusion complexes. Food Bioprocess Technol 10:1297–1309. https://doi.org/10.1007/s11947-017-1897-0

Salústio PJ, Pontes P, Conduto C et al (2011) Advanced technologies for oral controlled release: cyclodextrins for oral controlled release. AAPS PharmSciTech 12:1276–1292. https://doi.org/10.1208/s12249-011-9690-2

Salústio PJ, Pinto JF, Costa PC, Cabral-Marques HM (2013) Release profiles of indometacin in β-cyclodextrin complexes from HPMC capsules. J Incl Phenom Macrocycl Chem 75:101–109. https://doi.org/10.1007/s10847-012-0151-7

Sánchez-Maldonado AF, Schieber A, Gänzle MG (2011) Structure-function relationships of the antibacterial activity of phenolic acids and their metabolism by lactic acid bacteria. J Appl Microbiol 111:1176–1184. https://doi.org/10.1111/j.1365-2672.2011.05141.x

Santos EH, Kamimura JA, Hill LE, Gomes CL (2015) Characterization of carvacrol beta-cyclodextrin inclusion complexes as delivery systems for antibacterial and antioxidant applications. LWT Food Sci Technol 60:583–592. https://doi.org/10.1016/j.lwt.2014.08.046

Santos JS, Alvarenga Brizola VR, Granato D (2017) High-throughput assay comparison and standardization for metal chelating capacity screening: a proposal and application. Food Chem 214:515–522. https://doi.org/10.1016/j.foodchem.2016.07.091

Schieber M, Chandel NS (2014) ROS function in redox signaling and oxidative stress. Curr Biol 24:R453–R462. https://doi.org/10.1016/j.cub.2014.03.034

Shao P, Zhang J, Fang Z, Sun P (2014) Complexing of chlorogenic acid with β-cyclodextrins: inclusion effects, antioxidative properties and potential application in grape juice. Food Hydrocoll 41:132–139. https://doi.org/10.1016/j.foodhyd.2014.04.003

Sharif N, Golmakani MT, Niakousari M et al (2018) Active food packaging coatings based on hybrid electrospun gliadin nanofibers containing ferulic acid/hydroxypropyl-beta-cyclodextrin inclusion complexes. Nano 8(11):919. https://doi.org/10.3390/nano8110919

Sharma N, Baldi A (2016) Exploring versatile applications of cyclodextrins: an overview. Drug Deliv 23:739–757. https://doi.org/10.3109/10717544.2014.938839

Simsek T, Simsek S, Mayer C, Rasulev B (2019) Combined computational and experimental study on the inclusion complexes of β-cyclodextrin with selected food phenolic compounds. Struct Chem 30:1395–1406. https://doi.org/10.1007/s11224-019-01347-4

Son EJ, Kim JH, Kim K, Park CB (2016) Quinone and its derivatives for energy harvesting and storage materials. J Mater Chem A 4:11179–11202. https://doi.org/10.1039/c6ta03123d

Soto ML, Falqué E, Domínguez H (2015) Relevance of natural phenolics from grape and derivative products in the formulation of cosmetics. Cosmetics 2:259–276. https://doi.org/10.3390/cosmetics2030259

Stefanachi A, Leonetti F, Pisani L et al (2018) Coumarin: a natural, privileged and versatile scaffold for bioactive compounds. Molecules 23(2):250

Tao F, Hill LE, Peng Y, Gomes CL (2014) Synthesis and characterization of β-cyclodextrin inclusion complexes of thymol and thyme oil for antimicrobial delivery applications. LWT Food Sci Technol 59:247–255. https://doi.org/10.1016/j.lwt.2014.05.037

Teponno RB, Kusari S, Spiteller M (2016) Recent advances in research on lignans and neolignans. Nat Prod Rep 33(9):1044–1092

Thilakarathna SH, Vasantha Rupasinghe HP (2013) Flavonoid bioavailability and attempts for bioavailability enhancement. Nutrients 5:3367–3387. https://doi.org/10.3390/nu5093367

Tian B, Xiao D, Hei T et al (2020) The application and prospects of cyclodextrin inclusion complexes and polymers in the food industry: a review. Polym Int 69:597–603. https://doi.org/10.1002/pi.5992

Wu J, Sun X, Guo X et al (2018) Physicochemical, antioxidant, in vitro release, and heat sealing properties of fish gelatin films incorporated with β-cyclodextrin/curcumin complexes for apple juice preservation. Food Bioprocess Technol 11:447–461. https://doi.org/10.1007/s11947-017-2021-1

Xie J, Xu Y, Shishir MRI et al (2019) Green extraction of mulberry anthocyanin with improved stability using β-cyclodextrin. J Sci Food Agric 99:2494–2503. https://doi.org/10.1002/jsfa.9459

Ye Y, Zhu M, Miao K et al (2017) Development of antimicrobial gelatin-based edible films by incorporation of trans-anethole/β-cyclodextrin inclusion complex. Food Bioprocess Technol 10:1844–1853. https://doi.org/10.1007/s11947-017-1954-8

Yildirim S, Röcker B, Pettersen MK et al (2018) Active packaging applications for food. Compr Rev Food Sci Food Saf 17:165–199. https://doi.org/10.1111/1541-4337.12322

Yildiz ZI, Celebioglu A, Kilic ME et al (2018) Fast-dissolving carvacrol/cyclodextrin inclusion complex electrospun fibers with enhanced thermal stability, water solubility, and antioxidant activity. J Mater Sci 53:15837–15849. https://doi.org/10.1007/s10853-018-2750-1

Zhang H, Tsao R (2016) Dietary polyphenols, oxidative stress and antioxidant and anti-inflammatory effects. Curr Opin Food Sci 8:33–42. https://doi.org/10.1016/j.cofs.2016.02.002

Zhang JQ, Jiang KM, An K et al (2015) Novel water-soluble fisetin/cyclodextrins inclusion complexes: preparation, characterization, molecular docking and bioavailability. Carbohydr Res 418:20–28. https://doi.org/10.1016/j.carres.2015.09.013

Zhao J, Yang J, Xie Y (2019) Improvement strategies for the oral bioavailability of poorly water-soluble flavonoids: an overview. Int J Pharm 570:1–21. https://doi.org/10.1016/j.ijpharm.2019.118642

Zhu QY, Zhang QY, Cao J et al (2016) Cyclodextrin-assisted liquid-solid extraction for determination of the composition of jujube fruit using ultrahigh performance liquid chromatography with electrochemical detection and quadrupole time-of-flight tandem mass spectrometry. Food Chem 213:485–493. https://doi.org/10.1016/j.foodchem.2016.06.115

Chapter 7
Encapsulation of Essential Oils

Jaruporn Rakmai, Juan-Carlos Mejuto, Yaxin Sang, Seid Mahdi Jafari, Jianbo Xiao, and Jesus Simal-Gandara

7.1 Introduction to Essential Oils

Essential oils (EOs) are volatile compounds extracted from aromatic plants containing the secondary plant metabolites which are complex mixtures of many terpenes, terpenoids and their derivatives. Monoterpene and sesquiterpene derivatives consist of different classes of acyclic and cyclic compounds, such as alcohols, aldehydes, ketones, lactones, esters, phenols and oxides. Phenylpropanoids, other substantial components of EOs, are obtained from the carbon skeleton of phenylalanine (Kfoury et al. 2019). Figure 7.1 reports some examples of the chemical components of typical essential oils which can be subdivided into three groups of related to chemical constituents: terpenes, terpenoids and phenylpropanoids. The EOs are generally obtained by hydro or steam distillation or by a mechanical process. They usually

J. Rakmai
Biomass and Bio-energy Technology Division, Kasetsart Agricultural and Agro-Industrial Product Improvement Institute (KAPI), Kasetsart University, Bangkok, Thailand
e-mail: aapjor@ku.ac.th

J.-C. Mejuto
Department of Physical Chemistry, Faculty of Science, University of Vigo, Ourense, Spain
e-mail: xmejuto@uvigo.es

Y. Sang
College of Food Science and Technology, Hebei Agricultural University, Baoding, China

S. M. Jafari
Department of Food Materials and Process Design Engineering, Gorgan University of Agricultural Sciences and Natural Resources, Gorgan, Iran

Nutrition and Bromatology Group, Department of Analytical and Food Chemistry, Faculty of Food Science and Technology, University of Vigo, Ourense, Spain
e-mail: smjafari@gau.ac.ir; smjafari@uvigo.es

T. M. Ho et al. (eds.), *Functionality of Cyclodextrins in Encapsulation for Food Applications*, https://doi.org/10.1007/978-3-030-80056-7_7

consist of 20–60 components and some of these chemical components are responsible for the biological properties (Bilia et al. 2014).

EOs have been found various medicinal uses for human and animals; they are mostly used externally in the form of inhalations, mouthwashes and gargles. For oral application which is rarely used, EOs are designated as safe and placed under GRAS category for ingestion. In addition, the EOs (in diluted form) are generally considered safe for dermal application. However, some EOs such as citrus oils may cause skin irritation and darkening of skin upon exposure to sun. Some EOs may cause irritation to the eyes when used in inhalation form (Franz 2010). EOs afford interesting applications in food and drink as flavoring agents, pharmaceutical field, cosmetics and household cleaning products owing to their diverse and relevant biological activities (Lee and Choi 2012; Posadzki et al. 2012; Reeds 2000; Wadhwa et al. 2017).

Spices and herbs have been used for flavor, aroma, color, and preservation of foods and beverages for several hundred years (Embuscado 2015). They are acquiring importance in recent years as natural sources of food preservatives. Their phenolic compounds play an important role in antimicrobial and antioxidant effects (Martinez-Gracia et al. 2015). Application of EOs from spices and herbs is considered to be a better option than synthetic chemical additives. It has become popular among consumers who consume organic foods (Macwan et al. 2016).

EOs from herbs; including thyme, garlic, cinnamon, basil, tarragon, cumin, bay, black pepper, parsley and nutmeg, have been tested for antimicrobial activity. EOs form thyme, garlic and cinnamon showed strong antimicrobial activity against food-borne pathogens and spoilage microorganisms. EOs from cumin, bay, black pepper, parsley and nutmeg presented moderate antimicrobial activity with high minimum inhibitory concentration (MIC) value. However, EOs from orange, basil and tarragon had no inhibitory effect against tested microorganisms (Garcia-Diez et al. 2017). The application of various EOs as potential substitutes for synthetic antioxidants in the meat industry has been reported by Pateiro et al. (2018). The potential of oregano EO as a natural antioxidant was evaluated by Fasseas et al. (2007). During 12 days storage under refrigerated conditions, a significant reduction of the oxidation reactions in the samples, raw and cooked porcine and bovine ground meat, treated with 3% (w/w) of oregano EO was observed. The strong

J. Xiao
Nutrition and Bromatology Group, Department of Analytical and Food Chemistry, Faculty of Food Science and Technology, University of Vigo, Ourense, Spain

Institute of Chinese Medical Sciences, State Key Laboratory of Quality Research in Chinese Medicine, University of Macau, Taipa, Macau
e-mail: jianboxiao@uvigo.es

J. Simal-Gandara (✉)
Nutrition and Bromatology Group, Department of Analytical and Food Chemistry, Faculty of Food Science and Technology, University of Vigo, Ourense, Spain
e-mail: jsimal@uvigo.es

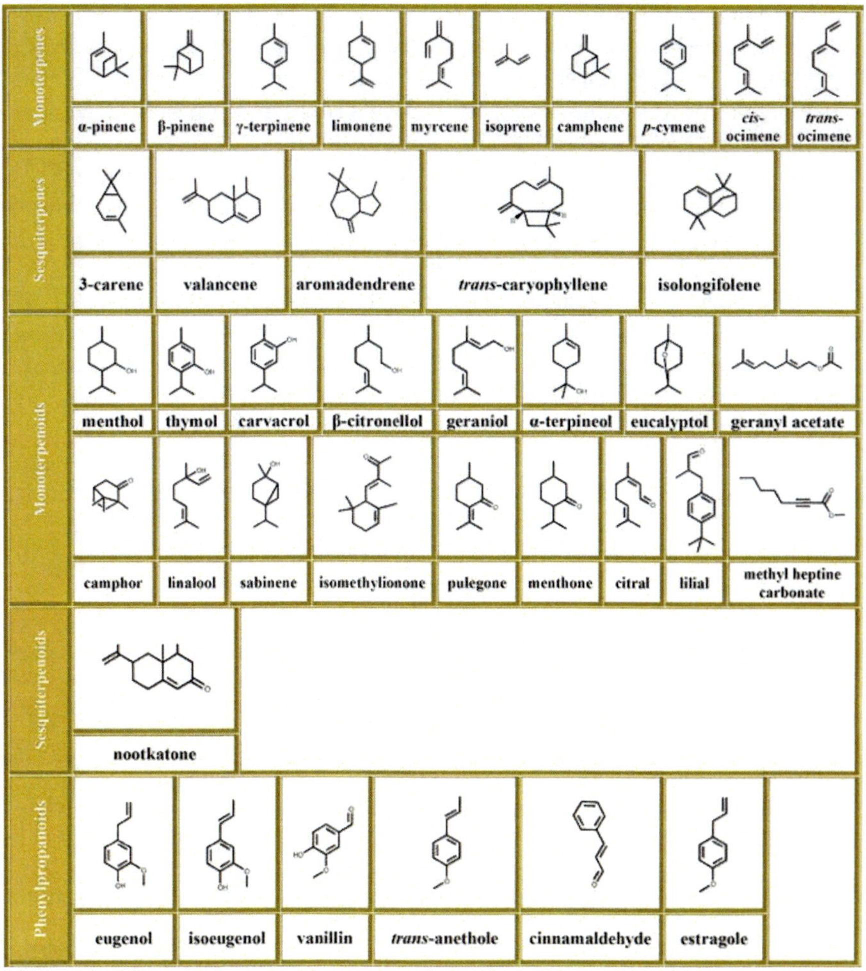

Fig. 7.1 Chemical components of typical essential oils which can be subdivided into three groups of related chemical constituents: terpenes, terpenoids and phenylpropanoids

antioxidant activity was attributed to the presence of thymol and carvacrol in oregano EO (Lagouri and Boskou 1995). EOs from basil (*Ocimum basilicum*) was also tested for antioxidant activity through lipid oxidation in order to apply as a natural food preservative in a meat product (Sharafati-Chaleshtori et al. 2015). This EO decreased lipid oxidation of beef burger during 12 days storage at 4 °C. EO has been reported to decrease the lipid peroxidation, reduce radical formation, avoid the deterioration of sarcoplasmic proteins, and decrease deterioration reactions extending the shelf-life of fresh chicken breast meat product (Fratianni et al. 2010). This EO can be used in small amount because of the presence of carvacrol, a powerful monoterpenoid with strong antioxidant properties. It has also been used to decrease lipid oxidation without affecting the proteolysis and sensory properties in dry fermented poultry meat sausage and fresh chicken sausages (El Adab and Hassouna 2016; Sharma et al. 2017). Rosemary and marjoram EOs were reported to be effective against lipid oxidation in frozen beef patties during 3 months storage at −18 °C. Besides, sensory evaluation also showed a positive effect of EOs incorporation among food panelist (Mohamed and Mansour 2012). Incorporation of EOs from ginger (*Zingiber officinale*) and basilica (*Ocimum gratissimum*) also enhanced the oxidative stability of the beef patties (Dzudie et al. 2004).

Citrus EOs derived from *Citrus*, a genus of lowering trees belongs to family Rutaceae, which produce citrus fruits, including important crops such as *Citrus sinensis* (Orange), *Citrus junos* (Yuzu), *Citrus reticulata* (Mandarin), *Citrus limon* (Lemon), *Citrus aurantifolia* (Limes), *Citrus bergamia* (Bergamot), *Citrus paradisi* (Grapefruit), and *Citrus japonica* (Kumquat) (Anwar et al. 2008; Dosoky and Setzer 2018). *Citrus* is beneficial for health promotion because they have high levels of elemental bioactive compounds which have numerous health benefits such as phenols, flavonoids, limonoids, and vitamins, particularly vitamin C and carotenoids (Duarte et al. 2016a, b). Several important secondary metabolites; such as phenolics, ascorbic acid, flavanones, and pectin, in citrus fruits are recognized to have antioxidants properties beneficial for human health (Chhikara et al. 2018). *Citrus* has also been reported to reduce the chances of developing mental health diseases, like anxiety and Alzheimer's, etc. (Ayaz et al. 2017). Many parts of *Citrus* are rich sources of EOs (Othman et al. 2016). Citrus EOs with their biological properties such as antioxidant, antidiabetic, insecticidal, antifungal, and antibacterial properties, have been applied in the food, agricultural, pharmaceutical, cosmetic and sanitary industries (Lawless 2002; Palazzolo et al. 2013; Wolffenbuttel et al. 2018). Citrus EOs have drawn more attention compared to other essential oils for wide applications in food including food formulations, packaging, and preservation to ensure food quality and safety (Bora et al. 2020; Calo et al. 2015).

Other edible plants such as tea tree and guava are also sources of EOs. Tea tree oil is obtained from leaves and terminal branchlets of the narrow-leaf tea tree (Melaleuca alternifolia) by distillation. This EO has been used as a remedy for skin diseases including acne, eczema, herpes simplex, warts, wounds, burns, insect bites, dandruff and nail mycoses. It has biological activities such as antibacterial, antifungal, antiviral, anti-inflammatory, anti-tumoral, analgesic, insecticidal and acaricidal activities (Barbosa et al. 2013; Benelli et al. 2013; Calcabrini et al. 2004; Carson et al. 2006; Gomez-Rincon et al. 2014; Pazyar et al. 2013; Satchell et al. 2002). In the market, it is known as natural antimicrobial and anti-inflammatory agent (Cox et al. 2001). The European Medicines Agency (European Medicines Agency 2013) has reported that tea tree EO can be applied for the treatment of small superficial wounds, insect bites, small boils, itching and irritation, minor inflammation of the oral mucosa. The EO products are present in various formulations, including pure oil, ointments, wart paint, (Bhushan and Beck 1997; Enshaieh et al. 2007; Hammer 2015), and household products such as cleansers, detergents and fabric softeners (Cox et al. 2001; Crawford et al. 2004; Hartford and Zug 2005). It is also used in cosmetic products, infectious diseases (such as colds, influenza and chickenpox), fever, aromatherapy for diseases of the respiratory system (such as asthma, tuberculosis, bronchitis, sinusitis and whooping cough) and genitourinary diseases (such as candidiasis, vaginitis, cystitis and genital pruritus) (Cox et al. 2001; Hartford and Zug 2005; Lawless 2014). Guava EO is extracted from guava leaf which has long been applied as a traditional medicine because of its biological properties (Jaiarj et al. 1999; Lozoya et al. 2002; Oh et al. 2005). It has antiproliferation, antioxidant and antimicrobial activities (Manosroi et al. 2006; Sacchetti et al. 2005). The major components of guava EO are limonene, β-caryophyllene, 1,8-cineole and α-pinene (Hsin-Chun et al. 2007; Ogunwande et al. 2003).

Flowering plants like lavender and yarrow which have been applied as traditional medicines for centuries are also sources of EOs. Lavender (Lavandula spp.) has been used as a therapeutic agent. It was widely used as an antibacterial in World War. It is traditionally believed to have some biological properties like sedative, carminative, anti-depressive, anti-inflammatory and antimicrobial effects. Lavender EO has been used as a traditional medicine and applied in cosmetic products (Cavanagh and Wilkinson 2002; Evelegh 1996; Gattefosse 1937; Grieve 1931). It has been found to be active against many bacterial species, including *Staphylococcus aureus*, *Streptococcus pyogenes*, *Enterobacter aerogenes*, *Pseudomonas aeruginosa*, *Citrobacter freundii*, *Escherichia coli*, *Proteus vulgaris*, *Shigella sonnei* and *Propionibacterium acne* (Hammer et al. 1999; Lis-Balchin et al. 1998; Moon et al. 2004a; Nelson 1997). Antifungal property also has been reported against *Aspergillus nidulans*, *Trichophyton mentagrophytes*, *Leptosphaeria maculans* and *Sclerotinia sclerotiorum* (Daferera et al. 2000; Moon et al. 2004b). Yarrow (*Achillea millefolium*) is widely applied in folk medicine because of its broad spectrum of pharmacological properties (Wichtl 1997). In Europe, it has been used as a remedy to treat digestive problems, amenorrhea, diabetes and hepatobiliary diseases. The consumption of yarrow also has been reported for antitumor and anti-inflammatory (Baretta et al. 2012; Cavalcanti et al. 2006; Jonsdottir et al. 2011). Chemical composition of yarrow EO comprises of carvacrol, camphor, thymol, linalool and 1,8-cineole. Carvacrol was found as a major component of yarrow EO (Alfatemi et al. 2014). Several biological studies of yarrow have been performed including antimicrobial, antioxidant, anti-inflammatory and antispasmodic properties (Al-Hindawi et al. 1989; Candan et al. 2003; Karamenderes and Apaydin 2003; Potrich et al. 2010; Skocibusic et al. 2004; Trumbeckaite et al. 2011).

EOs are mixtures of volatile organic compounds possess numerous biological properties. About 160 EOs are considered as Generally Recognized as Safe (GRAS) by the Food and Drug Administration (FDA). In food production, EOs have been used as flavoring agent, food grade preservative, bioactive components for functional food and active packaging for food products. Since EOs are not stable and can be volatilized during processing or long-time storage, the direct utilization of EOs is limited by their poor solubility, bioavailability and stability (in presence of oxygen, heat and light). Therefore, the EOs have to be encapsulated to enhance solubility, stability and bioavailability. Encapsulation also masks the objectionable taste and odor of volatile oxidation products. CDs are known to have adequate thermal and chemical stability. Although they are more susceptible to α-amylases (from saliva and pancreas), they are stable to high temperature and not be hydrolyzed by β-amylases and glucoamylases. Furthermore, α, β, γ-CDs have also been approved as additive for food industry in the European Union and granted GRAS status (Ciobanu et al. 2013; Costa et al. 2016; da Silveira et al. 2014; Hedges et al. 1995; Noichinda and Suppavorasatit 2019; Prakash et al. 2015; Rashed et al. 2017).

7.2 Preparation and Characteristics of Essential Oil-Cyclodextrin Complexes

7.2.1 Techniques for the Complex Preparation

The preparation technique employed for inclusion complex formation of cyclodextrins (CDs) and their guests plays an important role affecting morphometrical properties and performance of the products. Literature reports a variety of encapsulation techniques for preparing EO-CD complexes (Fig. 7.2).

Co-precipitation method has been applied for encapsulation of peppermint (*Mentha* × *villosa* Hudson) EO (Martins et al. 2007), β-caryophyllene (Liu et al.

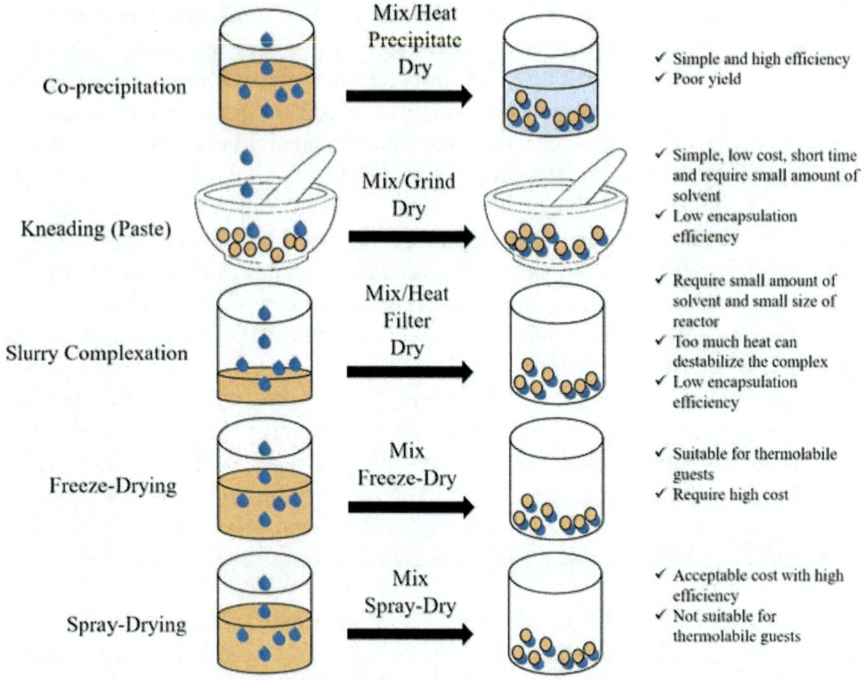

Fig. 7.2 Techniques for inclusion complex formation of cyclodextrins and essential oils

2013), *Xiangfu Siwu* decoction EO (Xi et al. 2015) and *Satureja montana* EO (Haloci et al. 2014) with β-CD. To prepare the complex, CD and water are added to a container at a constant temperature (40/50 °C) with stirring until obtain saturated solution. Ethanolic solution of EO is added dropwise to the saturated CD solution, with continuous stirring and the same temperature as mentioned above. Then the solution is cooled at room temperature and maintained overnight at 4 °C. The cold precipitated inclusion complex can be recovered by vacuum filtration and the filtrate is washed with suitable organic solvent before freeze dried in the last step. This technique is suitable for hydrophobic guests with high efficiency; however, it generates poor yield because of the competitive inhibition from organic solvents used as the precipitant (Hirayama and Uekama 1987; Saenger 1980).

Siqueira-Lima et al. (2014) used slurry method to prepare inclusion complex of β-CD and EO from *Lippia grata* leaf. In this method, slurry of certain amount of β-CD is prepared in water, then certain quantity of EO is added and stirred for about 36 h. The resultant product is filtered, and thereafter appropriately dried. The major advantage of this technique is the requirement of smaller amount of water and smaller size of reactor compared to some of other techniques. However, too much heat can destabilize the complex and complexation reaction may not be take place completely.

Kneading or paste method has been used for encapsulation of EOs form thyme, thymol (Tao et al. 2014) and *Lippia sidoides* (Fernandes et al. 2004). In this technique, β-CD is taken in porcelain motor and small amount of purified water is added and mixed until it forms paste. Then EO is incorporated into the paste through mixing. Finally, the solid is dried under vacuum at room temperature. It is a common, simple, low-cost and short-time method. However, it provides low encapsulation efficiency.

Super critical carbon dioxide has been used for preparation of β-CD inclusion complexes encapsulating carvacrol, eugenol and thymol (Locci et al. 2004). To prepare the complex, CD and EO are loaded into the autoclave thermostat and pressurized with carbon dioxide at constant pressure and contact time to form inclusion complex. Then a rapid drop in pressure is operated which allows carbon dioxide vaporization to separate solid inclusion complex from solvent (Nerome et al. 2013; Wadhwa et al. 2017).

Freeze-drying method has been used for encapsulation of EOs from cinnamon, clove with β-CD (Hill et al. 2013), black pepper (Rakmai et al. 2017a), guava (Rakmai et al. 2018), yarrow (Rakmai et al. 2017b), thymol and thyme (Tao et al. 2014). This technique is suitable for thermolabile or hydrophilic guests (Saenger 1980) and has been wildly applied for CD inclusion complex formation, especially water soluble hydroxypropyl-β-CD (Cheirsilp and Rakmai 2016); however, high cost is required for this technique. To prepare the complex, CD is dissolved in water and the required proportion of EO is slowly added to the CD solution with stirring to allow inclusion complex formation. Then the solution is freeze-dried and washed with organic solvent to remove unencapsulated EO. The powder is dried under vacuum (Hirayama and Uekama 1987; Junco et al. 2002).

Spray drying method has been applied for encapsulation of EOs from *lemongrass oil with β-CD (*Phunpee et al. 2017*)*. This technique is popular in industry because of its high efficiency with acceptable cost. The limitation of this technique is it not suitable for thermolabile guests. To encapsulate the guest, the required proportion of EO is slowly added to the CD solution with stirring to allow inclusion complex formation. The solution is homogenized and dried by spray-drying (Almeida and Marques 2004; Junco et al. 2002; Marreto et al. 2008).

7.2.2 Encapsulation Capacity

The interactions between CDs and their guests are not permanent, it is rather dynamic. The formation of inclusion complex depends on the equilibrium reached between complex and free substrates; CD or guest (Abarca et al. 2015). There are several factors appear to affect the interactions between CDs and guest molecules including type of CD which differ in cavity size (Astray et al. 2009), type of guest molecule (Fenyvesi et al. 2016), hydrophobicity level of guest molecule (Crini 2014) and preparation method (Abarca et al. 2015).

The percentage of a guest molecule encapsulated within CD is normally referred as encapsulation efficiency or entrapment efficiency. An extensive amount of research regarding inclusion complex formation of CDs and EOs or EO components, some of them were mentioned about encapsulation capacity (Table 7.1).

Encapsulation efficiency (EE) of the EOs is different from their pure compounds because of the presence of other minority components. For example, in the case of the encapsulation efficiency of black pepper EO, it was observed to be 50.55% (Rakmai et al. 2017a). This was less than its pure major compound, β-caryophyllene, which was found to be 85.30%. The explanation for this is because the presence of other components in black pepper EO such as limonene, pinene and δ-3-carene (Singh et al. 2004) can have high affinities for CD that compete for inclusion complex formations.

7.2.3 Release Profile of the Complex

The release profile of the guests out of inclusion complexes can be controlled using upfront dissolving and releasing mechanism's design. The guests can accurately release with a suitable release rate in the most appropriate time and a certain temperature (Babaoglu et al. 2017; Desai and Park 2005; Kytariolos et al. 2010; Ren et al. 2018). Several mathematical models can be applied to account for release kinetics of the guests. Release of EOs from CDs can be analyzed by zero-order, first-order and Avrami's equation kinetics. The zero-order model express extended release, the first-order represents an immediate mode, and Avrami's equation is mostly used to explain the controlled release kinetics (Costa and Lobo 2001; Dash

Table 7.1 Encapsulation capacity of cyclodextrins and essential oils (EO)

Major pure compound (MP) or EO	CD	Technique	EE (%)	References
Linalool (MP)	HPBCD	Microwave Irradiation	42.6–60.8	Rodríguez-López et al. (2020)
Linalool (MP)	HPBCD	Ultrasound	61.3–93.7	
Limonene (MP)	HPBCD	Freeze-drying	91.80	Rakmai et al. (2018)
Guava leaf oil (EO)	HPBCD	Freeze-drying	52.50	
β-Caryophyllene (MP)	HPBCD	Freeze-drying	85.30	Rakmai et al. (2017a)
Black pepper oil	HPBCD	Freeze-drying	50.55	
Carvacrol (MP)	HPBCD	Freeze-drying	86.59	Rakmai et al. (2017b)
Yarrow oil (EO)	HPBCD	Freeze-drying	45.05	
Carvacrol (MP)	BCD	Co-precipitation	66.13–73.63	Anaya-Castro et al. (2017)
Oregano oil (MP)	BCD	Co-precipitation	46.99–77.57	
Eugenol (MP)	BCD	Co-precipitation	96.32–97.94	
Clove oil (EO)	BCD	Co-precipitation	30.12–63.63	
Estragole (MP)	BCD	Freeze-drying	16–47	Kfoury et al. (2015)
Estragole (MP)	HPBCD	Freeze-drying	40–79	
Cinnamaldehyde (MP)	BCD	Freeze-drying	84.70	Hill et al. (2013)
Cinnamon oil (EO)	BCD	Freeze-drying	41.72	
Eugenol (MP)	BCD	Freeze-drying	90.15	
Clove oil (EO)	BCD	Freeze-drying	77.74	

et al. 2010; Gibaldi and Feldman 1967; Ren et al. 2018; Shiga et al. 2001, 2003). The release patterns can be divided and described by the following equations:

$$\text{Zero} - \text{order kinetics}: R_1 = k_1 t \tag{7.1}$$

where R_1 is the fraction of encapsulated compound released in time t and k_1 is the zero-order release constant.

$$\text{First} - \text{order kinetics}: \ln\left(1 - R_2\right) = k_2 t \tag{7.2}$$

where R_2 is the fraction of encapsulated compound released in time t and k_2 is the first order release constant.

$$\text{Avrami's equation}: R_3 = e^{\left[-(k_3 t)^n\right]} \tag{7.3}$$

where R_3 is the fraction of encapsulated compound released at time t, k_3 is the release rate constant and n represents the release parameter. The equation can be further rearranged by double logarithm in Eq. (7.4)

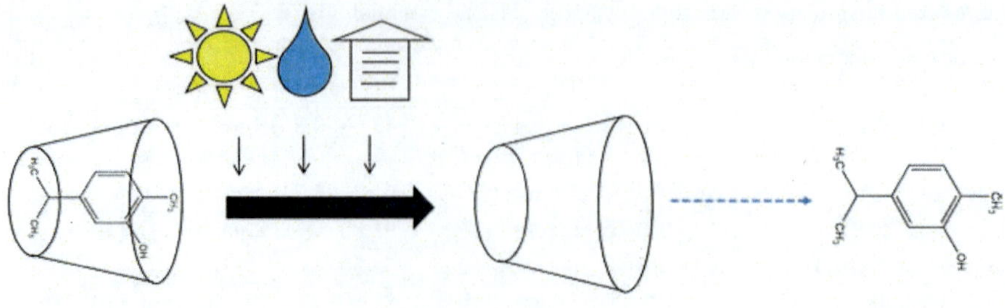

Fig. 7.3 Release of essential oils from complexes as a function of temperature, humidity and storage time

$$\ln(-\ln R_3) = n \ln k_3 + n \ln t \quad (7.4)$$

A linear plot of $\ln(-\ln R_3)$ against $\ln t$ allows the release rate constant k_3 (intercept) and release parameter n (slope) to be determined.

Environmental factors affecting release of EOs from inclusion complexes have been reported as temperature, humidity and storage time which can be adjusted to control the release of EOs (Fig. 7.3). Temperature impacts on the Brownian motion of EO molecules and CD's wall. With a rise of temperature, molecular Brownian motion is accelerated which improves the speed of EO escaping from inclusion complexes. High temperature can also damage the wall material and reduce the apparent activation energy, which cause an increase in number of activated molecules leading to easily release of EOs (Zhu et al. 2014). Ren et al. (2018) studied effect of temperature on release rate of eucalyptus EO from β-CD inclusion complex. The release rate was evaluated for 24 days at different temperature. It was found that at 5, 26 and 45 °C, the release of eucalyptus EO was 17.06%, 23.39% and 40.56%, respectively. Similar observation of CD-curcumin complex was reported by Mangolim et al. (2014).

Water participates releasing the guests from CD shells. The release rate of EOs from CD inclusion complexes increases with an increase of storage humidity. This is because the relative humidity causes wall suction and partial dissolution, destroying the wall structure and releasing EOs out of the complexes. In addition, encapsulated guests in the CD complexes can be replaced by water molecules during storage in high environmental humidity (Ayala-Zavala et al. 2008; Reineccius et al. 2002). Effect of humidity on release of EOs from inclusion complexes was reported. Kfoury et al. (2016) found that the deliverance of encapsulated trans-anethole, chemical component of EO, in β-CD increased with an increase in the humidity. Ren et al. (2018) investigated the release of eucalyptus EO from β-CD complex in different relative humidity values. They found that the retention rate of EO in the complex was gradually reduced to 67.17%, 55.40% and 38.83% with relative humidities of 52%, 71% and 94%, respectively.

Ren et al. (2018) investigated the retention rate of eucalyptus EO. After storage for 70 days, 64.71% of the initial EO remained. Among mathematical models, the release profile in this study showed better fit to the Avrami's equation with the highest coefficient of determination (R^2) of regression analysis. Similar observations on prolonged-release have been reported for EO from *Anemopsis californica* and usnic acid (Lira et al. 2009; Vlachou et al. 2017).

7.3 Applications of Essential Oil-Cyclodextrin Complexes in Food Production

7.3.1 Applications of Essential Oil-Cyclodextrin Complexes in Functional Foods, Natural Flavoring Agents and Natural Preservatives

There have been reports in recent years about nanoparticles or complexed molecules formed by various EOs with CDs (Matencio et al. 2016, 2017; Rakmai et al. 2018; Yildiz et al. 2018; Zhang et al. 2016). The complexation of bioactive compounds from EOs with CDs increases the capacity of bio-efficacies such as antioxidant, anti-aging, antimicrobial and anti-inflammatory, which can be used by the food industry to fortify products (Celebioglu et al. 2018; Das et al. 2019; Ho et al. 2017; Lopez-Nicolas et al. 2014).

Fragrant EOs consisting of aroma compounds have been used as natural flavors. They were added to enhance the aroma of a food or improve the original taste. EOs or their components including EOs from lemon, sweet orange, peppermint, lavender, *S. sclarea* L, and EO components including linalool and limonene have been reported to use as food additives for flavoring properties. Cinnamon EO was also used for desserts, chocolate, candies, liquor and tea as a flavor and condiment in cooking. Generally, the components of EOs are highly volatile compounds which may undergo oxidation or volatilization on exposure to light, heat or oxygen. Encapsulation of flavor compounds from EOs within CD has been reported. CDs, α, β or γ types were evaluated complexation efficiency for volatile flavor compounds from EOs including α-pinene, eucalyptol, linalool, limonene, camphene, β-pinene, ρ-cymene, myrcene, menthol, menthone, *trans*-anethole, camphor and pulegone (Ciobanu et al. 2013; Paliyath et al. 2010; Petrovic et al. 2013).

EOs from some spices such as cumin, thyme, clove, and cinnamon have been reported to have anti-rancidity effect and potential implications to prolong the induction time of corn oil and extend the shelf life of cooked turkey patties (Loizzo et al. 2015). EOs from cinnamon and garlic, or pure compounds from EOs such as estragole, linalool and eugenol, have been applied as food preservative. These EOs or some essential oil components were reported to possess antimicrobial, antifungal, antiviral and insecticidal properties. Some of them have also been approved by FDA as GRAS substances to be employed as food additives (Ayala-Zavala et al. 2008; Kfoury et al. 2015; Letizia et al. 2003).

Recent applications of EO-CD complexes in development of functional foods, natural flavoring agents and natural preservatives are summarized in Table 7.2.

Table 7.2 Applications of essential oil-cyclodextrin complexes in development of functional foods, natural flavoring agents, natural preservatives and food packaging

Cyclodextrin	Essential oil or pure compound	Applications	References
β-CD	Oregano	For food preservation; it preserved freshly cut purple yam, prevented browning and extended the shelf life of purple yam.	Huang et al. (2020)
β-CD	Orange	For food preservation; delivered antimicrobial against *Aspergillus* spp., it delayed the microbiological spoilage of cakes from 30 to 150 days.	Kringel et al. (2020)
β-CD	Basil	For food preservation; delivered antimicrobial against foodborne pathogens, it was incorporated into sachets to use as food preservatives and tested against the foodborne bacteria *S. aureus*, *E. coli*, *L. monocytogenes*, *P. aeruginosa*, *S. Enteritidis*, and the food-spoilage mold *B. nivea*.	Marques et al. (2019)
β-CD	Rosemary	For food preservation; delivered antimicrobial against *Salmonella* Typhimurim, *Listeria monocytogenes*, *Candida tropicalis*, and *Saccharomyces pastorianus* in pasteurized tomato juice.	Garcia-Sotelo et al. (2019)
α-CD, β-CD	Cinnamon	For food packaging; provided controlled-release active packaging and delivered antimicrobial against foodborne pathogens *E. coli* O157:H7 CECT 5947, *B. thermosphacta* CECT 847 and *L. monocytogenes* CECT 911.	Simionato et al. (2019)
β-CD	Cinnamon	For food packaging; provided active polyvinyl alcohol nanofibrous films for sustained release of antimicrobial against *Staphylococcus aureus* and *Escherichia coli*, it delayed decay of mushroom during storage.	Pan et al. (2019)
β-CD	Thymol, Carvacrol	For food packaging; combined essential oil-CD complexes with biodegradable polylactic acid, it gave fungus *Alternaria alternata* inhibition after 10 days, and it could be applied for agrofood.	Frine et al. (2019)
β-CD, HPBCD, α-CD	Coriander	For food packaging; provided controlled-release active packaging and delivered antimicrobial against foodborne pathogens *Campylobacter* spp.	Silva et al. (2019)
β-CD	Oregano	For functional food; developed a functional lipid-based nutrient supplement, provided targeted intestinal delivery of bioactive, reduce intestinal parasitic infections.	Gaur et al. (2018)

(continued)

Table 7.2 (continued)

Cyclodextrin	Essential oil or pure compound	Applications	References
β-CD	Cinnamon	For food packaging; provided polylactic acid active nanofilm and delivered antimicrobial against foodborne pathogens *Escherichia coli* and *Staphylococcus aureus*, it can prolong shelf life of pork.	Wen et al. (2016)
β-CD	Eugenol, cinnamon, clove, trans-cinnamaldehyde	For food preservation; delivered antimicrobial against foodborne pathogens, it was able to inhibit *Salmonella enterica* serovar, *Typhimurium* LT2 and *Listeria innocua*.	Hill et al. (2013)

Note: *β-CD* beta-cyclodextrin, *α-CD* alpha-cyclodextrin, *HPBCD* hydroxypropyl-beta-cyclodextrin

7.3.2 Applications of Essential Oil-Cyclodextrin Complexes in Active Food Packaging

As demand for higher quality food and food safety together with environmentally-friendly products increase with the continuous improvement of living standard, the development of natural active food packaging has been growing along with the development of those food products. Application of EOs in active packaging for food preservation has been considered as a key research and development priority in the twenty-first century. A number of researches reported that the major functions of EO-active packaging for food applications are antimicrobial and antioxidant properties (Ju et al. 2019; Ribeiro-Santos et al. 2017). EOs-active packaging system not only allows for slow release of EOs but also avoids the adverse effects of high temperature during processing and direct contact between EOs and food (Dai and Ying 2002). Solubility of antimicrobial agents in food is an important factor affecting the antimicrobial effects of EOs (Khaneghah et al. 2018).

Various EOs have been encapsulated in CDs and applied in food packaging. Frine et al. (2019) encapsulated EO active compounds, thymol or carvacrol, in β-CDs and applied in biodegradable packages polylactic acid. As the amount of encapsulation increased, the polymer surface showed a decrease in the number of cuts and folds, thereby reducing the stiffness of the packaging. The thermogravimetric analysis revealed that the temperature of degradation of the packaging slightly decreased as the concentration of the complexes increased. In addition, the inclusion complexes lowered the intermolecular forces of the polymer chains as plasticizers, thereby improving the breaking point. These active packages containing 2.5–5.0% complexes showed antifungal activity against *Alternaria alternata* after 10 days of incubation. Other applications of EO-CD complexes in development of food packaging are shown in Table 7.2.

7.4 Conclusion and Future Perspectives

EOs are mixtures of volatile organic compounds extracted from aromatic plants such as herbs and spices, citrus plants, edible plants and flowering plants. EOs possess numerous biological properties and many of them are considered as GRAS by the FDA. Chemical components of typical EOs can be subdivided into three groups of related chemical constituents: terpenes, terpenoids and phenylpropanoids. Application of EOs from spices and herbs is considered to be a better option than synthetic chemical additives. It has become popular among consumers who consume organic foods. However, the direct utilization of EOs is limited by their poor solubility, bioavailability and stability (in presence of oxygen, heat and light). Therefore, the EOs have to be encapsulated to overcome these limitations. Inclusion complex formation with CDs is widely used for several applications. The preparation technique employed for inclusion complex formation of CDs and EOs plays an important role on properties of the products. Several mathematical models can be applied to account for release kinetics of EOs from CDs such as zero-order, first-order and Avrami's equation kinetics. Environmental factors affecting release of EOs from complexes have been reported are temperature, humidity and storage time which can be adjusted to control the release of EOs. In food production, encapsulated EOs or complexes of CDs-EOs have been used as natural flavoring agents, natural preservatives, bioactive components for functional food and active packaging for food products. Further work is needed to reduce the cost of the encapsulated products.

References

Abarca RL, Rodriguez FJ, Guarda A et al (2015) Characterization of beta-cyclodextrin inclusion complexes containing an essential oil component. Food Chem 196:968–975

Alfatemi SMH, Rad JS, Rad MS et al (2014) Chemical composition, antioxidant activity and in vitro antibacterial activity of Achilleab wilhelmsii C. Koch essential oil on methicillin-susceptible and methicillin-resistant Staphylococcus aureus spp. 3. Biotech 5:39–44

Al-Hindawi MK, Al-Deen IH, Nabi MH et al (1989) Antiinflammatory activity of some Iraqi plants using intact rats. J Ethnopharmacol 26:163–168

Almeida R, Marques HC (2004) Pulmonary administration of beclomethasone: a-cyclodextrin complex. In: Duchene D (ed) Proceedings of the 12th international cyclodextrin symposium, Montpellier, France, 2004. APGI Publishing, Paris, pp 889–892

Anaya-Castro MA, Ayala-Zavala JF, Munoz-Castellanos L et al (2017) β-Cyclodextrin inclusion complexes containing clove (*Eugenia caryophyllata*) and Mexican oregano (*Lippia berlandieri*) essential oils: preparation, physicochemical and antimicrobial characterization. Food Packag Shelf Life 14:96–101

Anwar F, Naseer R, Bhanger MI et al (2008) Physico-chemical characteristics of citrus seeds and seed oils from Pakistan. J Am Oil Chem Soc 85:321–330

Astray G, Gonzalez-Barreiro C, Mejuto JC et al (2009) A review on the use of cyclodextrins in foods. Food Hydrocoll 23:1631–1640

Ayala-Zavala JF, Soto-Valdez H, Leon AG et al (2008) Microencapsulation of cinnamon leaf (*Cinnamomum zeylanicum*) and garlic (*Allium sativum*) oils in β-cyclodextrin. J Incl Phenom Macrocycl Chem 60:359–368

Ayaz M, Sadiq A, Junaid M et al (2017) Neuroprotective and anti-aging potentials of essential oils from aromatic and medicinal plants. Front Aging Neurosci 9:168

Babaoglu HC, Bayrak A, Ozdemir N et al (2017) Encapsulation of clove essential oil in hydroxypropyl beta-cyclodextrin for characterization, controlled release, and antioxidant activity. J Food Process Preserv 41:13202

Barbosa LCA, Silva CJ, Teixeira RR et al (2013) Chemistry and biological activities of essential oils from *Melaleuca* L. species. Agric Conspec Sci 78:11–23

Baretta IP, Felizardo RA, Bimbato VF et al (2012) Anxiolytic-like effects of acute and chronic treatment with *Achillea millefolium* L extract. J Ethnopharmacol 140:46–54

Benelli G, Canale A, Flamini G et al (2013) Biotoxicity of *Melaleuca alternifolia* (Myrtaceae) essential oil against the Mediterranean fruit fly, *Ceratitis capitata* (Diptera: Tephritidae), and its parasitoid *Psyttalia concolor* (Hymenoptera: Braconidae). Ind Crop Prod 50:596–603

Bhushan M, Beck MH (1997) Allergic contact dermatitis from tea tree oil in a wart paint. Contact Derm 36:117–118

Bilia AR, Guccione C, Isacchi B et al (2014) Essential oils loaded in nanosystems: a developing strategy for a successful therapeutic approach. Evid Based Complement Alternat Med 2014:651593

Bora H, Kamle M, Mahato DK et al (2020) Citrus essential oils (CEOs) and their applications in food: an overview. Plants 9:357

Calcabrini A, Stringaro S, Toccacieli L et al (2004) Terpinen-4-ol, the main component of *Melaleuca alternifolia* (tea tree) oil inhibits the in vitro growth of human melanoma cells. J Invest Dermatol 122:349–360

Calo JR, Crandall PG, O'Bryan CA et al (2015) Essential oils as antimicrobials in food systems—a review. Food Control 54:111–119

Candan F, Unlu M, Tepe B et al (2003) Antioxidant and antimicrobial activity of the essential oil and methanol extracts of *Achillea millefolium* subsp. *millefolium Afan.* (Asteraceae). J Ethnopharmacol 87:215–220

Carson CF, Hammer KA, Riley TV (2006) *Melaleuca alternifolia* (tea tree) oil: a review of antimicrobial and other medicinal properties. Clin Microbiol Rev 19:50–62

Cavalcanti AM, Baggio CH, Freitas CS et al (2006) Safety and antiulcer efficacy studies of *Achillea millefolium* L. after chronic treatment in Wistar rats. J Ethnopharmacol 107:277–284

Cavanagh HMA, Wilkinson JM (2002) Biological activities of lavender essential oil. Phytother Res 16:301–308

Celebioglu A, Yildiz ZI, Uyar T (2018) Thymol/cyclodextrin inclusion complex nanofibrous webs: enhanced water solubility, high thermal stability and antioxidant property of thymol. Food Res Int 106:280–290

Cheirsilp B, Rakmai J (2016) Inclusion complex formation of cyclodextrin with its guest and their applications. Biol Eng Med 2:1–6

Chhikara N, Kour R, Jaglan S et al (2018) Citrus medica: nutritional, phytochemical composition and health benefits—a review. Food Funct 9:1978–1992

Ciobanu A, Landy D, Fourmentin S (2013) Complexation efficiency of cyclodextrins for volatile flavor compounds. Food Res Int 53:110–114

Costa P, Lobo JMS (2001) Modeling and comparison of dissolution profiles. Eur J Pharm Sci 13:123–133

Costa G, Gidaro MC, Vulla D et al (2016) Active components of essential oils as anti-obesity potential drugs investigated by in silico techniques. J Agric Food Chem 64:5295–5300

Cox SD, Mann CM, Markham JL (2001) Interactions between components of the essential oil of *Melaleuca alternifolia.* J Appl Microbiol 91:492–497

Crawford GH, Sciacca JR, James WD (2004) Tea tree oil: cutaneous effects of the extracted oil of Melaleuca alternifolia. Dermatitis 15:59–66

Crini G (2014) Review: a history of cyclodextrins. Chem Rev 114:10940–10975

da Silveira E, Sa RDC, Andrade LN et al (2014) A review on anti-inflammatory activity of phenylpropanoids found in essential oils. Molecules 19:1459–1480

Daferera DJ, Ziogas BN, Polissiou MG (2000) GC-MS analysis of essential oils from some Greek aromatic plants and their fungitoxicity on *Penicillium digitatum*. J Agric Food Chem 48:2576–2581

Dai L, Ying L (2002) Infrared spectroscopic investigation of hydrogen bonding in EVOH containing PVA fibers. Macromol Mater Eng 287:509–514

Das S, Gazdag Z, Szente L et al (2019) Antioxidant and antimicrobial properties of randomly methylated β cyclodextrin-captured essential oils. Food Chem 278:305–313

Dash S, Murthy PN, Nath L et al (2010) Kinetic modeling on drug release from controlled drug delivery systems. Acta Pol Pharm 67:217–223

Desai KGH, Park HJ (2005) Recent developments in microencapsulation of food ingredients. Dry Technol 23:1361–1394

Dosoky NS, Setzer WN (2018) Biological activities and safety of *Citrus* spp. essential oils. Int J Mol Sci 19:1966

Duarte A, Carvalho C, Miguel G (2016a) Bioactive compounds of citrus as health promoters. In: Silva L, Silva B (eds) Natural bioactive compounds from fruits and vegetables as health promoters. Bentham Science Publishers, Sharjah, pp 29–97

Duarte A, Fernandes MJ, Bernardes JP et al (2016b) Citrus as a component of the Mediterranean diet. JSOD 4:289–304

Dzudie T, Kouebou CP, Essia-Ngang JJ et al (2004) Lipid sources and essential oils effects on quality and stability of beef patties. J Food Eng 65:67–72

El Adab S, Hassouna M (2016) Proteolysis, lipolysis and sensory characteristics of a Tunisian dry fermented poultry meat sausage with oregano and thyme essential oils. J Food Saf 36:19–32

Embuscado ME (2015) 11—Herbs and spices as antioxidants for food preservation. In: Shahidi F (ed) Handbook of antioxidants for food preservation, Woodhead publishing series in food science, technology and nutrition, 1st edn. Woodhead Publishing Ltd., Oxford, pp 251–283

Enshaieh S, Jooya A, Siadat AH et al (2007) The efficacy of 5% topical tea tree oil gel in mild to moderate acne vulgaris: a randomized, double-blind placebo-controlled study. Indian J Dermatol Venereol Leprol 73:22–25

European Medicines Agency (2013) Committee on herbal medicinal products (HMPC). http://www.ema.europa.eu. Accessed 13 Oct 2020

Evelegh T (1996) Lavender. Lorenz Books, Sydney

Fasseas MK, Mountzouris KC, Tarantilis PA et al (2007) Antioxidant activity in meat treated with oregano and sage essential oils. Food Chem 106:1188–1194

Fenyvesi E, Zemlenyi C, Orgovanyi J et al (2016) Can conversion mixture substitute beta-cyclodextrin in encapsulation of essential oils and their components? J Incl Phenom Macrocycl Chem 86:55–66

Fernandes LP, Ehen Z, Moura TF et al (2004) Characterization of *Lippia sidoides* oil extract-b-cyclodextrin complexes using combined thermoanalytical techniques. J Therm Anal Calorim 78:557–573

Franz CM (2010) Essential oil research: past, present and future. Flavour Fragr J 25:112–113

Fratianni F, de Martino L, Melone A et al (2010) Preservation of chicken breast meat treated with thyme and balm essential oils. J Food Sci 75:528–535

Frine VC, Hector AP, Manuel NDS et al (2019) Development and characterization of a biodegradable PLA food packaging hold monoterpene–cyclodextrin complexes against *Alternaria alternata*. Polymers 11:1720

Garcia-Diez J, Alheiro J, Falco V et al (2017) Chemical characterization and antimicrobial properties of herbs and spices essential oils against pathogens and spoilage bacteria associated to dry-cured meat products. J Essent Oil Res 29:117–125

Garcia-Sotelo D, Silva-Espinoza B, Perez-Tello M et al (2019) Antimicrobial activity and thermal stability of rosemary essential oil: β-cyclodextrin capsules applied in tomato juice. LWT Food Sci Technol 111:837–845

Gattefosse RM (1937) Gattefosse's aromatherapy. C.W. Daniel, Saffron Walden

Gaur S, Lopez EC, Ojha A et al (2018) Functionalization of lipid-based nutrient supplement with β-cyclodextrin inclusions of oregano essential oil. J Food Sci 83:1748–1756

Gibaldi M, Feldman S (1967) Establishment of sink conditions in dissolution rate determinations, theoretical considerations and application to nondisintegrating dosage forms. J Pharm Sci 56:1238–1242

Gomez-Rincon C, Langa E, Murillo P et al (2014) Activity of tea tree (*Melaleuca alternifolia*) essential oil against L3 larvae of Anisakis simplex. Bio Med Res Int 549510:6

Grieve M (1931) A modem herbal. Harcourt Brace & Co, New York

Haloci E, Toska V, Shkreli R et al (2014) Encapsulation of *Satureja montana* essential oil in β-cyclodextrin. J Incl Phenom Macrocycl Chem 80:147–153

Hammer KA (2015) Treatment of acne with tea tree oil (melaleuca) products: a review of efficacy, tolerability and potential modes of action. Int J Antimicrob Agents 45:106–110

Hammer K, Carson C, Riley T (1999) Antimicrobial activity of essential oils and other plant extracts. J Appl Microbiol 86:985–990

Hartford O, Zug KA (2005) Tea tree oil. Cutis 76:178–180

Hedges AR, Shieh WJ, Sikorski CT (1995) Use of cyclodextrins for encapsulation in the use and treatment of food products. In: Encapsulation and controlled release of food ingredients, ACS symposium series, vol 590. ASC Publications, Washington, DC, pp 60–71

Hill LE, Gomes C, Taylor TM (2013) Characterization of beta-cyclodextrin inclusion complexes containing essential oils (trans-cinnamaldehyde, eugenol, cinnamon bark, and clove bud extracts) for antimicrobial delivery applications. LWT Food Sci Technol 51:86–93

Hirayama F, Uekama K (1987) Methods of investigating and preparing inclusion compounds. In: Duchene D (ed) Cyclodextrins and their industrial uses. Les editions de Sante, Paris, pp 131–172

Ho S, Thoo YY, Young DJ et al (2017) Cyclodextrin encapsulated catechin: effect of pH, relative humidity and various food models on antioxidant stability. Food Sci Technol 85:232–239

Hsin-Chun C, Ming-Jen S, Li-Yun L et al (2007) Chemical composition of the leaf essential oil of *Psidium guajava* L. from Taiwan. J Essent Oil Res 19:345–347

Huang H, Huang C, Yin C et al (2020) Preparation and characterization of β-cyclodextrin–oregano essential oil microcapsule and its effect on storage behavior of purple yam. J Sci Food Agric 100:4849–4857

Jaiarj P, Khoohaswan P, Wongkrajang Y et al (1999) Anticough and antimicrobial activities of *Psidium guajava* Linn. leaf extract. J Ethnopharmacol 67:203–212

Jonsdottir G, Omarsdottird S, Vikingssona A et al (2011) Aqueous extracts from *Menyanthes trifoliate* and *Achillea millefolium* affect maturation of human dendritic cells and their activation of allogeneic CD4+ T cells in vitro. J Ethnopharmacol 136:88–93

Ju J, Chen X, Xie Y et al (2019) Application of essential oil as a sustained release preparation in food packaging. Trends Food Sci Technol 92:22–32

Junco S, Casimiro T, Ribeiro N et al (2002) A comparative study of naproxen–beta cyclodextrin complexes prepared by conventional methods and using supercritical carbon dioxide. J Incl Phenom Macrocycl Chem 44:117–121

Karamenderes C, Apaydin S (2003) Antispasmodic effect of *Achillea nobilis* L. subsp. sipylea (O. Schwarz) Bassler on rat isolated duodenum. J Ethnopharmacol 84:175–179

Kfoury M, Auezova L, Ruellan S et al (2015) Complexation of estragole as pure compound and as main component of basil and tarragon essential oils with cyclodextrins. Carbohydr Polym 118:156–164

Kfoury M, Auezova L, Greige-Gerges H et al (2016) Release studies of trans-anethole from β-cyclodextrin solid inclusion complexes by multiple headspace extraction. Carbohydr Polym 151:1245–1250

Kfoury M, Auezova L, Greige-Gerges H et al (2019) Encapsulation in cyclodextrins to widen the applications of essential oils. Environ Chem Lett 17:129–143

Khaneghah AM, Hashemi SMB, Limbo S (2018) Antimicrobial agents and packaging systems in antimicrobial active food packaging: an overview of approaches and interactions. Food Bioprod Process 111:1–19

Kringel DH, da Silva WMF, Biduski B et al (2020) Free and encapsulated orange essential oil into a β-cyclodextrin inclusion complex and zein to delay fungal spoilage in cakes. J Food Process Preserv 44:14411

Kytariolos J, Dokoumetzidis AP, Macheras P (2010) Power law IVIVC: an application of fractional kinetics for drug release and absorption. Eur J Pharm Sci 41:299–304

Lagouri V, Boskou D (1995) Screening for antioxidant activity of essential oils obtained from spices. In: Charalambous G (ed) Food flavors: generation, analysis and process influence. Elsevier, Amsterdam, pp 869–879

Lawless J (2002) The complete illustrated guide to aromatherapy. In: A practical approach to the use of essential oils for health and well-being. Harper Collins, London

Lawless J (2014) The encyclopedia of essential oils, 2nd edn. Harper Thorsons, London

Lee MS, Choi JC (2012) Aromatherapy for health care: an overview of systematic reviews. Maturitas 3:257–260

Letizia CS, Cocchiara J, Lalko J et al (2003) Fragrance material review on linalool. Food Chem Toxicol 41:943–964

Lira MCB, Ferraz MS, da Silva DGVC et al (2009) Inclusion complex of usnic acid with β-cyclodextrin: characterization and nanoencapsulation into liposomes. J Incl Phenom Macrocycl Chem 64:215–224

Lis-Balchin M, Deans SG, Eaglesham E (1998) Relationship between bioactivity and chemical composition of commercial essential oils. Flavour Fragr J 13:98–104

Liu H, Yang G, Tang Y et al (2013) Physicochemical characterization and pharmacokinetics evaluation of β-caryophyllene/β-cyclodextrin inclusion complex. Int J Pharm 450:304–310

Locci E, Lai S, Piras A et al (2004) 13C-CPMAS and 1H-NMR study of the inclusion complexes of β-cyclodextrin with carvacrol, thymol, and eugenol prepared in supercritical carbon dioxide. Chem Biodivers 1:1354–1366

Loizzo MR, Tundis R, Menichini F et al (2015) Anti-rancidity effect of essential oils, application in the lipid stability of cooked Turkey meat patties and potential implications for health. Int J Food Sci Nutr 66:50–57

Lopez-Nicolas JM, Rodriguez-Bonilla P, Garcia-Carmona F (2014) Cyclodextrins and antioxidants. Crit Rev Food Sci 54:251–276

Lozoya X, Reyes-Morales H, Chavez-Soto MA et al (2002) Intestinal anti-spasmodic effect of a phytodrug of *Psidium guajava* folia in the treatment of acute diarrheic disease. J Ethnopharmacol 83:19–24

Macwan SR, Dabhi BK, Aparnathi KD et al (2016) Essential oils of herbs and spices: their antimicrobial activity and application in preservation of food. Int J Curr Microbiol Appl Sci 5:885–901

Mangolim CS, Moriwaki C, Nogueira AC et al (2014) Curcumin-β-cyclodextrin inclusion complex: stability, solubility, characterisation by FT-IR, FT-Raman, X-ray diffraction and photoacoustic spectroscopy, and food application. Food Chem 153:361–370

Manosroi J, Dhumtanom P, Manosroi A (2006) Anti-proliferative activity of essential oil extracted from Thai medicinal plants on KB and P388 cell lines. Cancer Lett 235:114–120

Marques CS, Carvalho SG, Bertoli LD et al (2019) β-Cyclodextrin inclusion complexes with essential oils: obtention, characterization, antimicrobial activity and potential application for food preservative sachets. Food Res Int 119:499–509

Marreto RN, Almeida EECV, Alves PB et al (2008) Thermal analysis and gas chromatography coupled mass spectrometry analyses of hydroxypropyl-ß-cyclodextrin inclusion complex containing *Lippia gracilis* essential oil. Thermochim Acta 475:53–58

Martinez-Gracia C, Gonzalez-Bermudez CA, Cabellero-Valcarcel AM et al (2015) Use of herbs and spices for food preservation: advantages and limitations. Curr Opin Food Sci 6:38–43

Martins AD, Craveiro A, Machado M et al (2007) Preparation and characterization of Mentha × villosa Hudson oil–β-cyclodextrin complex. J Therm Anal Calorim 88:363–371

Matencio A, Garcia-Carmona F, Lopez-Nicolas JM (2016) Encapsulation of piceatannol, a naturally occurring hydroxylated analogue of resveratrol, by natural and modified cyclodextrins. Food Funct 7:2367–2373

Matencio A, Garcia-Carmona F, Lopez-Nicolas JM (2017) The inclusion complex of oxyresveratrol in modified cyclodextrins: a thermodynamic, structural, physicochemical, fluorescent and computational study. Food Chem 232:177–184

Mohamed HMH, Mansour HA (2012) Incorporating essential oils of marjoram and rosemary in the formulation of beef patties manufactured with mechanically deboned poultry meat to improve the lipid stability and sensory attributes. LWT Food Sci Technol 45:79–87

Moon T, Cavanagh HMA, Wilkinson JM (2004a) Lavender as an antibacterial essential oil—are all lavenders equal? In: AICA national conference 46 (abstracts)

Moon T, Chan YF, Wilkinson JM et al (2004b) Antifungal activity of Lava~zdula essential oil and oil volatiles. In: AICA national conference 46 (abstracts)

Nelson RRS (1997) In-vitro activities of five plant essential oils against methacilh-resistant Staphylococcus aureus and vancomycin-resistant *Enferococcus faecium.* J Antimicrob Chemother 40:305–306

Nerome H, Machmudah S, Fukuzato R et al (2013) Nanoparticle formation of lycopene/βcyclodextrin inclusion complex using supercritical antisolvent precipitation. J Supercrit Fluids 83:97–103

Noichinda W, Suppavorasatit I (2019) Enhancement of essential oil stability by cyclodextrins inclusion complex. J Food Technol (Siam University) 14:108–119

Ogunwande LA, Olawore NO, Adeleke KA et al (2003) Chemical composition of the leaf volatile oil of *Psidium guajava* L. growing in Nigeria. Flavour Fragr J 18:136–138

Oh WK, Lee CH, Lee MS et al (2005) Antidiabetic effects of extracts from *Psidium guajava.* J Ethnopharmacol 96:411–415

Othman M, Atiqah SN, Hassan MA et al (2016) Essential oils from the *Malaysian Citrus* (Rutaceae) medicinal plants. Medicines 3:13

Palazzolo E, Laudicina VA, Germana MA (2013) Current and potential use of citrus essential oils. Curr Org Chem 17:3042–3049

Paliyath G, Pessoa FL, Sidhu JS et al (2010) Handbook of fruit and vegetable flavors. Wiley, Hoboken

Pan J, Ai F, Shao P et al (2019) Development of polyvinyl alcohol/β-cyclodextrin antimicrobial nanofibers for fresh mushroom packaging. Food Chem 300:125249

Pateiro M, Barba FJ, Dominguez R et al (2018) Essential oils as natural additives to prevent oxidation reactions in meat and meat products: a review. Food Res Int 113:156–166

Pazyar N, Yaghoobi R, Bagherani N et al (2013) A review of applications of tea tree oil in dermatology. Int J Dermatol 52:784–790

Petrovic GM, Stojanovic GS, Radulovic NS (2013) Encapsulation of cinnamon oil in b-cyclodextrin. J Med Plant Res 4:1382–1390

Phunpee S, Ruktanonchai UR, Yoshii H et al (2017) Encapsulation of lemongrass oil with cyclodextrins by spray drying and its controlled release characteristics. Biosci Biotechnol Biochem 81:718–723

Posadzki P, Alotaibi A, Ernst E (2012) Adverse effects of aromatherapy: a systematic review of case reports and case series. Int J Risk Saf Med 24:147–161

Potrich FB, Allemand A, Silva LM et al (2010) Antiulcerogenic activity of hydroalcoholic extract of *Achillea millefolium* L.: involvement of the antioxidant system. J Ethnopharmacol 130:85–92

Prakash B, Kedia A, Mishra PK et al (2015) Plant essential oils as food preservatives to control moulds, mycotoxin contamination and oxidative deterioration of agri-food commodities-potentials and challenges. Food Control 47:381–391

Rakmai J, Cheirsilp B, Torrado-Agrasar A et al (2017a) Physico-chemical characterization and evaluation of bio-efficacies of black pepper essential oil encapsulated in hydroxypropyl-beta-cyclodextrin. Food Hydrocoll 65:157–164

Rakmai J, Cheirsilp B, Torrado-Agrasar A et al (2017b) Encapsulation of yarrow essential oil in hydroxypropyl-beta-cyclodextrin: physiochemical characterization and evaluation of bio-efficacies. CYTA J Food 15:409–417

Rakmai J, Cheirsilp B, Mejuto JC et al (2018) Antioxidant and antimicrobial properties of encapsulated guava leaf oil in hydroxypropyl-beta-cyclodextrin. Ind Crop Prod 111:219–225

Rashed AA, Mohd Nawi MN, Sulaiman K (2017) Assessment of essential oil as a potential anti-obesity agent: a narrative review assessment of essential oil as a potential anti-obesity agent: a narrative review. J Essent Oil Res 29:1–10

Reeds PJ (2000) Dispensable and indispensable amino acids for humans. J Nutr 130:1835–1840

Reineccius TA, Reineccius GA, Peppard TL (2002) Encapsulation of flavors using cyclodextrins: comparison of flavor retention in alpha, beta, and gamma types. Food Sci Nutr 67:3271–3279

Ren X, Yue S, Xiang H et al (2018) Inclusion complexes of eucalyptus essential oil with β-cyclodextrin: preparation, characterization and controlled release. J Porous Mater 25:1577–1586

Ribeiro-Santos R, Andrade M, Sanches-Silva A (2017) Application of encapsulated essential oils as antimicrobial agents in food packaging. Curr Opin Food Sci 14:78–84

Rodríguez-López MI, Mercader-Ros MT, Lucas-Abellan C et al (2020) Comprehensive characterization of linalool-HP-β-cyclodextrin inclusion complexes. Molecules 25:5069

Sacchetti G, Maietti S, Muzzoli MV et al (2005) Comparative evaluation of 11 essential oils of different origin as functional antioxidants: antiradicals and antimicrobials in foods. Food Chem 91:621–632

Saenger W (1980) Cyclodextrin inclusion compounds in research and industry. Angew Chem Int Ed 19:344–362

Satchell AC, Saurajen A, Bell C et al (2002) Treatment of dandruff with 5% tea tree oil shampoo. J Am Acad Dermatol 47:852–855

Sharafati-Chaleshtori R, Rokni N, Rafieian-Kopaei M et al (2015) Antioxidant and antibacterial activity of basil (*Ocimum basilicum* L.) essential oil in beef burger. J Agric Sci Technol 17:817–826

Sharma H, Mendiratta SK, Agrawal RK et al (2017) Use of various essential oils as bio preservatives and their effect on the quality of vacuum packaged fresh chicken sausages under frozen conditions. LWT Food Sci Technol 81:118–127

Shiga H, Yoshii H, Nishiyama T et al (2001) Flavor encapsulation and release characteristics of spray-dried powder by the blended encapsulant of cyclodextrin and gum Arabic. Dry Technol 19:1385–1395

Shiga H, Yoshii H, Taguchi R et al (2003) Release characteristics of flavor from spray-dried powder in boiling water and during rice cooking. Biosci Biotechnol Biochem 67:426–428

Silva F, Caldera F, Trotta F et al (2019) Encapsulation of coriander essential oil in cyclodextrin nanosponges: a new strategy to promote its use in controlled-release active packaging. Innov Food Sci Emerg Technol 56:102177

Simionato I, Domingues FC, Nerin C et al (2019) Encapsulation of cinnamon oil in cyclodextrin nanosponges and their potential use for antimicrobial food packaging. Food Chem Toxicol 132:110647

Singh G, Marimuthu P, Catalan C et al (2004) Chemical, antioxidant and antifungal activities of volatile oil of black pepper and its acetone extract. J Sci Food Agric 84:1878–1884

Siqueira-Lima PS, Araujo AA, Lucchese AM et al (2014) β-Cyclodextrin complex containing *Lippia grata* leaf essential oil reduces orofacial nociception in mice–evidence of possible involvement of descending inhibitory pain modulation pathway. Basic Clin Pharmacol Toxicol 114:188–196

Skocibusic M, Bezic N, Dunkic V et al (2004) Antibacterial activity of *Achillea clavennae* essential oil against respiratory tract pathogens. Fitoterapia 75:733–736

Tao F, Hill LE, Peng Y et al (2014) Synthesis and characterization of ß-cyclodextrin inclusion complexes of thymol and thyme oil for antimicrobial delivery applications. LWT Food Sci Technol 59:247–255

Trumbeckaite S, Benetis R, Bumblauskiene L et al (2011) *Achillea millefolium* L. s.l. herb extract: antioxidant activity and effect on the rat heart mitochondrial functions. Food Chem 127:1540–1548

Vlachou M, Papamichael M, Siamidi A et al (2017) Comparative in vitro controlled release studies on the chronobiotic hormone melatonin from cyclodextrins-containing matrices and cyclodextrin: melatonin complexes. Int J Mol Sci 18:1641

Wadhwa G, Kumar S, Chhabra L et al (2017) Essential oil–cyclodextrin complexes: an updated review. J Incl Phenom Macrocycl Chem 89:39–58

Wen P, Zhu DH, Feng K et al (2016) Fabrication of electrospun polylactic acid nanofilm incorporating cinnamon essential oil/β-cyclodextrin inclusion complex for antimicrobial packaging. Food Chem 196:996–1004

Wichtl M (1997) Teedrogen und phytopharmaka. Wissenschaftliche, Stuttgart, pp 395–399

Wolffenbuttel AN, Zamboni A, Becker G et al (2018) Citrus essential oils inhalation by mice: behavioral testing, GCMS plasma analysis, corticosterone, and melatonin levels evaluation. Phytother Res 32:160–169

Xi J, Qian D, Duan J et al (2015) Preparation, characterization and pharmacokinetic study of Xiangfu Siwu decoction essential oil/β-cyclodextrin inclusion complex. Molecules 20:10705–10720

Yildiz ZI, Celebioglu A, Kilic ME et al (2018) Fast-dissolving carvacrol/cyclodextrin inclusion complex electrospun fibers with enhanced thermal stability, water solubility, and antioxidant activity. J Mater Sci 53:15837–15849

Zhang L, Man S, Qiu H et al (2016) Curcumin-cyclodextrin complexes enhanced the anti-cancer effects of curcumin. Environ Toxicol Pharmacol 48:31–38

Zhu G, Xiao Z, Zhou R et al (2014) Study of production and pyrolysis characteristics of sweet orange flavor-β-cyclodextrin inclusion complex. Carbohydr Polym 105:75–80

Chapter 8
Encapsulation of Lipids

Afroza Sultana and Hidefumi Yoshii

8.1 Introduction

Lipids are very important for human health, as many lipids act as functional foods. An important group of lipids within unsaturated fatty acids is the long-chain (C16–C20) polyunsaturated fatty acids (PUFAs), including carotenoids, eicosapentaenoic acid (EPA), and docosahexenoic acid (DHA). EPA has antiatherosclerotic (Hay et al. 1982; Leaf and Weber 1988) and hypertriglyceridemic effects (Harris et al. 1983).

Daak et al. (2020) indicated biochemical and therapeutic effects of omega-3 fatty acids in sickle cell disease. The therapeutic effects of omega-3 fatty acids were attributed to their anti-inflammatory and anti-thrombotic eicosanoids, and the novel class of EPA- and DHA-derived lipid mediators, such as resolvins, protectins, and maresins. The chemical structures of EPA and DHA are shown in Fig. 8.1. Balk et al. (2006) reviewed the effects of omega-3 fatty acids on serum markers of cardiovascular disease risk and indicated the benefits of DHA and EPA supplementation. Maki et al. (2017) also showed that long-chain omega-3 polyunsaturated fatty acids (LC-OM3) supplementation was associated with a modest reduction in cardiac death. LC-OM3 affects blood pressure (Miller et al. 2014).

The decrease in atherogenic lipid levels by long-chain n-3 (LCn-3) PUFA is driven by a decrease in very low-density lipoprotein cholesterol (VLDL) secretion

A. Sultana
Department of Food Processing and Engineering, Faculty of Food Science and Technology, Chattogram Veterinary and Animal Sciences University, Chattogram, Bangladesh
e-mail: afroza-sultana@cvasu.ac.bd

H. Yoshii (✉)
Department of Food Science and Human Nutrition, Setsunan University, Osaka, Japan
e-mail: hidefumi.yoshii@setsunan.ac.jp

T. M. Ho et al. (eds.), *Functionality of Cyclodextrins in Encapsulation for Food Applications*, https://doi.org/10.1007/978-3-030-80056-7_8

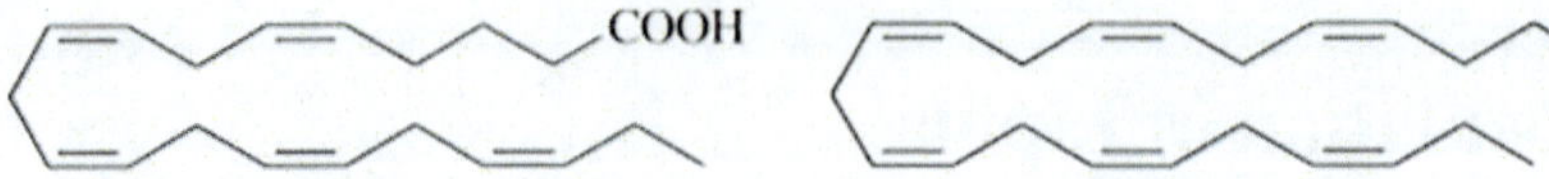

Eicosapentaenoic acid
(EPA, $C_{20}H_{30}O_2$)

Docosahexaenoic acid
(DHA, $C_{22}H_{32}O_2$)

Fig. 8.1 Chemical structures of EPA and DHA

by the liver, which typically coincides with a decrease in the triglycerides that are transported by this lipoprotein (Oscarsson and Hurt-Camejo 2017). PUFAs present in the phospholipids of the cell membrane are important inflammatory mediators involved in many molecular and cellular functions under physiological and pathological conditions (Meves 2008). Arachidonic acid (AA), EPA, and DHA play central roles in infant growth, neural development, and immune function. Lipids are generally difficult to disperse in food ingredients, and PUFAs are susceptible to autoxidation, which results in off-flavors and toxicity. To inhibit or retard the autoxidation of lipids, the encapsulation of lipids was investigated using wall material and lipid emulsion or molecular encapsulation with cyclodextrin (CD) (Imagi et al. 1992). Encapsulation refers to the isolation of a substance from the reactions of the surrounding matter. Matsuno and Adachi (1993) reviewed lipid encapsulation technology techniques and applications to foods, especially for conventional wall materials.

CDs are doughnut-shaped with an interior cavity and form specific inclusion complexes with many organic compounds. CD can therefore be used to stabilise labile compounds; emulsify oily compounds; mask taste, flavors, or odours; increase the solubility of materials; and convert oily compounds into powders. Szente et al. (1993) summarised the physicochemical properties and application of the fatty acid and CD complex. Imagi et al. (1992) showed that the oxidation of linoleic acid is completely prevented by entrapment into α-CD. Shimada et al. (1992) studied the structure of inclusion complexes of CDs with triglycerides on the vegetable oil/water interface. Schlenk and Sand (1961) indicated an increase in the water solubility of organic acids (C6–C12) with α- or β-CD and determined the molar ratio of organic acid to CD with crystal complexes. The structures of α- and β-CD complexes of linoleic and arachidonic acids were investigated by detailed spin-lattice relaxation time (T1) measurements and one-dimensional difference nuclear Overhauser enhancement (NOE) studies (Jyothirmayi et al. 1991). They indicated that the carboxyl arms of both fatty acids are present inside the cavity, and the olefinic group in linoleic acid is present at the edge of the wider end. The $-CH_2-$ groups (from positions 14 to 17) are located away from the cavity. Yoshii et al. (1995) investigated the quantitative analysis of α-CD inclusion complexes with fatty acids, methyl/ethyl esters, by X-ray diffractometry. Reichenbach and Min (1997) reported oxidation stability and nuclear magnetic resonance analyses of linoleic acid encapsulated in CD.

In this chapter, lipid encapsulation with CDs is summarized.

8.2 Preparation and Characteristics of Lipid-CD Complexes

8.2.1 Techniques for the Complex Preparation

PUFAs such as EPA and DHA have physiological functions and are used to form food supplements. The powder formation methods for inclusion complexes are spray drying, the co-precipitation method, the co-precipitation method based on phase solubility (the saturated aqueous method), heating in a sealed container, freeze-drying, and kneading. The co-precipitation method is not suitable for industrial-scale production with a large amount of water and is generally used for the formation on a small scale. On the other hand, the kneading method forms a complex powder with low water content. Yoshii et al. (1996) used a novel method for complexing PUFAs with CD as a dry powder by a twin-screw kneader, which was developed by Furuta et al. (1994). Hedges (1998) reviewed the industrial application of CD and summarised the production of complexes. He indicated the slurry method as a type of kneading method. For the slurry method, CD is suspended in water up to a 40–45% w/w concentration. The guest compound influences the viscosity of the slurry, and the concentrations are adjusted to allow mixing of the CD and guest. The paste method is the same method as the kneading method proposed by Furuta and Yoshii (1994) and uses a minimum amount of water (20–30% w/w). Frieler et al. (2019) encapsulated fish oil using a mixture of water and ethanol to induce complex crystallisation, and only γ-CD powder was able to form complexes with fish oil.

8.2.2 Encapsulation Capacity

Szente et al. (1993) reported that the solubility of lipids/CD complexes is generally very low (less than 0.1 mg/mL). Regiert (2007) indicated that linoleic acid forms the inclusion complex at 1:3 and 1:4 molar ratios to CD, as shown in Fig. 8.2.

Jyothirmayi et al. (1991) indicated that the carboxyl arms of linoleic acid and arachidonic acid are present inside the cavity, and the olefinic group in linoleic acid is present at the edge of the wider end. The $-CH_2-$ groups (from position 14 to 17) are located away from the cavity by 1-D difference NOE spectra at 270 MHz for the complex of arachidonic acid and α-CD. Yoshii et al. (1995) prepared α-CD inclusion complexes of fatty acid methyl/ethyl esters (FAME) at various molar ratios by the kneading method. The authors applied the twin screw kneading method to encapsulate FAME into α-CD. Figure 8.3 shows a plot of the amount of FAME included in α-CD against the initial amount of FAME added to α-CD. The dotted line in the figure indicates the theoretical value of the included FAME if all of it would be completely included in α-CD. The molar ratio of FAME to α-CD increased linearly with the amount of added FAME and reached a plateau, indicating the maximum inclusion value. These maximum values decreased with the length of the

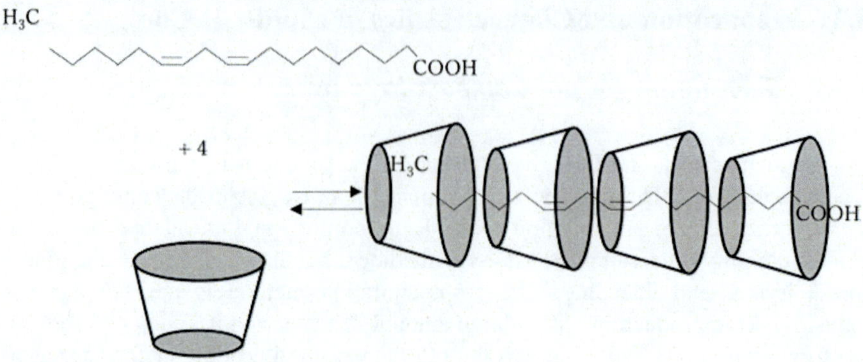

Fig. 8.2 Schematic diagram of the complex formation with α-CD and linoleic acid in a 1:4 molar ratio. (Adapted with permission from Regiert 2007)

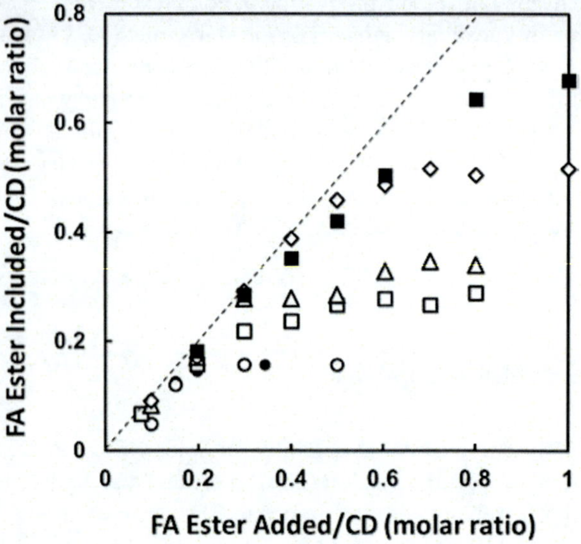

Fig. 8.3 Correlation of included fraction of α-CD complex of FAME to added molar ratio of various FAME to α-CD. ■: *n*-caproic acid methyl ester; ◇: *n*-caprylic acid ethyl ester; △: myristic acid methyl ester; □: linoleic acid methyl ester; ○: eicosapentaenoic acid ethyl ester; ●: docosahexaenoic acid ethyl ester. (Adapted with permission from Yoshii et al. 1995)

FAME. These values were similar to those for the fatty acids obtained by Schlenk and Sand (1961). From the plateau value in Fig. 8.3, Yoshii et al. (1995) estimated that the average number of α-CD molecules required for 1 mol of eicosapentaenoic acid ethyl ester (EPAEE) or docosahexaenoic acid ethyl ester (DHAEE) was about six. A molecule of α-CD covered 6.3 units of methylene chain in monophosphatidylcholine (Miyajima et al. 1985).

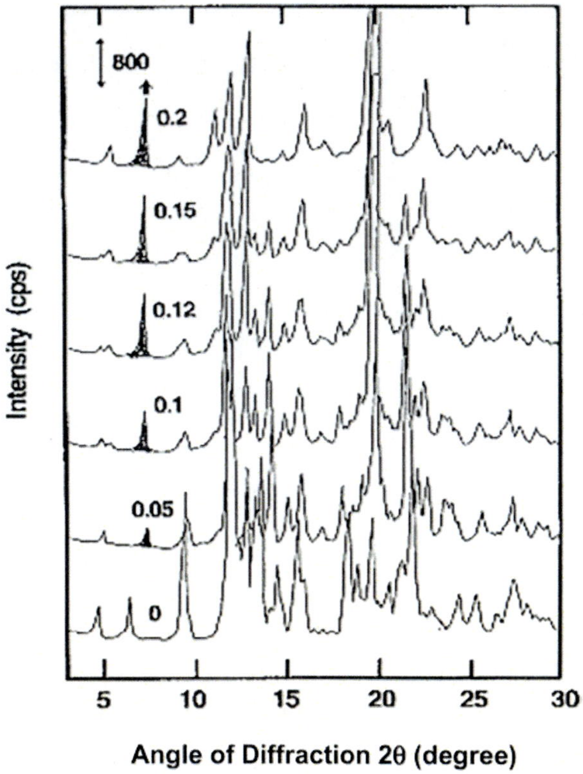

Fig. 8.4 X-ray diffractograms of complex powders between DHAEE and α-CD. (Adapted with permission from Yoshii et al. 1995)

Figure 8.4 illustrates the X-ray diffractograms of complex powders between DHAEE and α-CD at various amounts of DHAEE. The peak at 2θ = 7–8° increased as the molarity of DHAEE added to α-CD increased. The peak at 2θ = 22.5° also increased with added molar ratio of DHAEE to α-CD, but the quantifications of these peaks were difficult to evaluate in addition to the included amount of DHAEE into α-CD because of the irregular baseline. The refractive intensity of the peak occurred at 2θ = 7–8°. Refractive intensities for various FAME were plotted against the molar ratios of added FAME to α-CD and are shown in Fig. 8.5. The refractive intensities also reached the maximum plateau values at some constant molar ratios.

Using these plateau values of the included molar ratio of FAME to α-CD in Fig. 8.3 and the molar ratio of FAME to α-CD at the crick points in Fig. 8.4, the reciprocal values of the plateau values (the black dots using the data in Fig. 8.3 and open circles using the data of Fig. 8.5) were plotted against the carbon number of FAME without ethyl or methyl ester, as shown in Fig. 8.6. This figure indicates that 1 mol of EPAEE or DHAEE is included in about six α-CD. Schlenk and Sand (1961) reported that fatty acids with a carbon number of 18 and 12 were covered

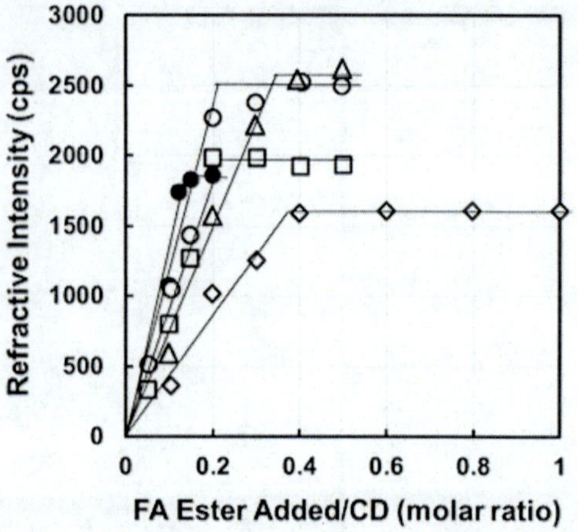

Fig. 8.5 Refractive peak intensities at 7–8° in X-ray diffractogram of FAME complex powders as function of initial molar ratio of added FAME to α-CD. (Adapted with permission from Yoshii et al. 1995)

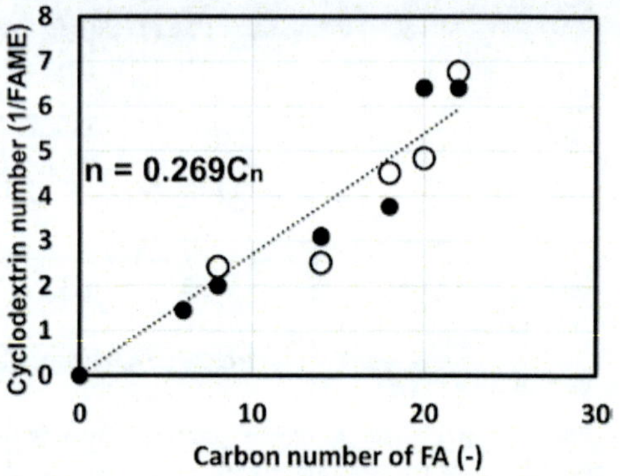

Fig. 8.6 Correlation of included CD number for FAME to the carbon number without ethyl or methyl ester of FAME using the maximum plateau values in Fig. 8.3 and the crick points in Fig. 8.5

with approximately 2.8 mol of α-CD according to the solubility data. These data suggest a tight fit if a hydrocarbon chain is inserted into the α-CD cavities.

8.3 Applications of Lipid-CD Complexes in Food Production

8.3.1 In Atlantic Salmon Oil

Atlantic salmon (*Salmo salar* L.) oil is very rich in monounsaturated and polyunsaturated fatty acids (MUFAs and PUFAs), with an average concentration of 63–79.5% (Rubio-Rodríguez et al. 2010; Wu and Bechtel 2008). PUFAs with a concentration of 25.4% were found in Atlantic salmon oil, where linoleic acid methyl ester (11.2%), EPA methyl ester (6.1%), and DHA methyl ester (4.1%) were the major fatty acid methyl esters (Hădărugă et al. 2016). Although EPA and DHA play a role in reducing cardiovascular disease risk by lowering blood cholesterol

(Balderas et al. 2010; Lopez-Huertas 2010), the unstable behaviour of these PUFAs essentially introduce protection against oxidation. Therefore, Atlantic salmon oil was complexed with β-CD by the co-crystallization and kneading method (Hădărugă et al. 2016). The authors used a molar ratio of β-CD to Atlantic salmon oil of 1:1 and 3:1 to make the complexed product. However, the molar ratio and complexation method significantly affected the recovered yield (percent ratio between the mass of the dried complex and the sum of the Atlantic salmon oil and hydrated β-CD masses). A molar ratio of 3:1 and the kneading method provided a higher recovered yield than the molar ratio of 1:1 and co-crystallization method.

8.3.2 In Menhaden Fish Oil

CDs are widely used in the pharmaceutical industry, and because of their inexpensive and non-toxic properties, they are commonly used to encapsulate functional compounds in the food industry. Fish oil, essential oil, flavors, and other bioactive compounds have been successfully encapsulated in CDs (Choi et al. 2010; Li et al. 2015; Pires et al. 2019; Nguyen and Yoshii 2018). The encapsulation mechanism of fish oil in β-CD was described by Choi et al. (2010). According to the authors, a partial inclusion complex between fish oil and β-CD is formed at the oil/water interface. The external agitation applied to the mixture crystallises CD molecules around the oily globules, which is caused by the interaction between fish oil and β-CD. They claimed that several parameters such as particle size, encapsulation efficiency, morphological shape, and membrane thickness influence β-CD encapsulation, which is determined by the mixing ratio of fish oil and β-CD. Because of the complicated make-up of fish oil with different components, such as linoleic acid and oleic acid, the molecular inclusion method does not allow for fish oil to be included as a core material. Therefore, a different technique, namely the self-assembly aggregation of β-CD, was used by Choi et al. (2010). For this technique, fish oil was dyed with 0.3% oil red (Choi et al. 2007), and 0.3% (w/w) β-CD was added to deionised water. Later, fish oil was added, and the mixture was placed in a shaking incubator at 250 rpm and 25 °C for 4 h for encapsulation by self-assembling aggregation. Grass carp (*Ctenopharyngodon idellus*) fish oil was encapsulated by a mixture of gum arabic/casein/β-CD as a barrier solution (Li et al. 2015). An optimal condition was found to be a solids concentration of 35%, ratio of oil to barrier material of 3:7, emulsifying temperature of 55 °C, and inlet temperature of spray drying at 220 °C. These parameters could affect the encapsulation efficiency of fish oil included in β-CD. The emulsion of fish oil in γ-CD showed better stability than that in β-CD when whey protein concentrates and xanthan gum were used to prepare the emulsion (Na et al. 2011). The authors also showed that CD could mask the odour of fish oil by up to 30% of the original intensity of non-encapsulated fish oil.

8.3.3 In Shrimp Oil

Shrimp oil not only consists of free astaxanthin, EPA, and DHA, but it is also rich in cholesterol (Gómez-Estaca et al. 2017). Due to the negative effects of cholesterol on human health (Jesch and Carr 2017), it is necessary to reduce the cholesterol level in this oil. In addition to the use of other components, β-CD is also used to lower the cholesterol level in these oils. β-CD not only reduces the cholesterol from shrimp oil but also reduces cholesterol in other foods (Dias et al. 2010). However, α- and γ-CD cannot reduce cholesterol because of the inner cavity size (Fenyvesi et al. 2016). The authors wrote an extensive review of the mobilisation and immobilisation of cholesterol into CDs. β-CD has a hydrophobic inner cavity, which includes cholesterol from food and thus creates lower cholesterol-containing food products, as shown in Fig. 8.7. To reduce the cholesterol from shrimp oil, 30 mL of ethyl acetate and 30 mL of distilled water were mixed with 3 g of oil (Raju and Benjakul 2020). Subsequently, β-CD was mixed with the solution with different oil and CD ratios. To remove the solvent, the supernatant was flushed with nitrogen, which was found after homogenisation and centrifugation. According to this study, cholesterol content was reduced by up to 95%, which was dependent on the addition of β-CD and homogenisation time. However, the authors concluded that the lower cholesterol-containing shrimp oil could be consumed as a nutraceutical food, as the total fatty acid and astaxanthin were increased by this method.

8.3.4 Encapsulation of Cholesterol in CDs

Cholesterol is a lipophilic molecule that is essential for human life. Watson and Meester (2016) edited Handbook of cholesterol. This book indicated that all animals possess cholesterol, which plays an important role and various functions in the

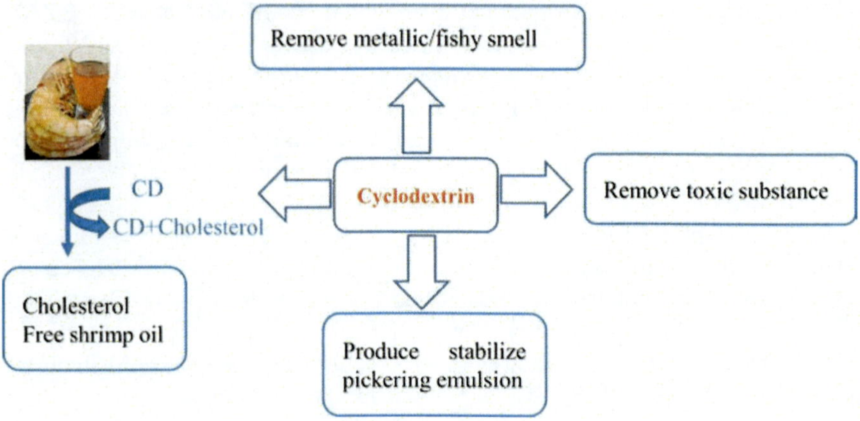

Fig. 8.7 CD to produce modified products (e.g., cholesterol-free shrimp oil)

body. There are 100–150 g of cholesterol in the human body. Cholesterol is one of the constituents of cell membranes and play role as modulator of membrane fluidity. A quarter of the cholesterol in the body is in the brain and most of it is in the myelin sheath. Another role of cholesterol is to be a precursor molecules of hormone synthesis. Hormones called steroid hormones are made from cholesterol in the adrenal cortex, testes, ovaries, and placenta. Cholesterol is also a constituent of bile acids. Atherosclerosis is likely to progress when the serum cholesterol concentration is 180 mg/dL or higher. However, excessive intake of high-cholesterol foods such as eggs and dairy products is often restricted. Therefore, it is necessary to cut-off cholesterol from food products. Ahn and Kwak (1999) studied the optimization of cholesterol removal in cream using β-CD and Lee et al. (1999) investigated cholesterol removal from homogenized milk with β-CD. Regardless other factors, the addition of β-CD concentration significantly increased the removal of cholesterol from the cream (Ahn and Kwak 1999). Using response surface methodology, the authors could optimize the cholesterol removal and found 97.99% removal at the condition of 15% addition of β-CD within 20 min stirring time. Lee et al. (1999) obtained the milk with 94.3% removal cholesterol by the addition of 1.5% β-CD at mixing temperature of 10 °C for 10 min. When β-CD was immobilized to chitosan beads, it could excellently adsorb cholesterol from egg yolk (Chiu et al. 2004). The maximum adsorption capacity of cholesterol was found 0.33 g cholesterol/g-adsorbent, which was calculated by Langmuir isotherm equation. Butter cholesterol removal using different complexation methods with β-CD was investigated by Dias et al. (2010). The authors found co-precipitation as the most effective method for removal of butter cholesterol rather kneading and physical mixture method. Phase solubility and stability of cholesterol/β-CD inclusion complexes were studied by dos Santos et al. (2011). Recently, Escobar Gianni et al. (2020) studied the effect of ultra-high-pressure homogenization combined with β-cyclodextrin in the development of a cholesterol reduced whole milk and obtained whole milk with reduced cholesterol of 87–89% using low amount of β-CD.

8.4 Conclusion and Further Perspectives

Various types of functional compounds are present in CD. Among the three types of CDs, β-CDs are usually used to encapsulate mono- and poly unsaturated fatty acids because of their inner cavity size and hydrophobicity. In addition to encapsulating the targeted compound, it also has additional benefits in food products such as decreasing cholesterol and toxic substances. CDs can extend the shelf life of thermally sensitive bioactive compounds by lowering their degradation. The release of functional compounds can vary with different humidity and temperature values. Thus, CD can be a good encapsulant for lipid compounds that can protect the physical barrier of these compounds.

By using CD, the stable lipid in the inclusion complex powder could be obtained. However, lipid contents in these CD powders are lower than those in the spray-dried powder of emulsified lipid oil. In order to overcome this demerit of low content, we have to investigate a method for producing a stable and high-content encapsulant by using a mixture of CD and other excipients in combination with inclusion complex and emulsified lipid.

References

Ahn J, Kwak HS (1999) Optimizing cholesterol removal in cream using β-cyclodextrin and response surface methodology. J Food Sci 64(4):629–632

Balderas C, Villasenor A, García A et al (2010) Metabolomic approach to the nutraceutical effect of rosemary extract plus Ω-3 PUFAs in diabetic children with capillary electrophoresis. J Pharm Biomed Anal 53(5):1298–1304

Balk EM, Lichtenstein AH, Chung M et al (2006) Effects of omega-3 fatty acids on serum markers of cardiovascular disease risk: a systematic review. Atherosclerosis 189(1):19–30

Chiu SH, Chung TW, Giridhar R et al (2004) Immobilization of β-cyclodextrin in chitosan beads for separation of cholesterol from egg yolk. Food Res Int 37(3):217–223

Choi MJ, Briançon S, Bazile D et al (2007) Effect of cryoprotectant and freeze-dried process on the stability of W/O/W emulsions. Dry Technol 25(5):809–819

Choi MJ, Ruktanonchai U, Min SG et al (2010) Physical characteristics of fish oil encapsulated by β-cyclodextrin using an aggregation method or polycaprolactone using an emulsion–diffusion method. Food Chem 119(4):1694–1703

Daak AA, Lopez-Toledano MA, Heeney MM (2020) Biochemical and therapeutic effects of Omega-3 fatty acids in sickle cell disease. Complement Ther Med 52:102482

Dias HMAM, Berbicz F, Pedrochi F et al (2010) Butter cholesterol removal using different complexation methods with beta-cyclodextrin, and the contribution of photoacoustic spectroscopy to the evaluation of the complex. Food Res Int 43(4):1104–1110

dos Santos C, Buera MP, Mazzobre MF (2011) Phase solubility studies and stability of cholesterol/β-cyclodextrin inclusion complexes. J Sci Food Agric 91(14):2551–2557

Escobar Gianni D, Jorcin S, Lema P et al (2020) Effect of ultra-high pressure homogenization combined with β-cyclodextrin in the development of a cholesterol-reduced whole milk. J Food Process Preserv 44(11):e14845

Fenyvesi E, Vikmon M, Szente L (2016) Cyclodextrins in food technology and human nutrition: benefits and limitations. Crit Rev Food Sci Nutr 56(12):1981–2004

Frieler L, Ho TM, Anthony A et al (2019) Crystallisation properties of amorphous cyclodextrin powders and their complexation with fish oil. J Food Sci Technol 56(3):1519–1529

Furuta T, Yoshii H (1994) Powdery encapsulation of *d*-limonene by kneading with mixed powders of β-cyclodextrin and maltodextrin at low water content. Biosci Biotechnol Biochem 58(5):847–850

Furuta T, Yoshii H, Miyamoto A et al (1994) Enhancement and inhibition by linear alcohol of the formation of inclusion complexes between *d*-limonene and cyclodextrins at low water content. Biosci Biotechnol Biochem 58(8):1381–1385

Gómez-Estaca J, Calvo MM, Álvarez-Acero I et al (2017) Characterization and storage stability of astaxanthin esters, fatty acid profile and α-tocopherol of lipid extract from shrimp (*L. vannamei*) waste with potential applications as food ingredient. Food Chem 216:37–44

Hădărugă DI, Ünlüsayin M, Gruia AT et al (2016) Thermal and oxidative stability of Atlantic salmon oil (*Salmo salar* L.) and complexation with β-cyclodextrin. Beilstein J Org Chem 12(1):179–191

Harris WS, Connor WE, McMurry MP (1983) The comparative reductions of the plasma lipids and lipoproteins by dietary polyunsaturated fats: salmon oil versus vegetable oils. Metabolism 32(2):179–184

Hay CRM, Durber AP, Saynor R (1982) Effect of fish oil on platelet kinetics in patients with ischaemic heart disease. Lancet 319(8284):1269–1272
Hedges AR (1998) Industrial applications of cyclodextrins. Chem Rev 98(5):2035–2044
Imagi J, Muraya K, Yamashita D et al (1992) Retarded oxidation of liquid lipids entrapped in matrixes of saccharides or proteins. Biosci Biotechnol Biochem 56(8):1236–1240
Jesch ED, Carr TP (2017) Food ingredients that inhibit cholesterol absorption. Prev Nutr Food Sci 22(2):67
Jyothirmayi N, Ramadoss CS, Divakar S (1991) Nuclear magnetic resonance studies of cyclodextrin complexes of linoleic acid and arachidonic acid. J Agric Food Chem 39(12):2123–2127
Leaf A, Weber PC (1988) Cardiovascular effects of n-3 fatty acids. N Engl J Med 318(9):549–557
Lee DK, Ahn J, Kwak HS (1999) Cholesterol removal from homogenized milk with β-cyclodextrin. J Dairy Sci 82(11):2327–2330
Li J, Xiong S, Wang F et al (2015) Optimization of microencapsulation of fish oil with gum Arabic/casein/beta-cyclodextrin mixtures by spray drying. J Food Sci 80(7):C1445–C1452
Lopez-Huertas E (2010) Health effects of oleic acid and long chain omega-3 fatty acids (EPA and DHA) enriched milks. A review of intervention studies. Pharmacol Res 61(3):200–207
Maki KC, Palacios OM, Bell M et al (2017) Use of supplemental long-chain omega-3 fatty acids and risk for cardiac death: an updated meta-analysis and review of research gaps. J Clin Lipidol 11(5):1152–1160
Matsuno R, Adachi S (1993) Lipid encapsulation technology-techniques and applications to food. Trends Food Sci Technol 4(8):256–261
Meves H (2008) Arachidonic acid and ion channels: an update. Br J Pharmacol 155(1):4–16
Miller PE, Van Elswyk M, Alexander DD (2014) Long-chain omega-3 fatty acids eicosapentaenoic acid and docosahexaenoic acid and blood pressure: a meta-analysis of randomized controlled trials. Am J Hypertens 27(7):885–896
Miyajima K, Tomita K, Nakagaki M (1985) Complex formation between di-and monophosphatidylcholines and cyclodextrins in water. Chem Pharm Bull 33(6):2587–2590
Na HS, Kim JN, Kim JM et al (2011) Encapsulation of fish oil using cyclodextrin and whey protein concentrate. Biotechnol Bioprocess Eng 16(6):1077–1082
Nguyen TVA, Yoshii H (2018) Release behavior of allyl sulfide from cyclodextrin inclusion complex of allyl sulfide under different storage conditions. Biosci Biotechnol Biochem 82(5):848–855
Oscarsson J, Hurt-Camejo E (2017) Omega-3 fatty acids eicosapentaenoic acid and docosahexaenoic acid and their mechanisms of action on apolipoprotein B-containing lipoproteins in humans: a review. Lipids Health Dis 16(1):149
Pires FQ, da Silva JKR, Sa-Barreto LL et al (2019) Lipid nanoparticles as carriers of cyclodextrin inclusion complexes: a promising approach for cutaneous delivery of a volatile essential oil. Colloids Surf B: Biointerfaces 182:110382
Raju N, Benjakul S (2020) Use of beta cyclodextrin to remove cholesterol and increase astaxanthin content in shrimp oil. Eur J Lipid Sci Technol 122(2):1900242
Regiert M (2007) Oxidation-stable linoleic acid by inclusion in α-cyclodextrin. J Incl Phenom Macrocycl Chem 57(1–4):471–474
Reichenbach WA, Min DB (1997) Oxidative stability and nuclear magnetic resonance analyses of linoleic acid encapsulated in cyclodextrins. J Am Oil Chem Soc 74(10):1329
Rubio-Rodríguez N, Beltrán S, Jaime I et al (2010) Production of omega-3 polyunsaturated fatty acid concentrates: a review. Innov Food Sci Emerg Technol 11(1):1–12
Schlenk H, Sand DM (1961) The association of α-and β-cyclodextrins with organic acids. J Am Chem Soc 83(10):2312–2320
Shimada KA, Kawano KI, Ishii JU et al (1992) Structure of inclusion complexes of cyclodextrins with triglyceride at vegetable oil/water interface. J Food Sci 57(3):655–656
Szente L, Szejtli J, Szemán J, Kató L (1993) Fatty acid-cyclodextrin complexes: properties and applications. J Incl Phenom Mol Recognit Chem 16(4):339–354

Watson RR, Meester FD (2016) Handbook of cholesterol: biology, function and role in health and diseases, vol 11. Wagenigen Academic Publishers, Wagenigen, p 494

Wu TH, Bechtel PJ (2008) Salmon by-product storage and oil extraction. Food Chem 111(4):868–871

Yoshii H, Furuta T, Kawasaki K et al (1995) Quantitative analysis of α-cyclodextrin inclusion complexes with fatty acid. Methyl/ethyl esters by X-ray diffractometry. J Appl Glycosci 42(3):243–249

Yoshii H, Furuta T, Yasunishi A et al (1996) Oxidation stability of eicosapentaenoic and docosahexaenoic acid included in cyclodextrins. J Incl Phenom Mol Recognit Chem 25(1–3):217–220

Chapter 9
Encapsulation of Nutraceuticals and Vitamins

Yukiko Uekaji and Keiji Terao

9.1 Introduction

Cyclodextrin (CD) is shaped like a bottomless bucket with hydrophilicity on the outside of the cavity and hydrophobicity (lipophilicity) on the inside. Fat-soluble substances can be incorporated inside the cavity through an action called "complexation". The water solubility of some fat-soluble substances can be improved by complexation with CD. However, there are cases in which complexation with CD forms a poorly soluble complex, and the water solubility is hardly changed or decreased. Typical fat-soluble substances are coenzyme Q_{10} (CoQ_{10}) and curcumin. Surprisingly, these γ-CD complexes have been reported to have significantly improved absorption when taken orally, even though their water solubility has not been improved (Purpura et al. 2018; Terao et al. 2006).

In general, fat-soluble substances ingested through the diet are taken up by micelles of bile acid in the intestine, dissolved in intestinal fluid, and absorbed into the body. However, highly fat-soluble substances, such as CoQ_{10} and curcumin, have lowered absorption and bioavailability because their molecules aggregate with each other, and they are difficult to take up by micelles of bile acid. As these γ-CD complexes have shown high bioavailability despite their low water solubility, we investigated the interaction of bile acid in intestinal fluid with γ-CD. We found that the solubilities of CoQ_{10} and curcumin in an aqueous solution containing bile acid

Y. Uekaji (✉)
CycloChem Bio Co., Ltd., Kobe, Hyogo, Japan
e-mail: yukiko.uekaji@cyclochem.com

K. Terao
CycloChem Bio Co., Ltd., Kobe, Hyogo, Japan

Department of Social/Community Medicine and Health Science, Food and Drug Evaluation Science, Graduate School of Medicine, Kobe University, Kobe, Hyogo, Japan
e-mail: keiji.terao@cyclochem.com

T. M. Ho et al. (eds.), *Functionality of Cyclodextrins in Encapsulation for Food Applications*, https://doi.org/10.1007/978-3-030-80056-7_9

hardly changed on their own, but improved significantly when their γ-CD complexes were used (Uekaji et al. 2012, 2013, 2017; Uekaji and Terao 2019). This result helped elucidate the mechanism of bioavailability improvement by complexation with γ-CD, as described in the next section.

Furthermore, when the solubility was tested with a surfactant that interacts with γ-CD, such as dipotassium glycyrrhizinate (GZK_2), instead of bile acid, the solubility was also improved with the γ-CD complex. This confirmed the applicability of this method to fat-soluble substances other than CoQ_{10} and curcumin that can be encapsulated with γ-CD. By using this technology, an aqueous solution can be prepared that contains a fat-soluble substance at high concentrations, which has potential applications to functional beverages and beauty essences. This is a breakthrough solution to improve the water solubility of fat-soluble substances, which was previously insufficient with CD alone (Terao et al. 2016).

In this chapter, we explain the mechanism for the improved absorption of fat-soluble substances through oral ingestion via the γ-CD complex, and we report on the improved absorption and bioavailability of various fat-soluble nutraceuticals and vitamins by complexation with γ-CD.

9.2 Mechanism of Improved Absorption of Fat-Soluble Nutraceuticals by Complexation with γ-Cyclodextrin

As indicated in Fig. 9.1, Uekaji et al. (2012, 2019) use CoQ_{10} as an example to show the mechanism of improvement of the absorption of fat-soluble substances by complexation with γ-CD. CoQ_{10} molecules are attracted to each other by intermolecular forces to form aggregates in water and hardly dissolve. In addition, as these CoQ_{10} aggregates are not dispersed finely in intestinal fluid, they are difficult for micelles of bile acid in the intestinal tract to take up, and their bioavailability is extremely low even when taken orally. In contrast, the CoQ_{10}–γ-CD complex is a fine particle in which CoQ_{10} molecules are incorporated in the γ-CD cavity one by one, and the inclusion complex disperses well in water. When the CoQ_{10}–γ-CD complex is orally ingested, it is replaced by the bile acid–γ-CD complex in the intestinal tract because bile acid has a higher affinity with γ-CD (Cabrer et al. 2000, 2003) than CoQ_{10} does (Fir et al. 2009). At the same time, CoQ_{10} is released from the complex on a molecular basis. The CoQ_{10} molecule released is taken up rapidly by micelles of bile acid and absorbed by the body. After γ-CD releases the CoQ_{10} molecule, it dissolves in the intestinal fluid as a bile acid–γ-CD complex, and γ-CD is decomposed by amylase and absorbed as glucose (Uekaji et al. 2013).

Thus, the mechanism can be described as follows:

1. The intermolecular forces of CoQ_{10} are cut by complexation with γ-CD, and each molecule is separated.
2. The CoQ_{10}–γ-CD complex is a fine particle that disperses well in the intestinal fluid, which improves the contact area with bile acid.

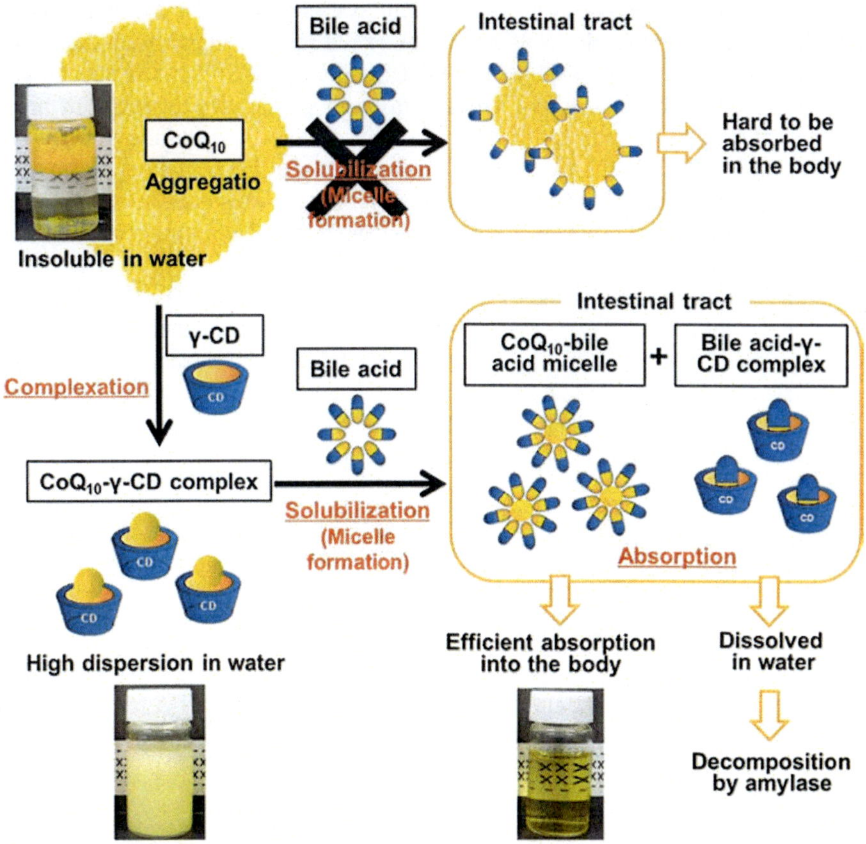

Fig. 9.1 Mechanism of the improved absorption of CoQ_{10} by complexation with γ-CD. (Adapted with permission from Uekaji et al. 2012)

3. Bile acid has a high affinity for γ-CD, and CoQ_{10} is released from γ-CD one molecule at a time as it is replaced as the guest molecule in the complex and is incorporated efficiently into the micelles of bile acid. This increases the absorption and bioavailability of CoQ_{10} and when taken orally. This is not a mechanism limited to CoQ_{10} and might be applicable to various fat-soluble substances that can be incorporated in the γ-CD cavity. Specific examples are presented in Sect. 9.4.

9.3 *In-Vitro* Evaluation Method of Absorption Using Artificial Intestinal Fluid

The mechanism of improving the absorption of fat-soluble substances with the γ-CD complex was used to conduct solubility tests of nutraceuticals and vitamins in artificial intestinal fluid containing bile acid. The results were used to develop an absorption index and verify the effects of complexation with γ-CD. Artificial intestinal fluid is an aqueous solution that reproduces the state of intestinal fluid during the fasted state (before meals) simulated intestinal fluid (FaSSIF) or fed state (after meals) simulated intestinal fluid (FeSSIF), in terms of the concentrations of bile acid and phospholipids and the pH (Vertzoni et al. 2004). A solubility test with FaSSIF or FeSSIF is a simple *in vitro* method that can be used to establish an absorption index. The results of previous studies have suggested a correlation between the results of this *in vitro* solubility test and those of *in vivo* absorption tests using actual animals and humans.

Taurocholic acid is a conjugate of cholic acid and taurine, and it is a type of bile acid. Sodium taurocholate (TCNa) is a commonly used reagent; its structure has hydrophilic cholanic acid sites and a hydrophobic steroid skeleton (Fig. 9.2a), and it acts as a surfactant. GZK_2 is used widely in cosmetics and pharmaceuticals because of its anti-inflammatory effect; it also has a hydrophobic triterpenoid skeleton as well as hydrophilic diglucuronic acid sites (Fig. 9.2b), and it can be

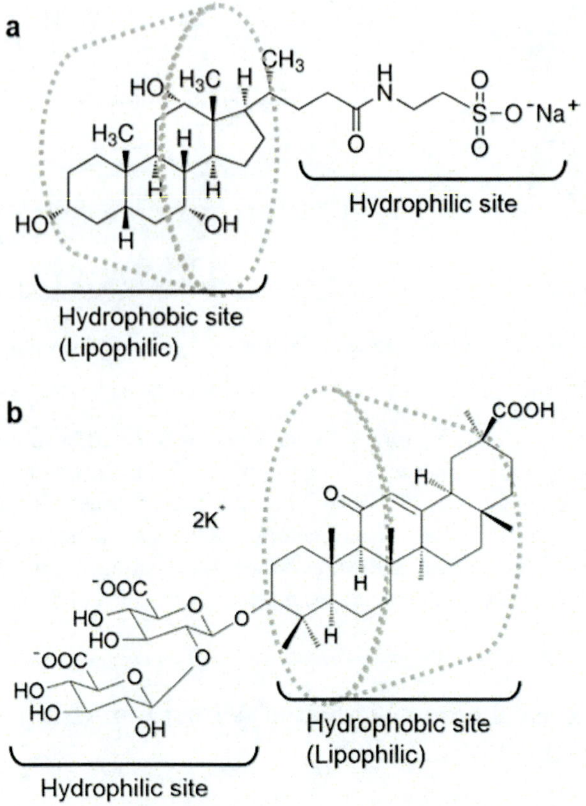

Fig. 9.2 Chemical structures of (**a**) TCNa and (**b**) GZK_2. (Adapted with permission from Uekaji et al. 2012)

used as a surfactant similar to TCNa. As a fat-soluble nutraceutical, the water solubility of GZK_2 (Koide et al. 1997) can be enhanced by the mechanism shown in Fig. 9.1, and the effect of the γ-CD complex can be verified by using GZK_2 instead of FeSSIF.

The next section introduces the improvements in the solubility, absorption, and bioavailability of various fat-soluble nutraceuticals and vitamins in FeSSIF containing TCNa or in GZK_2 aqueous solution: (1) CoQ_{10}, (2) curcumin, (3) ursolic acid, (4) caffeic acid phenethyl ester (CAPE; key effective ingredient in propolis), (5) vitamin D, (6) δ-tocotrienol, and (7) α-linolenic acid (key effective ingredient in perilla oil). Although the mechanism for absorption improvement is different, we also introduce interesting findings regarding (8) R-α-lipoic acid and (9) resveratrol.

9.4 Application of CD Complexes for Improved Bioavailability of Nutraceuticals

9.4.1 Solubility of CoQ_{10} in FeSSIF and Its Oral Absorption

CoQ_{10} (Fig. 9.3a) is a fat-soluble vitamin-like benzoquinone compound that functions as a cofactor in the production of adenosine triphosphate (ATP) by oxidative phosphorylation. CoQ_{10} has been reported to have energy-producing and antioxidant effects (Bhagavan and Chopra 2006; Crane 2001; Ernster and Dallner 1995), and it is used widely in health foods and cosmetics. However, as CoQ_{10} has poor water solubility and extremely low absorption and bioavailability, its functions are not fully exhibited. Therefore, a CoQ_{10}–γ-CD complex was prepared and tested for its solubility in FeSSIF as an absorption index, as well as its absorption in humans when taken orally.

When the solubility of CoQ_{10} in FeSSIF was evaluated at 37 °C, the CoQ_{10}–γ-CD complex showed a significantly higher CoQ_{10} concentration than CoQ_{10} alone (Fig. 9.3b) (Uekaji et al. 2019). This suggests that ingestion of the CoQ_{10}–γ-CD complex may improve the absorption of CoQ_{10}. An absorption test of the complex was conducted with healthy adult subjects. The CoQ_{10} concentration in the blood increased only slightly with the CoQ_{10}–microcrystalline cellulose mixture serving as the control, whereas the CoQ_{10} concentration in the blood increased significantly with the CoQ_{10}–γ-CD complex, which indicates excellent absorption (Fig. 9.3c) (Terao et al. 2006). This confirmed that the CoQ_{10}–γ-CD complex maintained a high CoQ_{10} concentration in the blood for a long time and showed persistence. The CoQ_{10}–γ-CD complex showed excellent pharmacokinetic properties, with a significantly higher area under the curve (AUC) for the CoQ_{10} concentration in blood plasma for 0–48 h and a higher maximum plasma concentration (C_{max}) for healthy adult subjects. Additionally, the CoQ_{10}–γ-CD complex showed significantly higher

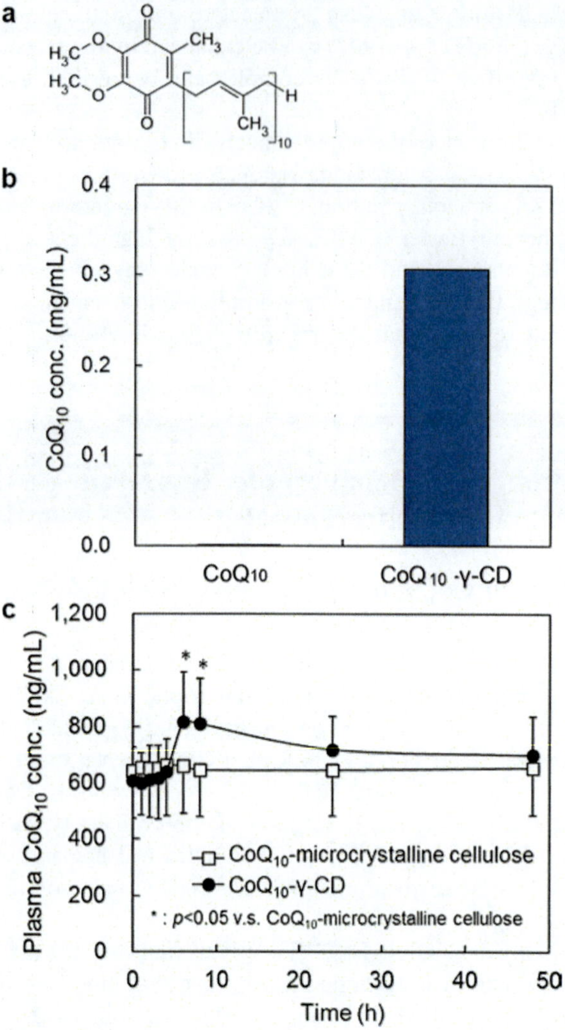

Fig. 9.3 (**a**) Chemical structure of CoQ_{10}; (**b**) solubility of the CoQ_{10}–γ-CD complex in FeSSIF; and (**c**) absorption when the CoQ_{10} preparations are taken orally. (Adapted with permission from Terao et al. 2006)

mean plasma levels than the control even after 24 and 48 h, and its half-life time ($T_{1/2}$) values were much longer.

Furthermore, the continuous ingestion of the CoQ_{10}–γ-CD complex with improved bioavailability maintained a high CoQ_{10} concentration in the blood, and the energy-producing activity and antioxidant effects of CoQ_{10} were improved. In addition, functionalities such as improved skin quality and endurance were enhanced (Terao et al. 2012; Terao and Uekaji 2015).

9.4.2 *Solubility of Curcumin in FeSSIF and Its Oral Absorption*

Curcumin (Fig. 9.4a) is a yellow pigment component and natural bioactive hydrophobic polyphenol isolated from the rhizome of turmeric (Curcuma longa). Curcumin has been reported to exhibit various pharmacological activities, such as antioxidant, anti-inflammatory, liver function improvement, and anti-Alzheimer's disease activities (Hatcher et al. 2008; Venigalla et al. 2015). Like CoQ_{10}, however, curcumin also has poor water solubility and extremely low absorption and bioavailability. The majority of orally ingested curcumin is excreted through the feces in a non-metabolized form (Ammon and Wahl 1991). Thus, even if curcumin is taken orally, its functions are not exhibited fully. A curcumin–γ-CD complex was prepared and tested for its solubility in FeSSIF as an absorption index and its absorption in humans when taken orally.

When the solubility of curcumin was evaluated in FeSSIF at 37 °C, the curcumin–γ-CD complex showed a curcumin concentration about five times higher than that of curcumin alone (Fig. 9.4b) (Nakata et al. 2017). This suggests that ingestion of the curcumin–γ-CD complex may improve the absorption of curcumin. Purpura et al. (2018) reported on the absorption of the curcumin–γ-CD complex in healthy subjects. They found that the curcumin–γ-CD complex increases the curcumin concentration in the blood significantly, compared with curcumin alone or two curcumin commercially available formulations (Fig. 9.4c). This suggests that the curcumin–γ-CD complex can provide the powerful benefits of curcumin to a much greater extent than existing commercial products.

9.4.3 *Solubility of Ursolic Acid in FeSSIF and Its Absorption*

Ursolic acid (Fig. 9.5a) is a pentacyclic triterpenoid carboxylic acid, contained in apple peel, kuding tea, basil, and rosemary. Ursolic acid has antihyperlipidemic and muscle-building effects (Li et al. 2013; Sakuma and Yamaguchi 2012; Woźniak et al. 2015). In recent years, it has been reported to prevent locomotive syndrome and sarcopenia. Thus, it has been attracting attention for many biological activities. However, as ursolic acid has poor water solubility and extremely low absorption and bioavailability, its use as a nutritional supplement has been limited. Many useful functions of ursolic acid are not exhibited sufficiently. Therefore, an ursolic acid–γ-CD complex was prepared, and its solubility in FeSSIF and its absorption in rats when taken orally were tested.

When the solubility of ursolic acid was evaluated in FeSSIF at 37 °C, the ursolic acid–γ-CD complex showed a significantly higher ursolic acid concentration than ursolic acid alone (Fig. 9.5b) (Ishida et al. 2018). This suggests that ingestion of the

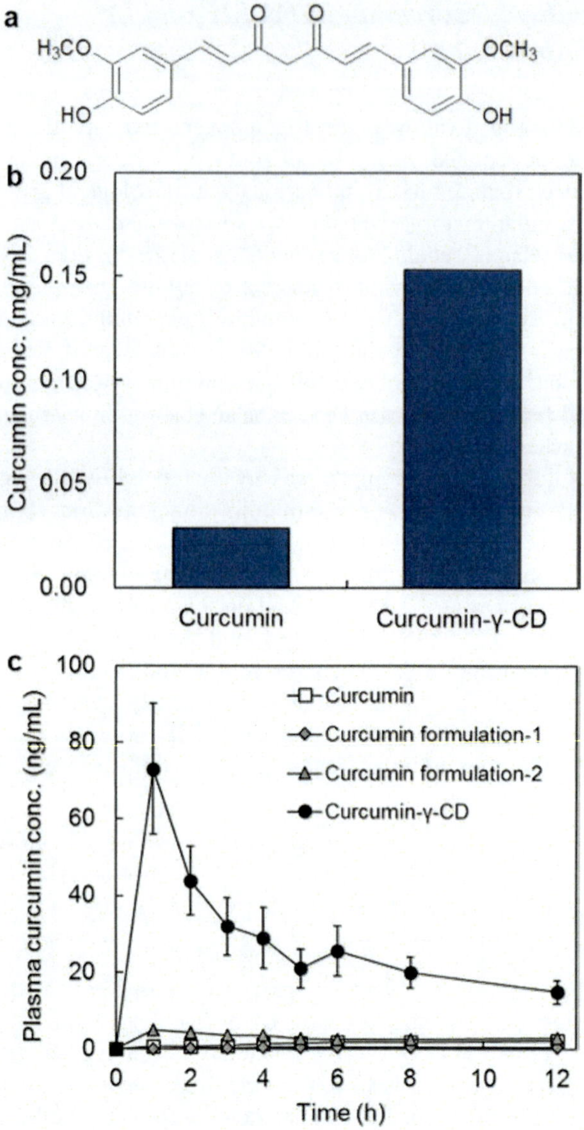

Fig. 9.4 (**a**) Chemical structure of curcumin; (**b**) solubility of the curcumin–γ-CD complex in FeSSIF; and (**c**) absorption when curcumin preparations are taken orally. (Adapted with permission from Purpura et al. 2018)

ursolic acid–γ-CD complex may improve the absorption of ursolic acid. Thus, an absorption test of the complex was conducted on rats. The ursolic acid concentration in the blood increased only slightly with ursolic acid alone as a control, whereas it increased significantly with the ursolic acid–γ-CD complex, which indicates excellent absorption (Fig. 9.5c).

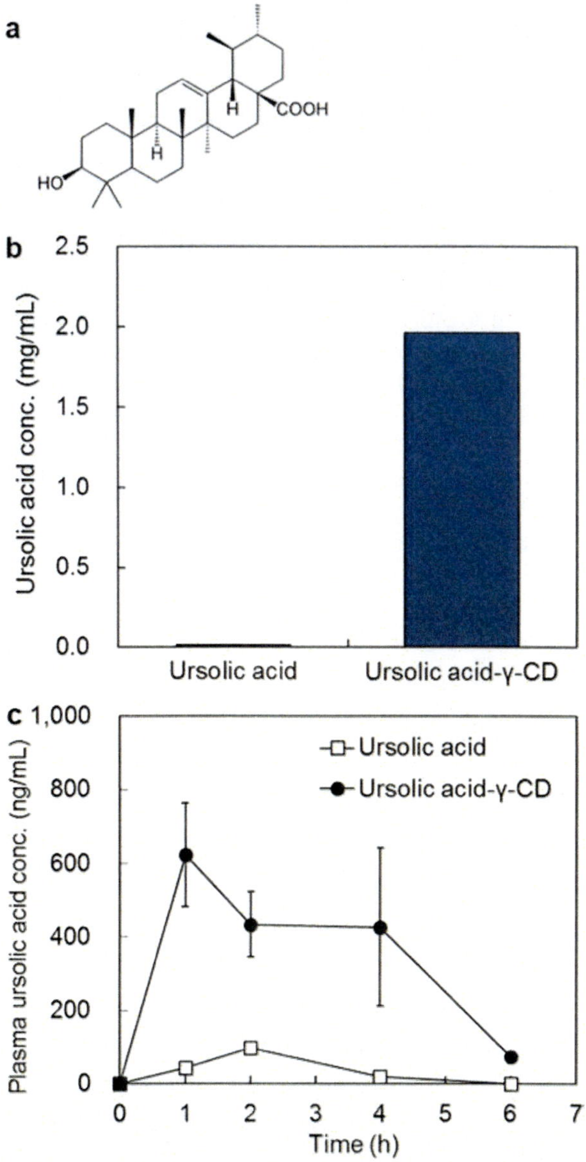

Fig. 9.5 (**a**) Chemical structure of ursolic acid; (**b**) solubility of the ursolic acid–γ-CD complex in FeSSIF; and (**c**) absorption when ursolic acid preparations are taken orally

As kuding tea extract contains ursolic acid, a kuding tea extract–γ-CD complex was also prepared. For FeSSIF at 37 °C, γ-CD was found to effectively improve the solubility of ursolic acid in kuding tea extract (Ishida et al. 2017). Therefore, the γ-CD complex was demonstrated to be effective even when the evaluation target was a mixture such as a plant extract.

9.4.4 Solubility of CAPE in TCNa Aqueous Solution and Its Anticancer Activity

Propolis is a natural nutraceutical that exhibits a wide range of biological activities, such as antibacterial, antiviral, and anti-inflammatory activities. The pharmacologically active substances in propolis are flavonoids, phenolic acids, and their esters (Castaldo and Capasso 2002). The active ingredient of propolis is characterized by its country of origin, such as New Zealand, Brazil, Europe, and China. CAPE (Fig. 9.6a) is a kind of cinnamic acid derivative and is characteristic of New Zealand propolis; it is attracting attention as a physiologically active substance that exhibits anti-inflammatory and anticancer effects (Grunberger et al. 1988; Song et al. 2002). However, as CAPE has poor water solubility and extremely low absorption and bioavailability, its functions are not fully exhibited. A CAPE–γ-CD complex was prepared and tested for its solubility in an aqueous solution of TCNa, which is a type of bile acid, and its anticancer activity was evaluated to assess its functionality.

When the solubility of CAPE was evaluated in the TCNa aqueous solution, the CAPE–γ-CD complex showed a higher CAPE concentration than CAPE alone (Fig. 9.6b) (Uekaji et al. 2013). This suggests that ingestion of the CAPE–γ-CD complex may improve the absorption of CAPE. Thus, an anticancer activity test was conducted on mice. In the non-administered control group, a marked increase in tumor volume was observed over time. In contrast, the increase in tumor volume was suppressed in the CAPE–γ-CD-administered mouse group when compared with the CAPE-administered group, and the growth of cancer cells was suppressed (Fig. 9.6c) (Wadhwa et al. 2016). Complexation with γ-CD was confirmed to improve the bioavailability of CAPE and enhance its anticancer activity.

9.4.5 Solubility of Vitamin D in FeSSIF

Vitamin D is a fat-soluble vitamin and generally refers to vitamin D_2 (ergocalciferol) (Fig. 9.7a) or D_3 (cholecalciferol) (Fig. 9.7b). Vitamin D is taken up from food and synthesized when the skin is exposed to the sun. As vitamin D controls bone metabolism, its deficiency can cause osteoporosis (Holick 2007; Lips 2006). However, as vitamin D has poor water solubility and extremely low absorption and bioavailability, its functions are not fully exhibited. Therefore, a vitamin D–γ-CD complex was prepared and tested for its solubility in FeSSIF as an absorption index.

When the solubility of vitamin D was evaluated in FeSSIF at 37 °C, the vitamin D–γ-CD complex showed a vitamin D concentration about 40 times higher than that of vitamin D alone (Fig. 9.7c) (Chikamoto et al. 2018). This suggests that ingestion of the vitamin D–γ-CD complex may improve the absorption of vitamin D.

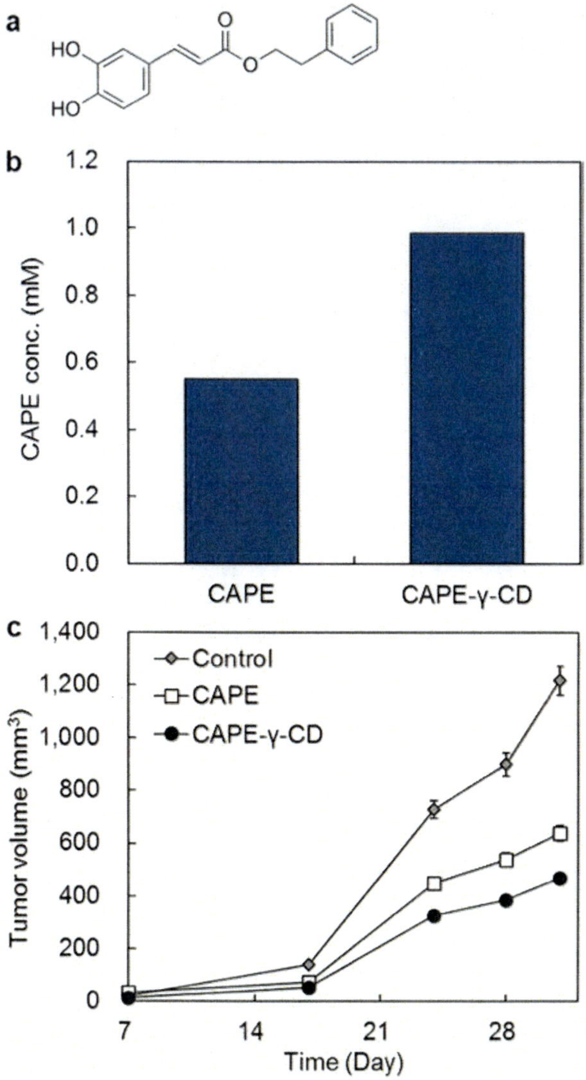

Fig. 9.6 (**a**) Chemical structure of CAPE; (**b**) solubility of the CAPE–γ-CD complex in TCNa aqueous solution; and (**c**) anticancer activity of CAPE preparations. (Adapted with permission from Wadhwa et al. 2016)

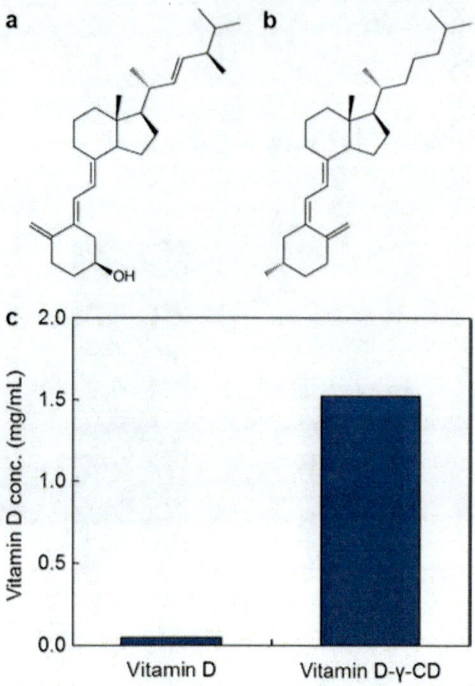

Fig. 9.7 Chemical structures of (**a**) vitamin D_2 (ergocalciferol), (**b**) D_3 (cholecalciferol); and (**c**) solubility of the vitamin D–γ-CD complex in FeSSIF

9.4.6 Solubility of δ-Tocotrienols in GZK_2 Aqueous Solution and Its Skin Absorption

In this section, we introduce interesting findings on the functionality and application of CD complexation with fat-soluble vitamins, although they are not food applications. Delta-tocotrienol (Fig. 9.8a), which is also called super vitamin E, exhibits various pharmacological activities, such as antioxidant, anti-inflammatory, cholesterol-lowering, and neuronal protection activities (Allen et al. 2017; Pervez et al. 2018; Qureshi et al. 2001). As it also inhibits tyrosinase activity (Ng et al. 2014), which has a whitening effect, it is useful not only in health foods but also in cosmetics. However, as δ-tocotrienol has poor water solubility and extremely low absorption and bioavailability, its functions are not fully exhibited. Therefore, a δ-tocotrienol–γ-CD complex was prepared and tested for its solubility and absorption. The solubility of the previous nutraceuticals was evaluated in FeSSIF to reflect absorption in the intestinal tract. However, as δ-tocotrienol is also used in cosmetics, the test was conducted with GZK_2 instead of TCNa.

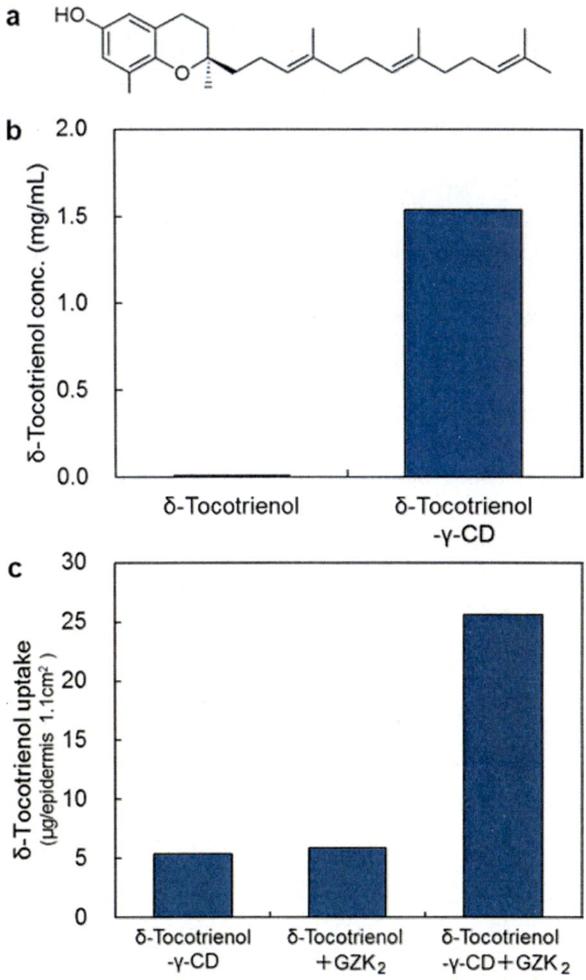

Fig. 9.8 (**a**) Chemical structure of δ-tocotrienol; (**b**) solubility of the δ-tocotrienol–γ-CD complex in GZK_2 aqueous solution; and (**c**) skin absorption of δ-tocotrienol preparations

In the solubility of δ-tocotrienol, effects of GZK_2, and γ-CD plus GZK_2 were shown in Fig. 9.8b (Nakata et al. 2014). This suggests that using the δ-tocotrienol–γ-CD complex with GZK_2 may improve the absorption of δ-tocotrienol. Therefore, the skin absorption of the δ-tocotrienol–γ-CD complex was tested by using a three-dimensional model of a cultured human epidermis. By using δ-tocotrienol–γ-CD complex with GZK_2, the uptake of δ-tocotrienol into the epidermal tissue was also four times higher than those of other samples (Fig. 9.8c). The results clarified that the above solubilization technique for TCNa can also be applied to GZK_2 for improved skin absorption.

9.4.7 Oral Absorption of α-Linolenic Acid

Alpha-linolenic acid (n-3 polyunsaturated fatty acid) (Fig. 9.9a) is abundant in perilla oil. It has been reported to prevent arteriosclerosis and improve serum lipids, and it is a food material useful for preventing lifestyle-related diseases (Ezaki et al. 1999; Ihara et al. 1998; Pan et al. 2012). However, as perilla oil is a liquid, it is difficult to handle. In addition, α-linolenic acid is extremely susceptible to oxidation, and it has low absorption and bioavailability. Therefore, Yoshikiyo et al. (2019) prepared a perilla oil–γ-CD complex powder and reported on the metabolism of n-3 fatty acids. They evaluated the absorption of the perilla oil–γ-CD complex in rats by examining the effect on the concentrations of α-linolenic acid and its metabolite eicosapentaenoic acid (EPA) in the blood. The α-linolenic acid (Fig. 9.9b) and EPA (Fig. 9.9c) concentrations in the blood increased significantly in the groups that ingested the liquid perilla oil and perilla oil–γ-CD complex. In particular, the α-linolenic acid concentration in the blood was significantly higher for the perilla oil–γ-CD complex group than for the liquid perilla oil group, which indicates that the complex led to excellent absorption properties. This result suggests that ingestion of the perilla oil–γ-CD complex may be a useful source of α-linolenic acid to increase n-3 fatty acids in plasma.

9.4.8 Stability of R-α-Lipoic Acid in Artificial Gastric Fluid and Its Oral Absorption

Here, we introduce another mechanism, that is, a stabilization technique for the improved absorption of nutraceuticals by complexation with γ-CD, where R-α-lipoic acid is taken as an example. Alpha-lipoic acid is a substance that acts as a coenzyme for energy production, and it has been reported to have antioxidant, anti-aging, and anti-glycation effects (Ghelani et al. 2018; Packer and Cadenas 2011; Patel and Packer 2008). There are two types of optical isomers of α-lipoic acid: R-form and S-form. Only the R-form exists in the living body (Fig. 9.10a). Industrially, racemates containing equal amounts of R-form and S-form are produced, and many commercial health foods containing α-lipoic acid use racemates. As many events harmful to the living body have been reported for the S-form in the racemates, the R-form is desirable. However, the R-form is extremely unstable, and it has the drawback of being easily polymerized by acid or heat. Therefore, even if the R-α-lipoic acid is ingested, it is decomposed by gastric acid when it reaches the stomach, and the amount of absorption decreases. This instability of R-α-lipoic acid leads to a reduction of its bioavailability. Therefore, an R-α-lipoic acid–γ-CD complex was prepared and tested for its stability in artificial gastric fluid and its absorption in humans when taken orally.

In the stability of R-α-lipoic acid in artificial gastric fluid, R-α-lipoic acid alone had a low residual rate of 50% or less, whereas the R-α-lipoic acid–γ-CD complex

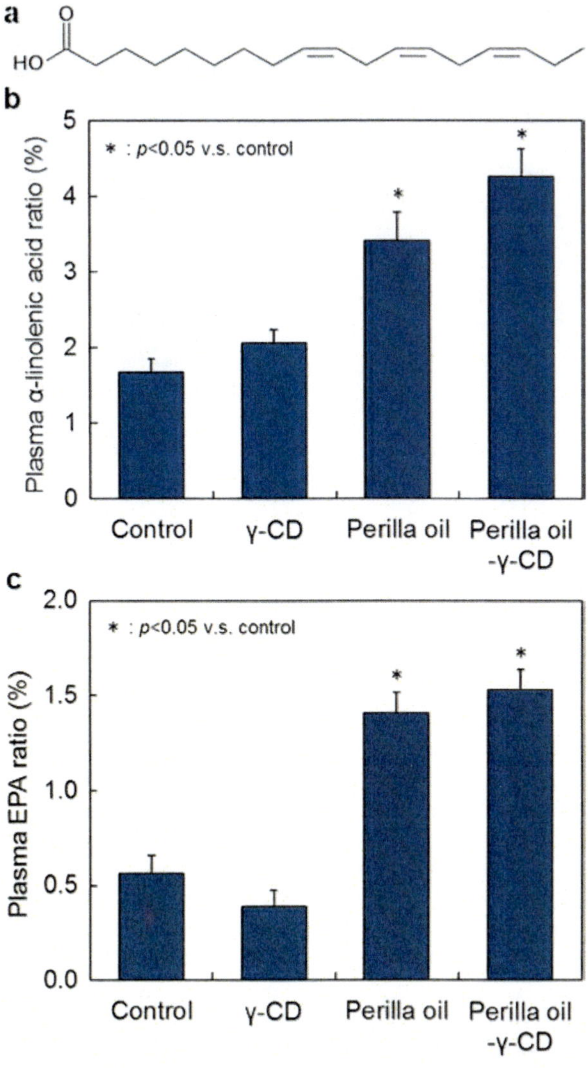

Fig. 9.9 (**a**) Chemical structure of α-linolenic acid; and absorption of (**b**) α-linolenic acid and (**c**) EPA when perilla oil preparations are taken orally. (Adapted with permission from Yoshikiyo et al. 2019)

had a residual rate of 100%, and it showed high stability against acid (Fig. 9.10b) (Ikuta et al. 2013). These suggest that ingestion of the R-α-lipoic acid–γ-CD complex may improve the absorption of R-α-lipoic acid. Thus, an absorption test of its complex was conducted in healthy adult subjects. The R-α-lipoic acid concentration in the blood was increased significantly with the R-α-lipoic acid–γ-CD complex as compared with R-α-lipoic acid alone, and the absorption was about 2.5 times higher (Fig. 9.10c) (Ikuta et al. 2016; Uekaji et al. 2017).

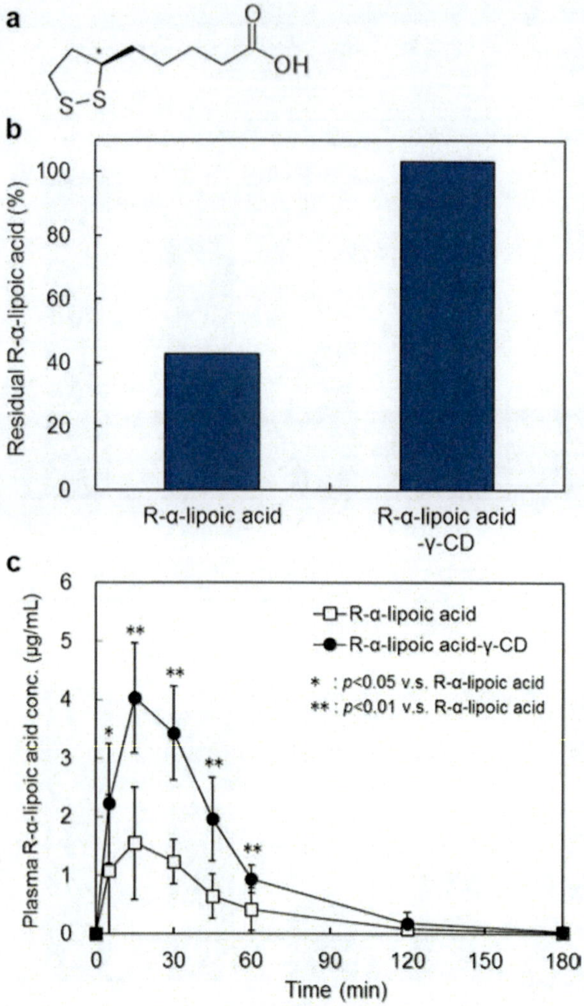

Fig. 9.10 (**a**) Chemical structure of R-α-lipoic acid; (**b**) stability of the R-α-lipoic acid–γ-CD complex in artificial gastric fluid; and (**c**) absorption when R-α-lipoic acid preparations are taken orally. (Adapted with permission from Ikuta et al. 2013, 2016)

These results confirmed that complexation of R-α-lipoic acid with γ-CD improved its stability in gastric fluid and its absorption efficiency into the living body. Furthermore, ingestion of the R-α-lipoic acid–γ-CD complex increased the bioavailability of R-α-lipoic acid and enhanced its original functionality. The increased bioavailability of R-α-lipoic acid with complexation can help prevent nonalcoholic steatohepatitis by improving the antidiabetic effects and insulin resistance (Terao and Ikuta 2014).

9.4.9 Solubility of Resveratrol in Water and Its Absorption

Complexation with α-CD in addition to γ-CD has also been found to be effective at improving absorption (Inagaki et al. 2016). Resveratrol is a kind of polyphenol contained in grapes and has been attracting attention for its antioxidant and anti-arteriosclerosis effects as an anti-aging ingredient (Alarcon De La Lastra and Villegas 2005; Baur and Sinclair 2006; Fremont 2000). However, resveratrol is not absorbed efficiently by the body because of its low water solubility. In general, the bioavailability of fat-soluble substances can be improved by adding solubilizers to increase the water solubility. Complexation with α-CD was found to markedly improve the water solubility of resveratrol (Onishi et al. 2013). An absorption test with rats revealed that the resveratrol concentration in the blood was increased significantly with the resveratrol–α-CD complex compared with resveratrol alone. These results confirmed that the absorption of fat-soluble substances taken orally can be improved by improving the water solubility through complexation with α-CD.

9.5 Conclusion and Further Perspectives

Some nutraceuticals and vitamins may not be able to exert their original functionality because poor water solubility and chemical instability do not allow for sufficient absorption and bioavailability. CD complexation is an extremely useful approach to solving such problems; in this chapter, we introduce how it improves the absorption and bioavailability of nutraceuticals and vitamins, such as CoQ_{10}, curcumin, δ-tocotrienol and R-α-lipoic acid. The solubilization and stabilization technologies using α-CD and γ-CD are useful in improving absorption and bioavailability, and those CD are expected to be widely applicable to fat-soluble functional components other than the nutraceuticals and vitamins introduced here.

References

Alarcon De La Lastra C, Villegas I (2005) Resveratrol as an anti-inflammatory and anti-aging agent: mechanisms and clinical implications. Mol Nutr Food Res 49(5):405–430

Allen L, Ramalingam L, Menikdiwela K et al (2017) Effects of delta-tocotrienol on obesity-related adipocyte hypertrophy, inflammation and hepatic steatosis in high-fat-fed mice. J Nutr Biochem 48:128–137

Ammon HP, Wahl MA (1991) Pharmacology of *Curcuma longa*. Planta Med 57(1):1–7

Baur JA, Sinclair DA (2006) Therapeutic potential of resveratrol: the *in vivo* evidence. Nat Rev Drug Discov 5(6):493–506

Bhagavan HN, Chopra RK (2006) Coenzyme Q_{10}: absorption, tissue uptake, metabolism and pharmacokinetics. Free Radic Res 40(5):445–453

Cabrer PR, Alvarez-Parrilla E, Meijide F et al (2000) Complexation of bile salts with α-, β- and γ-cyclodextrins. Effects of the cavity size in the complexation of steroids. In: Proceedings of the 10th International cyclodextrin symposium, pp 170–176

Cabrer PR, Alvarez-Parrilla E, Al-Soufi W et al (2003) Complexation of bile salts by natural cyclodextrins. Supramol Chem 15(1):33–43

Castaldo S, Capasso F (2002) Propolis, an old remedy used in modern medicine. Fitoterapia 73:S1–S6

Chikamoto K, Okamoto H, Ishida Y et al (2018) Evaluations of the stability and bioavailability of vitamin D-γ-cyclodextrin complex. In: Abstracts of the 35th Cyclodextrin symposium in Japan, pp 170–171

Crane FL (2001) Biochemical functions of coenzyme Q_{10}. J Am Coll Nutr 20(6):591–598

Ernster L, Dallner G (1995) Biochemical, physiological and medical aspects of ubiquinone function. Biochim Biophys Acta 1271(1):195–204

Ezaki O, Takahashi M, Shigematsu T et al (1999) Long-term effects of dietary α-linolenic acid from perilla oil on serum fatty acids composition and on the risk factors of coronary heart disease in Japanese elderly subjects. J Nutr Sci Vitaminol 45(6):759–772

Fir MM, Milivojevic L, Prosek M et al (2009) Property studies of coenzyme Q_{10}-cyclodextrins complexes. Acta Chim Slov 56(4):885–891

Fremont L (2000) Biological effects of resveratrol. Life Sci 66(8):663–673

Ghelani H, Razmovski-Naumovski V, Pragada RR et al (2018) (R)-α-Lipoic acid inhibits fructose-induced myoglobin fructation and the formation of advanced glycation end products (AGEs) *in vitro*. BMC Complement Altern Med 18:13

Grunberger D, Banerjee R, Eisinger K et al (1988) Preferential cytotoxicity on tumor cells by caffeic acid phenethyl ester isolated from propolis. Experientia 44(3):230–232

Hatcher H, Planalp R, Cho J et al (2008) Curcumin: from ancient medicine to current clinical trials. Cell Mol Life Sci 65(11):1631–1652

Holick MF (2007) Vitamin D deficiency. N Engl J Med 357(3):266–281

Ihara M, Umekawa H, Takahashi T et al (1998) Comparative effects of short- and long-term feeding of safflower oil and perilla oil on lipid metabolism in rats. Comp Biochem Physiol B Biochem Mol Biol 121(2):223–231

Ikuta N, Sugiyama H, Shimosegawa H et al (2013) Analysis of the enhanced stability of R(+)-alpha lipoic acid by the complex formation with cyclodextrins. Int J Mol Sci 14(2):3639–3655

Ikuta N, Okamoto H, Furune T et al (2016) Bioavailability of an R-α-lipoic acid/γ-cyclodextrin complex in healthy volunteers. Int J Mol Sci 17(6):949

Inagaki H, Ito R, Setoguchi Y et al (2016) Administration of piceatannol complexed with α-cyclodextrin improves its absorption in rats. J Agric Food Chem 64(18):3557–3563

Ishida Y, Nakata D, Terao K (2017) Development of ku-ding-tea extract-γ-cyclodextrin complex containing ursolic acid. In: Abstracts of the 34th Cyclodextrin symposium in Japan, pp 280–281

Ishida Y, Mori A, Okamoto H et al (2018) Development of ku-ding-tea extract-γ-cyclodextrin complex powder. In: Abstracts of the 35th Cyclodextrin symposium in Japan, pp 164–165

Koide M, Ishii R, Ukawa J et al (1997) Structural features of inclusion complexes between glycyrrhizinates and cyclodextrins. Nippon Kagaku Kaishi 1997(7):489–496

Li L, Xu LJ, Ma GZ et al (2013) The large-leaved Kudingcha (*Ilex latifolia* Thunb and *Ilex kudingcha* C.J. Tseng): a traditional Chinese tea with plentiful secondary metabolites and potential biological activities. J Nat Med 67(3):425–437

Lips P (2006) Vitamin D physiology. Prog Biophys Mol Biol 92(1):4–8

Nakata D, Okamoto H, Onishi M et al (2014) Bioavailability enhancement of δ-tocotrienol by γ-cyclodextrin complexation. In: Abstracts of the 31st Cyclodextrin symposium in Japan, pp 208–209

Nakata D, Okamoto H, Furune T et al (2017) Evaluation of properties of curcumin-cyclodextrin complex. In: Abstracts of the 34th Cyclodextrin symposium in Japan, pp 298–299

Ng LT, Lin LT, Chen CL et al (2014) Anti-melanogenic effects of δ-tocotrienol are associated with tyrosinase-related proteins and MAPK signaling pathway in B16 melanoma cells. Phytomedicine 21(7):978–983

Onishi M, Uekaji Y, Nakata D et al (2013) Solubility enhancements of functional ingredients by cyclodextrins. In: Abstracts of the 30th Cyclodextrin symposium in Japan, pp 280–281
Packer L, Cadenas E (2011) Lipoic acid: energy metabolism and redox regulation of transcription and cell signaling. J Clin Biochem Nutr 48(1):26–32
Pan A, Chen M, Chowdhury R et al (2012) α-Linolenic acid and risk of cardiovascular disease: a systematic review and meta-analysis. Am J Clin Nutr 96(6):1262–1273
Patel MS, Packer L (eds) (2008) Lipoic acid: energy production, antioxidant activity and health effects. CRC Press, Boca Raton
Pervez MA, Khan DA, Ijaz A et al (2018) Effects of delta-tocotrienol supplementation on liver enzymes, inflammation, oxidative stress and hepatic steatosis in patients with nonalcoholic fatty liver disease. Turk J Gastroenterol 29(2):170–176
Purpura M, Lowery RP, Wilson JM et al (2018) Analysis of different innovative formulations of curcumin for improved relative oral bioavailability in human subjects. Eur J Nutr 57(3):929–938
Qureshi AA, Salser WA, Parmar R et al (2001) Novel tocotrienols of rice bran inhibit atherosclerotic lesions in C57BL/6 ApoE-deficient mice. J Nutr 131(10):2606–2618
Sakuma K, Yamaguchi A (2012) Novel intriguing strategies attenuating to sarcopenia. J Aging Res 2012:251217
Song YS, Park EH, Hur GM et al (2002) Caffeic acid phenethyl ester inhibits nitric oxide synthase gene expression and enzyme activity. Cancer Lett 175(1):53–61
Terao K, Ikuta N (2014) Scientific evidence of R(+)-α-lipoic acid/γ-cyclodextrin inclusion complex: enhanced bioavailability and health improvement effect of R(+)-α-lipoic acid/γ--cyclodextrin complex. Food Process Ingredients 49(12):81–84
Terao K, Uekaji Y (2015) Health promotion effects by combination with absorption type coenzyme Q_{10} (γ-CD inclusion complex) and L-ascorbic acid. Food Process Ingredients 50(6):83–90
Terao K, Nakata D, Fukumi H et al (2006) Enhancement of oral bioavailability of coenzyme Q_{10} by complexation with γ-cyclodextrin in healthy adults. Nutr Res 26(10):503–508
Terao K, Okamoto H, Furune T et al (2012) Coenzyme Q_{10}-γ-cyclodextrin complex for sports nutrition. Food Process Ingredients 47(8):80–83
Terao K, Fukumi H, Nakata D et al (2016) Method for producing aqueous solution containing fat-soluble substance. JP5938587B2
Uekaji Y, Terao K (2019) Bioavailability enhancement of hydrophobic nutraceuticals using γ-cyclodextrin. J Incl Phenom Macrocycl Chem 93(1–2):3–15
Uekaji Y, Jo A, Ohnishi M et al (2012) A new generation of nutra-ceuticals and cosme-ceuticals complexing lipophilic bioactives with γ-cyclodextrin. Process Eng 36:540–550
Uekaji Y, Jo A, Urano A et al (2013) Application of γ-cyclodextrin in nanomedicinal foods and cosmetics. In: Bagchi D, Bagchi M, Moriyama H et al (eds) Bio-nanotechnology: a revolution in food, biomedical and health sciences. Wiley, Chichester, pp 179–211
Uekaji Y, Ikuta N, Rimbach G et al (2017) Enhancement of oral bioavailability of functional ingredients by complexation with cyclodextrin. J Drug Des Res 4:1043
Uekaji Y, Ishida Y, Nakata D et al (2019) Enhanced oral absorption/bioavailability of functional compounds by complexation with cyclodextrins. Food Process Ingredients 54(8):4–9
Venigalla M, Gyengesi E, Münch G (2015) Curcumin and apigenin – novel and promising therapeutics against chronic neuroinflammation in Alzheimer's disease. Neural Regen Res 10(8):1181–1185
Vertzoni M, Fotaki N, Kostewicz E et al (2004) Dissolution media simulating the intralumenal composition of the small intestine: physiological issues and practical aspects. J Pharm Pharmacol 56(4):453–462
Wadhwa R, Nigam N, Bhargava P et al (2016) Molecular characterization and enhancement of anticancer activity of caffeic acid phenethyl ester by γ cyclodextrin. J Cancer 7(13):1755–1771
Woźniak Ł, Skąpska S, Marszałek K (2015) Ursolic acid – a pentacyclic triterpenoid with a wide spectrum of pharmacological activities. Molecules 20(11):20614–20641
Yoshikiyo K, Yoshioka Y, Narumiya Y et al (2019) Thermal stability and bioavailability of inclusion complexes of perilla oil with γ-cyclodextrin. Food Chem 294:56–59

Chapter 10
Encapsulation of Antimicrobial Compounds

Adrián Matencio, Silvia Navarro-Orcajada, Francisco García-Carmona, and José Manuel López-Nicolás

10.1 Introduction to Antimicrobial Compounds

Globalization enables a food produced in Spain to be sold in China, for example. To make this possible, besides suitable means of transport, additives are needed to preserve the external appearance of food stuffs because of the importance of this characteristic for consumers. The food packaging industry has developed many petrochemical materials to control the shelf life of foods (Han et al. 2018). However, the society demands novel biodegradable strategies to decrease impact on the environment (Arrieta et al. 2014; Atarés and Chiralt 2016).

Apart from different strategies such as reducing the oxygen supply to prevent microbial growth, the use of antimicrobial compounds may also be considered. An antimicrobial compound is an agent that kills microorganisms or stops their growth (Madigan et al. 2015). For example, antibiotics are used against bacteria and antifungals are used against fungi. Antibacterials can be subdivided into two groups: bactericidal agents, which kill bacteria, and bacteriostatic agents, which slow down or stall bacterial growth (Madigan et al. 2015). Among the elements of the periodic table, some atoms have antimicrobial effect, for example incorporating copper in work surfaces (Grass et al. 2011; Parra et al. 2018).

A. Matencio (✉)
Department of Biochemistry and Molecular Biology-A, Faculty of Biology, University of Murcia, Murcia, Spain

Department of Chemistry, University of Turin, Torino, Italy
e-mail: adrian.matencio@um.es

S. Navarro-Orcajada · F. García-Carmona · J. M. López-Nicolás (✉)
Department of Biochemistry and Molecular Biology-A, Faculty of Biology, University of Murcia, Murcia, Spain
e-mail: silvia.navarro6@um.es; gcarmona@um.es; josemln@um.es

T. M. Ho et al. (eds.), *Functionality of Cyclodextrins in Encapsulation for Food Applications*, https://doi.org/10.1007/978-3-030-80056-7_10

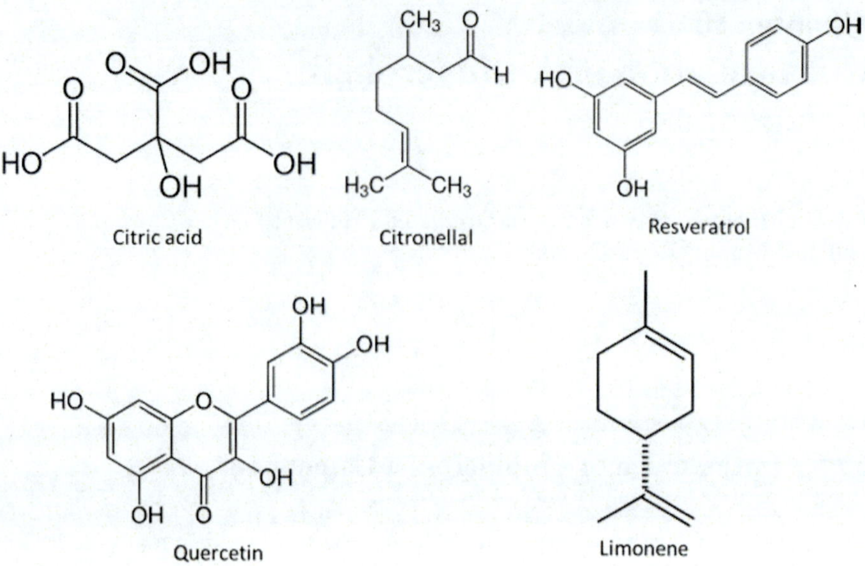

Fig. 10.1 Examples of natural antimicrobials

In foods, the use of synthetic antimicrobials is limited due to legal requirements as it is causing antimicrobial resistance in farmed food products (Xiong et al. 2018). In addition, the opinions of consumers concerning the possible undesirable effects not only for humans but also for the environment, encouraged us to look for different biological agents (Fig. 10.1). A wide number of antimicrobial agents are present in both animal and plant tissues, and also in different species of the same. Substances that are naturally occurring and directly derived from biological systems without alteration or modification in a laboratory setting are recognized as natural antimicrobials. These can be considered ideal due to the possibility of obtaining them in large quantities as products of a given organism that is Generally Recognised As Safe (GRAS) for having formed part of our diet for many years.

Organic acids (lactic acid, citric acid, acetic acid, etc.) are widely used in food products and are used as preservatives due to their antimicrobial capacity among others. In addition, many essential oils, such as bay, cinnamon or clove oils have antimicrobial capacity. Other interesting biomolecules with antimicrobial effects are polyphenols (e.g. stilbenes), or other secondary plant-metabolites. Sometimes these types of substances evaporate easily and have low stability and/or solubility so that their properties usually need to be improved. In this case, complexation with cyclodextrins (CDs) can be considered a good strategy to achieve this target.

CDs are torus-shaped oligosaccharides made up of α-(1→4) linked glucopyranoside units. The most common CDs are α, β and γ-CD, which contain six, seven and eight glucose units, respectively (Matencio et al. 2020a; Astray et al. 2020). These types of natural CDs have GRAS status and appear in the list of additives approved for alimentary use with the E-numbers E-457, E-459 and E-458 for α-, β- and γ-CD, respectively. Moreover, a derivative called hydroxypropyl-β-CD (HPβ-CD) is used as an orphan drug for Niemann Pick disease type C (Matencio et al. 2018a, 2020b). Although CDs can form inclusion complexes with organic molecules their salts and neutral molecules (Fourmentin et al. 2018), they are better known for being used to encapsulate poorly water-soluble compounds and hydrophobic moieties of amphiphilic molecules, which are also highly water-soluble. In the last 10 years, the number of publications in the Web of Science (WOS) involving CD complexes with antimicrobial compounds in foods has increased more than ninefold, and this tendency is expected to continue in forth coming years.

Bearing the above in mind, in this chapter we discuss the strategies that can be used to complex antimicrobials in CDs for food applications and append a list of studies on the subject.

10.2 Preparation and Characteristics of the Antimicrobial Complexes

10.2.1 Techniques for Preparation of Complexes

An inclusion complex is formed when the antimicrobial compound enters the inner cavity of a CD and is totally or partially included by the CD in a type of non-covalent interaction. The stability of the inclusion complex can be described in terms of its complexation constant (K_c) or dissociation constant (K_D):

$$CD + Antimicrobial \xrightarrow{K_C} CD - Antimicrobial \tag{10.1}$$

$$CD - Antimicrobial \xrightarrow{K_D} CD + Antimicrobial \tag{10.2}$$

The most commonly used inclusion complexes are prepared by several methods (Gharib et al. 2015) such as kneading, co-evaporation (even co-precipitation sometimes) and freeze drying methods (Fig. 10.2).

- Kneading: CDs and aqueous solution are mixed in a mortar until a homogeneous paste, then, the antimicrobial is added while grinding to dissolve the antimicrobial. The product is ground with an appropriate quantity of water. The complex can be obtained by filtration, or rotary evaporation, among other techniques.
- Co-evaporation or solvent evaporation method: The organic phase containing the antimicrobial compound and an aqueous solution of CDs are mixed to achieve molecular dispersion. The resulting suspension is centrifuged to remove the undissolved antimicrobial and the solvent is evaporated.
- Complex preparation in aqueous phase: The CDs are dissolved in aqueous phase and then the antimicrobial compound is added. The water-soluble complex is formed after a routine period at the desired temperature.

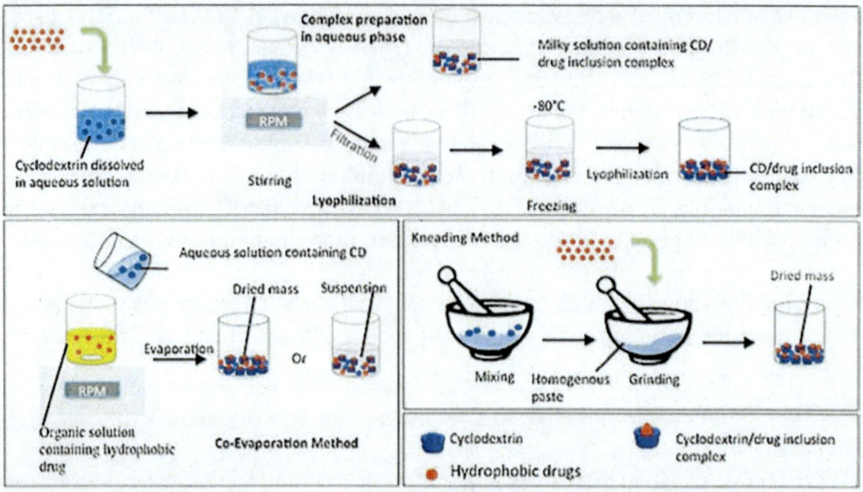

Fig. 10.2 Preparation methods of CD/antimicrobial compound inclusion complexes. (Adapted with permission from Gharib et al. 2015)

- Freeze-drying (lyophilisation): After stirring the antimicrobial/CD solution, the sample is frozen and lyophilized under reduced pressure.
- Spray-drying: After stirring the antimicrobial/CD solution, a spray-drying machine is used to form the inclusion complexes.

10.2.2 Encapsulation Mechanism

The selection of the most suitable complexation technique depends on the physical and chemical stability and concentration of the compounds used in the encapsulation process, the required particle size, the release mechanism, etc. However, the retention of the antimicrobial compound is related to its chemical nature, including its molecular weight, chemical functionality, polarity, and volatility, and by the properties of the selected CD.

Among the different processes, steric hindrance, and the substitution of unfavorable interactions by favorable new ones usually determine the energy state of the inclusion complex. Interactions that may contribute to the formation of the complex include electrostatic/coulomb forces, van der Waals interactions, hydrogen bonding, and hydrophobic effects, among others, although their impact depends on several factors such as type of guest molecule, the solvent used, etc. (Jansook et al. 2018; Schönbeck and Holm 2019). Such changes are related to the classic thermodynamic parameters, enthalpy, entropy and binding free energy. Although it is not general, the encapsulation is usually an exothermic process (Matencio et al. 2017;

Schönbeck and Holm 2019). So, the increase in temperature results in weakening of the complex.

In general, attractive electrostatic and van der Waals forces are assumed to result in enthalpy-driven binding, while hydrophobic attraction seems to be characterized by a large gain in entropy. Several CDs can be used to complex a guest molecule, and so, the different cavities of CDs and any modifications necessary must be analyzed. For the food industry, only the three natural CDs can be used, the cavity constituting the main difference between them, although β-CD and its derivatives such as HP-β-CD are the most widely used (López-Nicolás et al. 2009a; Hill et al. 2013; Wadhwa et al. 2017). As regards the effect of polarity, the most important factors to consider are structure, the solvent and the temperature. The solvent is important, since the presence of organic solvents might decrease the complexation constant due to the higher solubility of the molecule in question (Charumanee et al. 2016). On the other hand, the presence of a minimal quantity of water is necessary to form the inclusion complex (Kamihira et al. 1990).

When an organic acid is used as antimicrobial agent, the encapsulation capacity and release is influenced by the electronic state of the acid (the ionized or charged form of a molecule shows a poorer binding compared to the non-ionized or neutral forms) (Matencio et al. 2016, 2019a). This decrease in stability has been attributed to the overall increased hydrophilicity of the substrate upon ionization which reduces the substrates interaction with the more hydrophobic cavity of the CD. Therefore, conditions favoring the ionization of antimicrobial compounds will increase the fraction of non-complexed molecules in solution. However, the addition of different salts to neutralize the charge can optimize encapsulation in these cases. If the cyclodextrin is itself charged, this effect may be even more apparent when the charge of the antimicrobial compound and CD are similar, resulting in coulombic repulsion. If the charge on the antimicrobial compound and the CD are opposite, there may be a coulombic attraction component to the binding.

Usually, complexation constants of between 100 and 5000 M^{-1} are used in practical applications. In the case of antimicrobials of essential oil nature, a strong complexation constant is desirable to reduce the volatility of the antimicrobial essential oil. In addition, some recommend complexing the most hydrophobic guest first when several guests are present in the solution (Marques 2010).

10.2.3 Release Properties and Dissolution Models of the Antimicrobial/CD Complexes

Complexation in CDs slows down the release of a antimicrobial compound (Abril-Sánchez et al. 2019), although there are several aspects that could accelerate or decrease the rate. In general, the dissolution of the solid inclusion complex and hence its dilution promote the release of the antimicrobial compound from the CD system. This subsection focuses on several cases of this behavior: In the case of dilution in the presence of a competing agent (e.g. plasma administration) (Stella et al. 1999), the antimicrobial compound is closely dependent on the degree of dilution and on the complexation constant of the competing agent (the higher the complexation constant, the faster the release of our antimicrobial compound), and also on its concentration. This effect is mitigated when the complexation constant of the antimicrobial compound is higher than the complexation constant of the competing antimicrobial compound. However, if the solution also presents another agent with an affinity for the antimicrobial compound (e.g. plasma proteins) the release is faster than without the agent. As a conclusion to these experiments, the authors said that a simple dilution could affect in the case of weak CD complexes (when the encapsulation constant is low). When the complexation constant is higher (the authors tested 10,000 M^{-1}), the additional effect of competitive displacement, plasma or protein binding may also play an important role.

Koester and co-workers (2004) made a complete evaluation of the release of carbamazepine/β-CD complexes from hydroxypropylmethylcellulose matrix tablets. The release of the complexes prepared using three different techniques (spray-drying, a freeze-drying and a physical mixture as control) was fitted against the equations of different dissolution models: zero-order, first-order, Higuchi and Weibull. A complete description of these models can be found in the review by Dash et al. (2010). The results showed that the kinetic of the complexes is different from the simple physical mixture. Weibull was considered a good model since it has parameters that are sensitive to the ranges of dissolution profiles. In another study, Weibull was also shown to be the best model (in this study, Higuchi was not used) in the case of amlodipine/HPβ-CD complexes (Dalmolin et al. 2019). However, the transport of furosemide/HPβ-CD complexes across CaCo-2 monolayer was better fitted to the Higuchi model (in this study, Weibull was not used) (Sadighi et al. 2012). These data suggest that different model should be tested according to the conditions and parameters of the study.

In short, although the dilution factor is a critical aspect, the use of different direct (antimicrobial compound-protein binding, direct antimicrobial compound participation in certain reactions) or indirect (temperature, pH, volatility, competition for the inner cavity of the CD) agents could accelerate the release of the antimicrobial compound from the complexes.

10.3 Applications of the Complexes in Food Production

The use of free antibacterial or antifungal compounds in foods can give rise to certain problems, such as low solubility in aqueous systems, limited stability in the face of degradation or inactivation, modification of sensory qualities or the uncontrolled release of the compound (Fu et al. 2016). These problems could be overcome by complexing the antimicrobial compound in CDs before adding them to food (Fig. 10.3).

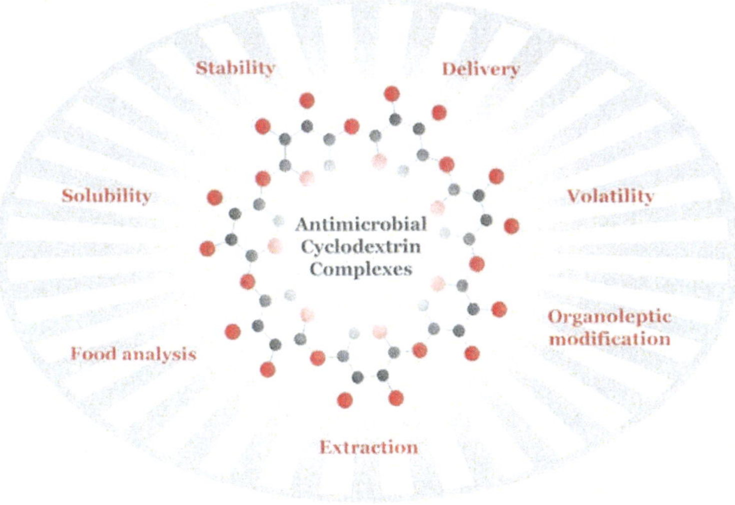

Fig. 10.3 Applications of CDs with antimicrobial compounds, an overview

10.3.1 Enhancement of Aqueous Solubility

Enhancement of aqueous solubility of the antimicrobials is one of the best-known applications of these complexes. As CDs have an outer surface which is mainly hydrophilic, they are relatively soluble in water, and can increase the aqueous solubility of some lipophilic antimicrobial compounds that enter their inner hydrophobic cavity. This is the case of some stilbenes (López-Nicolás and García-Carmona 2008; López-Nicolás et al. 2009a, b; Matencio et al. 2016, 2019b), carotenoids (Chen et al. 2007; Yuan et al. 2008, 2013) and essential oils and their components (Kfoury et al. 2019), among others.

Analysis of *trans*-anethole/CD complexes to resolve the low solubility of the antimicrobial compound revealed that CDs enhanced its aqueous solubility as well as its thermal and UV-light stability (Kfoury et al. 2014). Curcumin is also a poorly water soluble antimicrobial compound whose encapsulation in HP-β-CD contributed to enhancing its solubility and creating more active core-shell electrospun nanofibres to deliver the molecule (Aytac and Uyar 2017).

Another example is the bioactive molecule thymol (from thyme essential oil) whose inclusion in β-CD improved its solubility and were enabled grafting in gelatine films for sustained release (Li et al. 2018).

An increase in solubility after CD complexation could also improve the antimicrobial activity of a bioactive compound by increasing the amount in the solution. For instance, Hill et al. (2013) and Kamimura et al. (2014) associated the increase in activity of essential oils and their main antimicrobial components complexed in β-CD or HP-β-CD against *Salmonella enterica* serovar typhimurium LT2, *Listeria innocua* and/or *Escherichia coli* K12, with the increase in solubility after encapsulation, because of the increased contact between the pathogens and the antimicrobial agents. Matencio et al. (2020a) observed increased antioxidant activity of the antimicrobial compound oxyresveratrol complexed with β-CD due to by the higher amount of soluble bioactive compound. These complexes increased the doubling time of *Escherichia coli* at an oversaturated concentration of oxyresveratrol. However, the fact that this did not happen at lower concentrations was attributed to a balance between the increase in the carbon source with β-CD and the antimicrobial activity, as previously reported by Abril-Sánchez et al. (2019). The water solubility of triclosan was better in inclusion complexes with β-CD and γ-CD, also leading to a higher antibacterial effect against *Staphylococcus aureus* and *Escherichia coli* (Kayaci et al. 2013b).

Piperine solubility was improved after β-CD complexation (Quilaqueo et al. 2019), and the authors found that CDs were able to enhance the bioaccessibility of the molecule after *in vitro* gastrointestinal digestion and simulated colonic fermentation.

10.3.2 Increased Through Protecting Against Degradation or Inactivation

The protective cover that the CD molecule offers to the guest molecule can prevent the degradation or inactivation of the antimicrobial compound by thermal processes, oxidation, light or interaction with other components of the food (Castro-Rosas et al. 2017).

For example, neochlorogenic acid, an antimicrobial isomer of chlorogenic acid, is very susceptible to enzymatic oxidation due to its o-diphenol structure, however, CD inclusion complexes with this acid have been demonstrated to maintain up to 94% of the antimicrobial compound after a polyphenol oxidase reaction, compared to 50% when CDs were not used (Navarro-Orcajada et al. 2021). These authors also observed an increase in aqueous solubility due to encapsulation. Eugenol is a fragrant agent with antibacterial and antifungal properties but is also very thermosensitive. Kayaci et al. (2013a) enhanced its thermal stability and durability by complexation with β-CD and γ-CD. Similarly, the thermal stability of the antimicrobial and the fragrant compound menthol in polymethyl methacrylate nanofibres improved after encapsulation in natural CDs, the binding strength increasing as the CD cavity size increased (Uyar et al. 2009).

This protection can lead to increased antibacterial or antifungal activity after storage due to increased stability of the molecule. In this sense, Rakmai et al. (2017, 2018) demonstrated that the encapsulation of black pepper and guava leaf oils, both rich in antimicrobial compounds, in HP-β-CD provided 2–4 times more antibacterial activity against *Staphylococcus aureus* and *Escherichia coli* than without CDs. It has been suggested that CDs increase the antimicrobial activity of essential oils because they can reduce the interactions between the bioactive compounds and other hydrophobic constituents of the matrix (lipids and proteins) that usually affect their activity (Kfoury et al. 2019). Arana-Sánchez et al. (2010) observed that the antimicrobial activity of oregano essential oils (high in carvacrol)/β-CD complexes against *Staphylococcus aureus*, *Escherichia coli* and *Pseudomonas aeruginosa* was retained and the antioxidant activity increased compared with the observed with the compounds in free form.

10.3.3 Delivery and Controlled Release of Antimicrobial Compounds

The sustained release of antimicrobial molecules is another interesting application of CDs in foods. If the antimicrobial compound is free in the food or the food packaging, its effect is likely to be lost during storage. However, if the antimicrobial compound forms a complex with CDs, it better protected during storage and released in a controlled way in food, prolonging its effect.

For instance, a higher preservative effect (due to the decrease in the growth of *Pseudomonas* spp.) was observed in grass carp fillets stored at 4 °C when coated with fish gelatin and curcumin/β-CD complexes (Sun et al. 2019). Paun et al. (2014) monitored the release behavior with pH of gallic acid/HP-β-CD/chitosan membranes in sodium phosphate buffer, and concluded that the composite membranes with the antimicrobial compound provided a desirable sustained slow release rate at basic pH values.

Moreover, if the antimicrobial compound is in the packaging containing fresh-cut fruits or vegetables, the increase in humidity in the headspace (which is usually associated with spoilage) can trigger the release of the bioactive molecules, for which the CDs act as carriers (Ayala-Zavala et al. 2008a; Matencio et al. 2020a). This effect was observed in fresh-cut tomatoes contained in packaging with a filter paper with garlic oil/β-CD complexes (rich in the antimicrobial allyl disulphide) (Ayala-Zavala and González-Aguilar 2010). These complexes improved the shelf life of the tomatoes by inhibiting the growth of mesophilic, mould and yeast. In a more recent study (Buendía-Moreno et al. 2020), β-CD complexes with a blend of carvacrol, oregano and cinnamon essential oils were applied to active cardboard boxes containing fresh tomatoes. The quality of the tomatoes stored with the complexes was better than that of the controls as a result of the lower growth of mesophilic, psychrotrophic, enterobacteria, yeast and mould.

Citral and *trans*-cinnamaldehyde complexes with β-CD have been incorporated in food packages to improve the shelf life of beef by inhibiting bacteria growth (Chen et al. 2019). The controlled release of the antimicrobial compounds was checked in aqueous, acid, alcoholic and fatty food simulants, consisting of aqueous solutions with (1) nothing else, (2) 3% acetic acid, (3) 10% ethanol or (4) 95% ethanol, respectively. Canales et al. (2019) designed fungicide films that contained carvacrol and *trans*-cinnamaldehyde complexed with β-CD against *Botrytis cinerea*.

The gradual release of the antifungal compounds from the complexes resulted in a slower response than the free compounds, which is a desirable characteristic for improving food storage. *Trans*-anethole/β-CD complexes were applied in a gelatine edible coating to provide microbial safety and were found to extend shelf life of foods (Ye et al. 2017), the authors stressed the fact that the inclusion complexes gave UV light barrier properties to the coatings too.

Sodium benzoate is a commonly used food preservative that inhibits bacterial and fungal growth. Yang et al. (2019) and Birck et al. (2016) developed films that delivered the molecule in β-CD and HP-β-CD complexes, respectively, and found that they provided more activity against *Staphylococcus aureus* than without CD. Birck et al. (2016) also performed the assay against *Escherichia coli* and *Candida* sp. and obtained similar outcomes.

10.3.4 Reducing Volatilization Ratio of Antimicrobials

Some antimicrobial compounds are volatile molecules and may evaporate during storage or food processing, especially in the case of heat treatments, which can cause organoleptic modifications or diminishing their antimicrobial activity. Their encapsulation in CDs provides a protective cover to the molecule, preventing volatilization and loss. For example, Abril-Sánchez et al. (2019) observed that HP-β-CD can successfully regulate the volatility of citronellal, a bioactive agent widely used for its antimicrobial, insect repellent and aromatic properties. Kayaci et al. (2014) designed polymers with geraniol/γ-CD complexes embedded in polyvinyl alcohol using an electrospinning technique, and claimed that encapsulation in CDs was essential for preserving the antimicrobial compound during polymer manufacturing or during food storage. Somewhat similar observations were made by Aytac et al. (2014) using allyl isothiocyanate/β-CD complexes in the same type of polymer: lower volatility and consequently, higher antibacterial activity against *Escherichia coli* and *Staphylococcus aureus*.

Ayala-Zavala et al. (2008a, b) encapsulated cinnamon leaf and garlic oils (which are rich in cinnamaldehyde and sulphur-containing compounds, respectively) in β-CD to overcome the volatility of these antimicrobial compounds. The inclusion complexes showed a dose-dependent antifungal effect against *Alternaria alternata*, influenced, perhaps, by the possibility that fungi may use β-CD as a carbohydrate source. Another example is methyl salicylate which can be used as a food additive for its activity against pathogens. However, it is highly volatile, and its effectiveness could be lost during food processing, a problem that was solved by its complexation in β-CD, which permitted the controlled release of the molecule (Lee et al. 2020). D-limonene evaporation was also minimised when it formed β-CD inclusion complexes in polybutylene succinate films (Mallardo et al. 2016). Moreover, the aromatic and antimicrobial compound 2-nonanone from strawberries was encapsulated in β-CD to control its volatilization from active packages with antifungal activity against *Botrytis cinerea* (Abarca et al. 2017).

An exhaustive study (Ciobanu et al. 2013) to identify the most appropriate CD to encapsulate 13 antimicrobial volatile compounds commonly used in the food industry (α-pirene, β-pirene, camphene, camphor, eucalyptol, linalool, myrcene, menthol, menthone, pulegone, limonene, trans-anethole and *p*-cymene) was performed by measuring the stability constant using static headspace gas chromatography. It was found that β-CD was the most versatile CD and that it led to greater complexation efficiency.

It was found that the antifungal and bacteriostatic agent 2-phenyl ethanol could not be directly integrated into chitosan films because it evaporated rapidly during film processing, then only 8% of the compound retained. However, Zarandona et al. (2020) managed to preserve more than 90% of the antimicrobial agent using β-CD inclusion complexes. The volatility of the antifungal propionic acid was also reduced after β-CD encapsulation and incorporation into chitosan/gelatine polymers (Poverenov et al. 2013). The authors declared that β-CD significantly increased the loading capacity of the film and helped to inhibit the release. *Trans*-hexenal/β-CD complexes minimised the rapid evaporation of the antimicrobial compound from a polylactic acid matrix and did not affect its activity against the food spoiling fungus *Alternaria solani* (Joo et al. 2012).

10.3.5 Organoleptic Modification

While the incorporation of antimicrobial compounds from essential oils and other natural sources may increase the shelf life of foods by inhibiting microbial growth, the same compounds may also give rise to an undesirable change in sensory quality. This side effect can sometimes be mitigated by the use of CD inclusion complexes capable of masking the strong flavor of these molecules, and also the off-flavors from volatile oxidation products (Wadhwa et al. 2017).

For example, allyl isothiocyanate is an antimicrobial compound with a pungent taste and odor that can alter the sensory characteristics of food when used. However, β-CD inclusion complexes can effectively mask this flavor, while also controlling its release. Such complexes have demonstrated antimicrobial activity against *Listeria monocytogenes* in packaged fresh-cut onions (Piercey et al. 2012).

CD inclusion complexes can also mask the bitterness of some natural occurring antimicrobial compounds in green coffee beans, such as chlorogenic acid, caffeic acid and quinic acid, improving in the organoleptic profile of the final product (Aree 2019).

10.3.6 Extraction of Antimicrobial Compounds

Due to their ability to trap hydrophobic molecules, CDs can be used as extraction agents by forming inclusion complexes with antimicrobial compounds from natural sources and, so enable their separation. Polyphenols were extracted from olive leaves and pomegranates with the aid of methyl-β-CD (M-β-CD), β-CD or HP-β-CD (Diamanti et al. 2017; Athanasiadis et al. 2018). Catechin and epicatechin were also recovered from red grape pomace by using β-CD and HP-β-CD as extraction agents (López-Miranda et al. 2016).

Malapert et al. (2018) studied the β-CD complexation of hydroxytyrosol-O-glucoside, tyrosol, caffeic, and *p*-coumaric acids from alperujo (a highly valuable olive oil by-product), and concluded that β-CD was a suitable extraction and storage agent for the antimicrobial, and did not alter their *in vitro* bioavailability in Caco-2 cells.

Furthermore, CDs can be used as elicitors in the production of antimicrobial compounds from plant cell cultures (Matencio et al. 2020a). On the other hand, HP-β-CD and/or M-β-CD were used in the extraction of stilbenes and flavonoids from *Morus alba* and *Bryophyllum* sp. cell cultures, respectively (Komaikul et al. 2019; García-Pérez et al. 2019). In other organisms, such as cyanobacteria, the use of CDs increased the production of antifungal compounds (anabaenolysins) against *Candida albicans* (Shishido et al. 2015).

10.3.7 Food Analysis

CDs can be useful to detect and quantify antimicrobial molecules present in foods, acting as nanosensors due to their ability to amplify signals and chiral recognition isomers (Matencio et al. 2020a). For instance, Matencio et al. (2018b) amplified the fluorescent signal of ellagic acid in a ternary complex with CDs and borax, which enabled this antimicrobial compound to be quantitatively analyzed when found in low concentrations in foods. Moreover, several terpenoid isomers from the essential oils of *Mentha spicata* were separated in gas chromatography by the use of modified CDs (Pragadheesh et al. 2015).

The complexes formed by the inclusion of antimicrobial compounds in CDs have been shown to be capable of overcoming the problems of solubility, degradation and volatility that this type of bioactive agent may present during food processing or storage, while also leading to an increase in biological activity in some cases. Conversely, since CDs can also act as a carbon source, the antibacterial or antifungal activity may be diminished, which must be taken into consideration when optimizing the concentrations of guest and host molecules. For this reason, it is important to evaluate the effectiveness of the complexes in complete food systems in order to confirm that the activity remains. If such benefits cannot be demonstrated, the encapsulation process might not even unnecessary. Furthermore, these complexes can be used to deliver and control the release of the antimicrobial compound from the food or food packaging, to mask bitterness or strong flavors, to extract it from its natural source or from cell cultures, or to improve qualitative and quantitative determinations in food analysis.

10.4 Conclusion and Future Perspectives

The formation of inclusion complexes can obtain the capacity to control the release antimicrobials. In general, the complexation strength will depend on the nature, selection of the best strategy (spray drying, freeze drying, etc.) and conditions (temperature, pH). The ability of CDs to (1) increase the solubility of antimicrobials, (2) increase their stability, (3) improve the release and delivery, (4) prevent evaporation, (5) control their organoleptic properties, (6) extract their antimicrobial from its source and (7) analyze its presence in foods has been demonstrated, confirming the potential of these molecules for use in several stages of food production.

As mentioned above, research into novel antimicrobials is increasing, as in the research into their complexation in order to increase their medicinal bioactivity. Bearing the above in mind, CD/antimicrobial complexation can be considered a good strategy not only for use in food products, but also for example, for decreasing antibiotic resistance.

Acknowledgments This research was funded by the Spanish Ministry of Science and Innovation, project AGL2017-86526-P (MCI/AEI/FEDER, UE) and by the Programa de Ayudas a Grupos de Excelencia de la Región de Murcia, Fundación Séneca, Agencia de Ciencia y Tecnología de la Región de Murcia (Spain) (Project 19893/GERM/15). This work is the result of a predoctoral contract for the training of research staff (Silvia Navarro-Orcajada, number 21269/FPI/19) and also an aid to postdoctoral training and improvement abroad (for Adrián Matencio, number 21229/PD/19) financed by the Consejería de Empleo, Universidades, Empresa y Medio Ambiente of the CARM, through the Fundación Séneca-Agencia de Ciencia y Tecnología de la Región de Murcia.

References

Abarca RL, Rodríguez FJ, Guarda A et al (2017) Application of β-cyclodextrin/2-nonanone inclusion complex as active agent to design of antimicrobial packaging films for control of Botrytis cinerea. Food Bioprocess Technol 10:1585–1594. https://doi.org/10.1007/s11947-017-1926-z

Abril-Sánchez C, Matencio A, Navarro-Orcajada S et al (2019) Evaluation of the properties of the essential oil citronellal nanoencapsulated by cyclodextrins. Chem Phys Lipids 219:72–78. https://doi.org/10.1016/j.chemphyslip.2019.02.001

Arana-Sánchez A, Estarrón-Espinosa M, Obledo-Vázquez EN et al (2010) Antimicrobial and antioxidant activities of Mexican oregano essential oils (Lippia graveolens H. B. K.) with different composition when microencapsulated in β-cyclodextrin. Lett Appl Microbiol 50:585–590. https://doi.org/10.1111/j.1472-765X.2010.02837.x

Aree T (2019) Understanding structures and thermodynamics of beta-cyclodextrin encapsulation of chlorogenic, caffeic and quinic acids: implications for enriching antioxidant capacity and masking bitterness in coffee. Food Chem 293:550–560. https://doi.org/10.1016/j.foodchem.2019.04.084

Arrieta MP, López J, Hernández A, Rayón E (2014) Ternary PLA–PHB–limonene blends intended for biodegradable food packaging applications. Eur Polym J 50:255–270. https://doi.org/10.1016/j.eurpolymj.2013.11.009

Astray G, Mejuto JC, Simal-Gandara J (2020) Latest developments in the application of cyclodextrin host-guest complexes in beverage technology processes. Food Hydrocoll 106:105882. https://doi.org/10.1016/j.foodhyd.2020.105882

Atarés L, Chiralt A (2016) Essential oils as additives in biodegradable films and coatings for active food packaging. Trends Food Sci Technol 48:51–62. https://doi.org/10.1016/j.tifs.2015.12.001

Athanasiadis V, Grigorakis S, Lalas S et al (2018) Methyl β-cyclodextrin as a booster for the extraction for Olea europaea leaf polyphenols with a bio-based deep eutectic solvent. Biomass Conv Bioref 8:345–355. https://doi.org/10.1007/s13399-017-0283-5

Ayala-Zavala JF, González-Aguilar GA (2010) Optimizing the use of garlic oil as antimicrobial agent on fresh-cut tomato through a controlled release system. J Food Sci 75:M398–M405. https://doi.org/10.1111/j.1750-3841.2010.01723.x

Ayala-Zavala JF, Del-Toro-Sánchez L, Alvarez-Parrilla E et al (2008a) High relative humidity in-package of fresh-cut fruits and vegetables: advantage or disadvantage considering microbiological problems and antimicrobial delivering systems? J Food Sci 73:R41–R47. https://doi.org/10.1111/j.1750-3841.2008.00705.x

Ayala-Zavala JF, Soto-Valdez H, González-León A et al (2008b) Microencapsulation of cinnamon leaf (Cinnamomum zeylanicum) and garlic (Allium sativum) oils in β-cyclodextrin. J Incl Phenom Macrocycl Chem 60:359–368. https://doi.org/10.1007/s10847-007-9385-1

Aytac Z, Uyar T (2017) Core-shell nanofibers of curcumin/cyclodextrin inclusion complex and polylactic acid: enhanced water solubility and slow release of curcumin. Int J Pharm 518:177–184. https://doi.org/10.1016/j.ijpharm.2016.12.061

Aytac Z, Dogan SY, Tekinay T et al (2014) Release and antibacterial activity of allyl isothiocyanate/β-cyclodextrin complex encapsulated in electrospun nanofibers. Colloids Surf 120:125–131. https://doi.org/10.1016/j.colsurfb.2014.04.006

Birck C, Degoutin S, Maton M et al (2016) Antimicrobial citric acid/poly(vinyl alcohol) crosslinked films: effect of cyclodextrin and sodium benzoate on the antimicrobial activity. LWT 68:27–35. https://doi.org/10.1016/j.lwt.2015.12.009

Buendía-Moreno L, Sánchez-Martínez MJ, Antolinos V et al (2020) Active cardboard box with a coating including essential oils entrapped within cyclodextrins and/or halloysite nanotubes. A case study for fresh tomato storage. Food Control 107:106763. https://doi.org/10.1016/j.foodcont.2019.106763

Canales D, Montoille L, Rivas LM et al (2019) Fungicides films of low-density polyethylene (LDPE)/inclusion complexes (Carvacrol and Cinnamaldehyde) against Botrytis Cinerea. Coatings 9:795. https://doi.org/10.3390/coatings9120795

Castro-Rosas J, Ferreira-Grosso CR, Gómez-Aldapa CA et al (2017) Recent advances in microencapsulation of natural sources of antimicrobial compounds used in food - a review. Food Res Int 102:575–587. https://doi.org/10.1016/j.foodres.2017.09.054

Charumanee S, Okonogi S, Sirithunyalug J et al (2016) Effect of cyclodextrin types and co-solvent on solubility of a poorly water soluble drug. Sci Pharm 84:694–705. https://doi.org/10.3390/scipharm84040694

Chen X, Chen R, Guo Z et al (2007) The preparation and stability of the inclusion complex of astaxanthin with β-cyclodextrin. Food Chem 101:1580–1584. https://doi.org/10.1016/j.foodchem.2006.04.020

Chen H, Li L, Ma Y et al (2019) Development of active packaging film containing bioactive components encapsulated in β-cyclodextrin and its application. Food Hydrocoll 90:360–366. https://doi.org/10.1016/j.foodhyd.2018.12.043

Ciobanu A, Landy D, Fourmentin S (2013) Complexation efficiency of cyclodextrins for volatile flavor compounds. Food Res Int 53:110–114. https://doi.org/10.1016/j.foodres.2013.03.048

Dalmolin MC, da Silva CE, Lunelli CE et al (2019) Modified β-cyclodextrin/amlodipine inclusion complexes: preparation and application in aqueous systems. J Mol Liq 276:531–540. https://doi.org/10.1016/j.molliq.2018.11.116

Dash S, Murthy PN, Nath L, Chowdhury P (2010) Kinetic modeling on drug release from controlled drug delivery systems. Acta Pol Pharm 67:217–223

Diamanti AC, Igoumenidis PE, Mourtzinos I et al (2017) Green extraction of polyphenols from whole pomegranate fruit using cyclodextrins. Food Chem 214:61–66. https://doi.org/10.1016/j.foodchem.2016.07.072

Fourmentin S, Crini G, Lichtfouse E (eds) (2018) Cyclodextrin fundamentals, reactivity and analysis. Springerplus, Berlin

Fu Y, Sarkar P, Bhunia AK, Yao Y (2016) Delivery systems of antimicrobial compounds to food. Trends Food Sci Technol 57:165–177. https://doi.org/10.1016/j.tifs.2016.09.013

García-Pérez P, Losada-Barreiro S, Gallego PP et al (2019) Cyclodextrin-elicited bryophyllum suspension cultured cells: enhancement of the production of bioactive compounds. Int J Mol Sci 20:5180. https://doi.org/10.3390/ijms20205180

Gharib R, Greige-Gerges H, Fourmentin S et al (2015) Liposomes incorporating cyclodextrin–drug inclusion complexes: current state of knowledge. Carbohydr Polym 129:175–186. https://doi.org/10.1016/j.carbpol.2015.04.048

Grass G, Rensing C, Solioz M (2011) Metallic copper as an antimicrobial surface. Appl Environ Microbiol 77:1541–1547. https://doi.org/10.1128/AEM.02766-10

Han J-W, Ruiz-Garcia L, Qian J-P et al (2018) Food packaging: a comprehensive review and future trends. Compr Rev Food Sci Food Saf 17:860–877. https://doi.org/10.1111/1541-4337.12343

Hill LE, Gomes C, Taylor TM (2013) Characterization of beta-cyclodextrin inclusion complexes containing essential oils (trans-cinnamaldehyde, eugenol, cinnamon bark, and clove bud extracts) for antimicrobial delivery applications. LWT 51:86–93. https://doi.org/10.1016/j.lwt.2012.11.011

Jansook P, Ogawa N, Loftsson T (2018) Cyclodextrins: structure, physicochemical properties and pharmaceutical applications. Int J Pharm 535:272–284. https://doi.org/10.1016/j.ijpharm.2017.11.018

Joo MJ, Merkel C, Auras R et al (2012) Development and characterization of antimicrobial poly(l-lactic acid) containing trans-2-hexenal trapped in cyclodextrins. Int J Food Microbiol 153:297–305. https://doi.org/10.1016/j.ijfoodmicro.2011.11.015

Kamihira M, Asai T, Yamagata Y et al (1990) Formation of inclusion complexes between cyclodextrins and aromatic compounds under pressurized carbon dioxide. J Ferment Bioeng 69:350–353. https://doi.org/10.1016/0922-338X(90)90242-O

Kamimura JA, Santos EH, Hill LE et al (2014) Antimicrobial and antioxidant activities of carvacrol microencapsulated in hydroxypropyl-beta-cyclodextrin. LWT 57:701–709. https://doi.org/10.1016/j.lwt.2014.02.014

Kayaci F, Ertas Y, Uyar T (2013a) Enhanced thermal stability of eugenol by cyclodextrin inclusion complex encapsulated in electrospun polymeric nanofibers. J Agric Food Chem 61:8156–8165. https://doi.org/10.1021/jf402923c

Kayaci F, Umu OCO, Tekinay T et al (2013b) Antibacterial electrospun poly(lactic acid) (PLA) nanofibrous webs incorporating triclosan/cyclodextrin inclusion complexes. J Agric Food Chem 61:3901–3908. https://doi.org/10.1021/jf400440b

Kayaci F, Sen HS, Durgun E et al (2014) Functional electrospun polymeric nanofibers incorporating geraniol–cyclodextrin inclusion complexes: high thermal stability and enhanced durability of geraniol. Food Res Int 62:424–431. https://doi.org/10.1016/j.foodres.2014.03.033

Kfoury M, Auezova L, Greige-Gerges H et al (2014) Cyclodextrin, an efficient tool for trans-anethole encapsulation: chromatographic, spectroscopic, thermal and structural studies. Food Chem 164:454–461. https://doi.org/10.1016/j.foodchem.2014.05.052

Kfoury M, Auezova L, Greige-Gerges H et al (2019) Encapsulation in cyclodextrins to widen the applications of essential oils. Environ Chem Lett 17:129–143. https://doi.org/10.1007/s10311-018-0783-y

Koester LS, Ortega GG, Mayorga P et al (2004) Mathematical evaluation of in vitro release profiles of hydroxypropylmethylcellulose matrix tablets containing carbamazepine associated to β-cyclodextrin. Eur J Pharm Biopharm 58:177–179. https://doi.org/10.1016/j.ejpb.2004.03.017

Komaikul J, Kitisripanya T, Likhitwitayawuid K et al (2019) Improvement of stilbenoid production by 2-hydroxypropyl-β-cyclodextrin in white mulberry (Morus alba L.) callus cultures. Nat Prod Res 33:2762–2769. https://doi.org/10.1080/14786419.2018.1499643

Lee M, Dey KP, Lee YS (2020) Complexation of methyl salicylate with β-cyclodextrin and its release characteristics for active food packaging. Food Sci Biotechnol. https://doi.org/10.1007/s10068-020-00749-z

Li M, Zhang F, Liu Z et al (2018) Controlled release system by active gelatin film incorporated with β-cyclodextrin-thymol inclusion complexes. Food Bioprocess Technol 11:1695–1702. https://doi.org/10.1007/s11947-018-2134-1

López-Miranda S, Serrano-Martínez A, Hernández-Sánchez P et al (2016) Use of cyclodextrins to recover catechin and epicatechin from red grape pomace. Food Chem 203:379–385. https://doi.org/10.1016/j.foodchem.2016.02.100

López-Nicolás JM, García-Carmona F (2008) Rapid, simple and sensitive determination of the apparent formation constants of trans-resveratrol complexes with natural cyclodextrins in aqueous medium using HPLC. Food Chem 109:868–875. https://doi.org/10.1016/j.foodchem.2008.01.022

López-Nicolás JM, Rodríguez-Bonilla P, García-Carmona F (2009a) Complexation of Pinosylvin, an analogue of resveratrol with high antifungal and antimicrobial activity, by different types of cyclodextrins. J Agric Food Chem 57:10175–10180. https://doi.org/10.1021/jf902519d

López-Nicolás JM, Rodríguez-Bonilla P, Méndez-Cazorla L et al (2009b) Physicochemical study of the complexation of Pterostilbene by natural and modified cyclodextrins. J Agric Food Chem 57:5294–5300. https://doi.org/10.1021/jf900285e

Madigan MT, Martinko JM, Bender KS et al (2015) Brock biology of microorganisms, Global Edition. Pearson Education Limited, London

Malapert A, Tomao V, Margier M et al (2018) β-Cyclodextrin does not alter the bioaccessibility and the uptake by Caco-2 cells of olive by-product phenolic compounds. Nutrients 10:1653. https://doi.org/10.3390/nu10111653

Mallardo S, De Vito V, Malinconico M et al (2016) Poly(butylene succinate)-based composites containing β-cyclodextrin/d-limonene inclusion complex. Eur Polym 79:82–96. https://doi.org/10.1016/j.eurpolymj.2016.04.024

Marques HMC (2010) A review on cyclodextrin encapsulation of essential oils and volatiles. Flavour Fragr J 25:313–326. https://doi.org/10.1002/ffj.2019

Matencio A, García-Carmona F, López-Nicolás JM (2016) Encapsulation of piceatannol, a naturally occurring hydroxylated analogue of resveratrol, by natural and modified cyclodextrins. Food Funct 7:2367–2373. https://doi.org/10.1039/c6fo00557h

Matencio A, García-Carmona F, López-Nicolás JM (2017) Aggregation of t10,c12 conjugated linoleic Acid in presence of natural and modified cyclodextrins. A physicochemical, thermal and computational analysis. Chem Phys Lipids 204:57–64. https://doi.org/10.1016/j.chemphyslip.2017.03.008

Matencio A, Alcaráz-Gómez MA, García-Carmona F et al (2018a) Application of a simple methodology to analyze Hydroxypropyl-β-Cyclodextrin in urine using HPLC–LS in early Niemann–Pick disease type C patient. J Chromatogr B 1093–1094:47–51. https://doi.org/10.1016/j.jchromb.2018.06.051

Matencio A, Navarro-Orcajada S, Garcia-Carmona F et al (2018b) Ellagic acid-borax fluorescence interaction. Application to a novel cyclodextrin-borax nanosensor for analyzing ellagic acid in food samples. Food Funct. https://doi.org/10.1039/C8FO00906F

Matencio A, Hernández-García S, García-Carmona F et al (2019a) A way to increase the bioaccesibility and photostability of roflumilast, a COPD treatment, by cyclodextrin monomers. Polymers (Basel) 11:801. https://doi.org/10.3390/polym11050801

Matencio A, Navarro-Orcajada S, Conesa I et al (2019b) Evaluation of juice and milk "food models" fortified with oxyresveratrol and β-Cyclodextrin. Food Hydrocoll 98:105250. https://doi.org/10.1016/j.foodhyd.2019.105250

Matencio A, Navarro-Orcajada S, García-Carmona F et al (2020a) Applications of cyclodextrins in food science. A review. Trends Food Sci Technol. https://doi.org/10.1016/j.tifs.2020.08.009

Matencio A, Navarro-Orcajada S, González-Ramón A et al (2020b) Recent advances in the treatment of Niemann Pick disease type C: a mini-review. Int J Pharm 584:119440. https://doi.org/10.1016/j.ijpharm.2020.119440

Navarro-Orcajada S, Matencio A, Vicente-Herrero C et al (2021) Study of the fluorescence and interaction between cyclodextrins and neochlorogenic acid, in comparison with chlorogenic acid. Sci Rep 11(1):3275

Parra A, Toro M, Jacob R et al (2018) Antimicrobial effect of copper surfaces on bacteria isolated from poultry meat. Braz J Microbiol 49:113–118. https://doi.org/10.1016/j.bjm.2018.06.008

Paun G, Neagu E, Tache A, Radu GL (2014) New type of chitosan/2-hydroxypropyl-β-cyclodextrin composite membrane for gallic acid encapsulation and controlled release. Acta Chim Slov 61:27–33

Piercey MJ, Mazzanti G, Budge SM et al (2012) Antimicrobial activity of cyclodextrin entrapped allyl isothiocyanate in a model system and packaged fresh-cut onions. Food Microbiol 30:213–218. https://doi.org/10.1016/j.fm.2011.10.015

Poverenov E, Granit R, Gabai S (2013) Encapsulation and controlled release of antifungal propionic acid utilizing biodegradable active films based on natural polymers. Eur Food Res Technol 237:19–26. https://doi.org/10.1007/s00217-013-2011-0

Pragadheesh VS, Yadav A, Chanotiya CS (2015) Role of substituents in cyclodextrin derivatives for enantioselective gas chromatographic separation of chiral terpenoids in the essential oils of Mentha spicata. J Chromatogr B 1002:30–41. https://doi.org/10.1016/j.jchromb.2015.07.034

Quilaqueo M, Millao S, Luzardo-Ocampo I et al (2019) Inclusion of piperine in β-cyclodextrin complexes improves their bioaccessibility and in vitro antioxidant capacity. Food Hydrocoll 91:143–152. https://doi.org/10.1016/j.foodhyd.2019.01.011

Rakmai J, Cheirsilp B, Mejuto JC et al (2017) Physico-chemical characterization and evaluation of bio-efficacies of black pepper essential oil encapsulated in hydroxypropyl-beta-cyclodextrin. Food Hydrocoll 65:157–164. https://doi.org/10.1016/j.foodhyd.2016.11.014

Rakmai J, Cheirsilp B, Mejuto JC et al (2018) Antioxidant and antimicrobial properties of encapsulated guava leaf oil in hydroxypropyl-beta-cyclodextrin. Ind Crop Prod 111:219–225. https://doi.org/10.1016/j.indcrop.2017.10.027

Sadighi A, Ostad SN, Rezayat SM et al (2012) Mathematical modelling of the transport of hydroxypropyl-β-cyclodextrin inclusion complexes of ranitidine hydrochloride and furosemide loaded chitosan nanoparticles across a Caco-2 cell monolayer. Int J Pharm 422:479–488. https://doi.org/10.1016/j.ijpharm.2011.11.016

Schönbeck C, Holm R (2019) Exploring the origins of enthalpy–entropy compensation by calorimetric studies of cyclodextrin complexes. J Phys Chem B 123:6686–6693. https://doi.org/10.1021/acs.jpcb.9b03393

Shishido TK, Jokela J, Kolehmainen C-T et al (2015) Antifungal activity improved by coproduction of cyclodextrins and anabaenolysins in cyanobacteria. Proc Natl Acad Sci U S A 112:13669–13674. https://doi.org/10.1073/pnas.1510432112

Stella VJ, Rao VM, Zannou EA, Zia V (1999) Mechanisms of drug release from cyclodextrin complexes. Adv Drug Deliv 36:3–16. https://doi.org/10.1016/S0169-409X(98)00052-0

Sun X, Guo X, Ji M et al (2019) Preservative effects of fish gelatin coating enriched with CUR/βCD emulsion on grass carp (Ctenopharyngodon idellus) fillets during storage at 4 °C. Food Chem 272:643–652. https://doi.org/10.1016/j.foodchem.2018.08.040

Uyar T, Nur Y, Hacaloglu J, Besenbacher F (2009) Electrospinning of functional poly(methyl methacrylate) nanofibers containing cyclodextrin-menthol inclusion complexes. Nanotechnology 20:125703. https://doi.org/10.1088/0957-4484/20/12/125703

Wadhwa G, Kumar S, Chhabra L et al (2017) Essential oil–cyclodextrin complexes: an updated review. J Incl Phenom Macrocycl Chem 89:39–58. https://doi.org/10.1007/s10847-017-0744-2

Xiong W, Sun Y, Zeng Z (2018) Antimicrobial use and antimicrobial resistance in food animals. Environ Sci Pollut Res Int 25:18377–18384. https://doi.org/10.1007/s11356-018-1852-2

Yang Y-C, Mei X-W, Hu Y-J et al (2019) Fabrication of antimicrobial composite films based on xylan from pulping process for food packaging. Int J Biol Macromol 134:122–130. https://doi.org/10.1016/j.ijbiomac.2019.05.021

Ye Y, Zhu M, Miao K et al (2017) Development of antimicrobial gelatin-based edible films by incorporation of trans-anethole/β-cyclodextrin inclusion complex. Food Bioprocess Technol 10:1844–1853. https://doi.org/10.1007/s11947-017-1954-8

Yuan C, Jin Z, Xu X et al (2008) Preparation and stability of the inclusion complex of astaxanthin with hydroxypropyl-β-cyclodextrin. Food Chem 109:264–268. https://doi.org/10.1016/j.foodchem.2007.07.051

Yuan C, Du L, Jin Z, Xu X (2013) Storage stability and antioxidant activity of complex of astaxanthin with hydroxypropyl-β-cyclodextrin. Carbohydr Polym 91:385–389. https://doi.org/10.1016/j.carbpol.2012.08.059

Zarandona I, Barba C, Guerrero P et al (2020) Development of chitosan films containing β-cyclodextrin inclusion complex for controlled release of bioactives. Food Hydrocoll 104:105720. https://doi.org/10.1016/j.foodhyd.2020.105720

Chapter 11
Encapsulation for Packaging

Éva Fenyvesi, István Puskás, and Lajos Szente

11.1 Roles of Packaging in Food Production

The changing market needs require continuous development in food production also including the design of packaging which should not only grip the consumer's interest but also has a function to maintain the physical, chemical, and microbiological characteristics of food from production to final consumption.

According to definition by the Commission of European Communities (2009) active packaging systems "are intended to extend the shelf life or to maintain or improve the condition of packaged food; they are designed to deliberately incorporate components that would release or absorb substances into or from the packaged food or the environment surrounding the food".

One of the main roles of active packaging is to control or preserve microbial contamination. For instance, it contains antimicrobial compounds released slowly or only upon a trigger in order to reduce, delay or hinder the growth of microbes on the surface of the food without compromising the sensory and nutritional qualities. The antimicrobial packaging makes possible to reduce the preservatives concentration added to the food and thus the consumption of these—often not harmless—chemicals.

The rising demand for ready-to-eat fresh foods, only slightly processed if at all, not containing synthetic preservatives but still having reasonable shelf life created an emerging market for active packaging growing from approximately USD 20 billion in 2019 worldwide to over USD 30 billion in 2025 (Research and Market 2020). The market report identified the increasing health-related concerns as a major reason for the growing demand for active packaging. Further reasons are the rising complexity of the products to be stored or transported to distant places, and the

É. Fenyvesi (✉) · I. Puskás · L. Szente
Cyclodextrin Research & Development Laboratory Ltd., Budapest, Hungary
e-mail: fenyvesi.e@cyclolab.hu; puskas@cyclolab.hu; szente@cyclolab.hu

T. M. Ho et al. (eds.), *Functionality of Cyclodextrins in Encapsulation for Food Applications*, https://doi.org/10.1007/978-3-030-80056-7_11

increasing E-commerce delivering food to the door of consumers. On the other hand, it should be also considered that extending the shelf life could result in a significant economic benefit for food companies.

The environmental aspects are also getting in the focus. In 2015, out of the 381 million metric tons worldwide production of polymer resin and fiber (mostly polyethylene, polyethylene terephthalate and polypropylene), 146 million tons were used for every kind of packaging (Geyer et al. 2017) and over 91 million tons was estimated for the use in packaging of food and beverages. To reduce the accumulation of non-biodegradable plastics of petrochemical origin recycling and replacement by biopolymers from natural sources are encouraged. Polysaccharide (chitosan, starch, agar, cellulose, carrageenan, etc.) and protein (gelatin, alginate, collagen, gluten, whey protein, etc.) films are proper alternatives of synthetic packaging (Jamróz and Kopel 2020). Bio-based synthetic polymers, such as polyethylene, polypropylene, polyvinylchloride, produced from renewable biomass slowly decrease the share of fossil-based plastics. The worldwide production of bio-based and biodegradable plastics was approximately 2.1 million metric tons in 2019 (European Bioplastics 2020).

The interest in natural preservatives is growing as people want to avoid chemically synthetized compounds in foods (Oswell et al. 2018; Baptista et al. 2019). Materials used for food packaging should not contain any potentially toxic substances or at least their migration into the food should be hindered.

Various essential oils and plant extracts were proved to exert antimicrobial effect against typical foodborne bacteria and molds, some of them also have antioxidant and insect-repellent effects (Jajasena and Jo 2014; Marques et al. 2019; Marsin et al. 2020). These beneficial effects are utilized in active packaging (Siró 2012). The natural antioxidants and antimicrobials such as plant-derived essential oils, however, are volatile and light- and heat-sensitive, often released too fast from the packaging material, so it was obvious to improve these properties via inclusion complex formation with cyclodextrins. These cyclic oligosaccharides built up from 6, 7 or 8 glucopyranose units are of great interest due to their low toxicity, biodegradability, thermal stability and ability for including other molecules such as components of essential oils and plant extracts into their cavities (Szejtli 2004; Crini 2014; Crini et al. 2018). The included molecules are protected from environmental effects like heat and oxidation; their diffusion and volatility are reduced but wettability is increased due to the hydrophilic wrapping around them. Complexation can control the release of these bioactives. All these properties are beneficial in food packaging. Not only the inclusion complexes can be useful in active packaging but cyclodextrins themselves as they act as traps for volatile and easily migrating compounds thus hindering that harmful components can get into food or valuable flavor components escape during transport and storage (Szente and Fenyvesi 2018).

Cyclodextrins are widely used in food industry and these applications have been reviewed from time to time (Szente and Szejtli 2004; Astray et al. 2009, 2020; Martina et al. 2013; Fenyvesi et al. 2016; Tian et al. 2020; Matencio et al. 2020). The native cyclodextrins are safe to consume and therefore can be incorporated into food packaging without any risk; their regulatory status in various countries

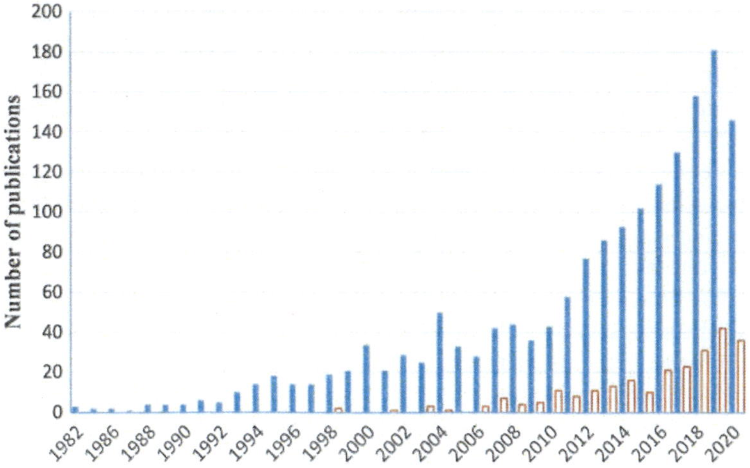

Fig. 11.1 Number of publications in Scopus database for the keywords: cyclodextrin AND food (filled columns) and cyclodextrin AND food AND packaging (empty columns) (Scopus 2020)

concerning food application has recently been overviewed by da Silva (2018). The growing interest toward such applications are shown in the increasing number of publications shown in Fig. 11.1.

Figure 11.1 also shows the number of publications on cyclodextrin's application in food packaging indicating that although this field appeared in the literature much later than the use in food but has been developing dynamically in the last 10–15 years. A comprehensive review on cyclodextrin-enabled packaging was compiled by Szente and Fenyvesi (2018).

As related to patents, the use of cyclodextrins in packaging started already in the mid of 1970s by a Japanese patent application on polypropylene film coated with ethyl cellulose mixed with β-cyclodextrin/iodine complex to extend the shelf life of seafood (Yoshitomi et al. 1976). The number of hits on keywords (cyclodextrin + food + packaging) has reached 116 out of 1311 hits for (cyclodextrin + food) in the database of European Patent Office in October 2020 (Espacenet 2020).

Out of the native cyclodextrins mostly β-cyclodextrin is used in packaging materials partly because of economic reason (this is the cheapest cyclodextrin) and due to its versatility in complexing various bioactive compounds. There are, however, a few examples for using α-cyclodextrin for complexing small molecules such as allyl isothiocyanate (Plackett et al. 2007), carbon dioxide (Ho et al. 2018) and 1-methyl cyclopropene (Daly and Kourelis 2001). The larger γ-cyclodextrin found rarely in the packaging-related papers has been usually suggested for complexation of larger bioactives like quercetin, eugenol, geraniol, α-tocopherol (Koontz et al. 2009, 2010; Kayaci et al. 2013a, 2014; Aytac et al. 2017a; Figuera-Lopez et al. 2020). In some studies also hydroxypropyl derivatives were used successfully, e.g. carvacrol and gallic acid and resveratrol complexes with hydroxypropyl-β-cyclodextrin (Higueras et al. 2014; Aytac et al. 2016) and hydroxypropyl-γ-cyclodextrin (Silva et al. 2016), respectively, were incorporated into packaging.

Random methylated β-cyclodextrin enhanced both antioxidant and antimicrobial activities of essential oils (Das et al. 2018). Sulfobutylether β-cyclodextrin was applied for catechin encapsulation to be used in antimicrobial packaging (Jin et al. 2019). Acetylated, octanyl succinate and trimethyl silyl cyclodextrin derivatives were suggested by Wood and Beaverson (1998) to improve compatibility with hydrophobic packaging. Citrate derivative of β-cyclodextrin improved the interaction with chitosan (Adel et al. 2019). None of these derivatives, however, have been approved as food additives so far. Although some of them are officially accepted excipients in pharmaceutical formulations, their safety should be studied prior to their application in food packaging.

The native cyclodextrins are readily biodegradable under composting conditions while the derivatives are degraded slower by soil bacteria depending on the type and number of substituents on the cyclodextrin ring (Verstichel et al. 2004). The biodegradability is not significantly influenced by the presence of antimicrobial agents, such as allylisothiocyanate and imazalil within the cavity. When incorporated into biodegradable packaging films, e.g. into L-poly(lactide) and L-poly(lactide)-poly(caprolactone) co-polymer films, complete biodegradation was observed under controlled composting conditions (Plackett et al. 2006).

The cyclodextrin complexes are usually stable in solid state but dissociate in solution to reach association–dissociation equilibrium (Szejtli 2004). The release of the included guest molecules starts only on the effect of humidity. This property makes the cyclodextrin complexes especially useful for active packaging of moisture-containing food products.

11.2 Techniques to Incorporate Cyclodextrins into Packaging Materials

There are many ways to improve the performance of packaging by applying either cyclodextrins or their complexes. The simplest method patented as early as in 1978 was to put a sachet, similarly to silica gel sachets for moisture absorption, into the package beside the food (Edo and Yoshida 1978). The sachet contained cyclodextrin-complexed ethanol as preservative. Ethanol was released slowly as a vapor preventing the growth of mold. This method can be used independently of the chemical nature of packaging material. Similar sachets with essential oil complexes such as cinnamon oil or garlic oil have been suggested even nowadays for controlled release of these volatile natural compounds to reduce microbial growth on the packaged fruits and vegetables (Ayala-Zavala and González-Aguilar 2010; Wang et al. 2020).

Another simple method is to spray the solution of a preservative inclusion complex on the packaging material preferably on paper board, able to absorb or adsorb the solution. Iodine spray stabilized by methyl β-cyclodextrin gave antimicrobial film (Hara et al. 2002). The packaging material, e.g. paper or unwoven cloth can be also impregnated with a solution of cyclodextrin/preservative complex. For instance, cloths immersed in β-cyclodextrin/hinokitiol complex solution were used for wrapping freshly harvested lemons to be stored without cooling for 60 days with no significant quality reduction (Fukazawa and Sato 1986, 1987). Filter paper impregnated with a solution of garlic oil inclusion complex, when placed into polystyrene food containers, extended the shelf life of fresh-cut tomato (Ayala-Zavala and González-Aguilar 2010). Another application is aromatized tea bags made of filter paper impregnated with a solution of aroma compounds complexed by cyclodextrin (Holobradi et al. 1986).

The permanent fixing of cyclodextrins or their complexes on the surface or within the packaging material will be discussed in the following parts of this subchapter.

11.2.1 Incorporating Cyclodextrin into Thermoplastic Polymers

Thermoplastic materials such as polyethylene, polypropylene, polybutylene, poly(vinyl chloride) and polyesters are widely used in food packaging in the form of films, trays, sheets, cups, bottles, containers, etc. The packaging films or containers are usually made by melt extrusion at elevated temperature high enough to cause considerable loss through volatilization and thermal decomposition of the bioactive components. This loss can be significantly reduced when the bioactive components are complexed by cyclodextrin.

Cyclodextrin complexes are usually prepared in aqueous systems. It is important to dry the complexes carefully to remove the strongly bound crystal water, otherwise it will escape at the melting temperature and cause foaming which results in uneven surface and bad mechanical properties (Szente and Fenyvesi 2018).

The hydrophilic cyclodextrin is not well compatible with most of the hydrophobic thermoplastic polymers, such as polyethylene. When incorporated into the packaging material agglomerations are formed resulting in reduced elasticity and barrier capacity (Fig. 11.2a, b) (Fenyvesi et al. 2007; López-de-Dicastillo et al. 2011a, b). Better compatibility was observed when less hydrophilic acetylated, octanyl succinate and trimethyl silyl cyclodextrin derivatives were used (Wood and Beaverson 1998). Applying triacetyl derivative no agglomeration of cyclodextrin was observed and also the mechanical and barrier properties were better than those of the polyethylene films with native β-cyclodextrin (Shin et al. 2019). These derivatives, however, are not approved for application in food.

It was observed that the inclusion complexes have better compatibility than the hydrophilic "empty" cyclodextrins because the part of guest molecule protruding from the complex can be less hydrophilic and gives an amphiphilic character to the

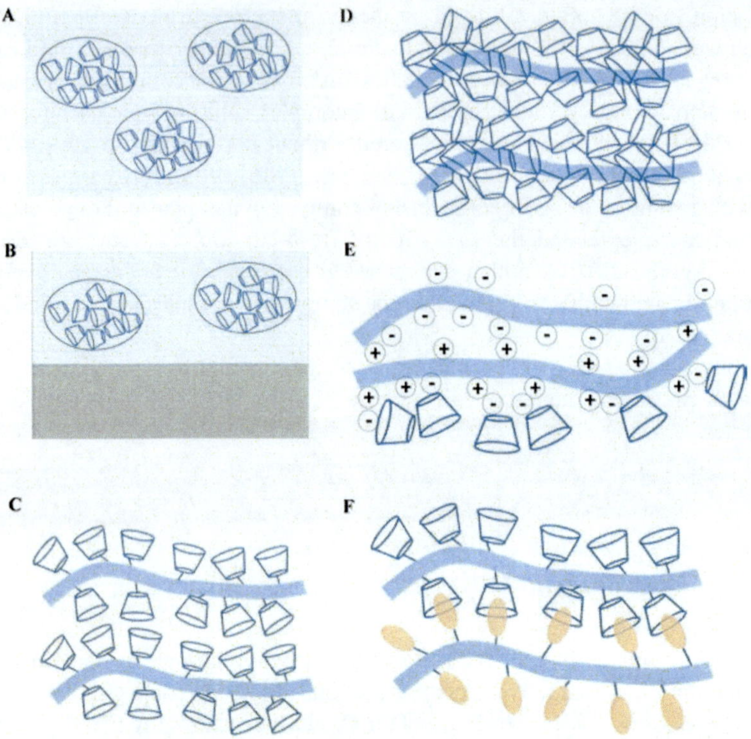

Fig. 11.2 Scheme of various methods for fixing cyclodextrin on packaging materials. (**a**) Cyclodextrin embedded in thermoplasts by melting and extrusion; (**b**) Cyclodextrin embedded in coatings by melting and extrusion; (**c**) Grafting activated cyclodextrin on packaging material able to react with it; (**d**) Crosslinking cyclodextrin on the surface of fibrous packaging using appropriate coupling agent; (**e**) Layer-by-layer assembly based on electrostatic interactions; (**f**) Layer-by-layer assembly based on host-guest interactions

clathrate making it more compatible (Szente and Fenyvesi 2018). Applying appropriate adjuvants, such as glycerol also improves the compatibility (Fenyvesi et al. 2007; López-de-Dicastillo et al. 2011a, b).

Homogeneity of cyclodextrin distribution within the polymer matrix can be improved by applying masterbatch, a preformed mixture of cyclodextrin or complexes with the polymer by extruding and forming pellets. Addition of cyclodextrin in the form of a masterbatch reduced the negative effects on mechanical properties and color and improved compatibility with the thermoplasts such as poly(lactic acid) (Joo et al. 2011).

Low-density polyethylene granules were mixed with cyclodextrin/hinokitiol complex and blown into film by extrusion then reinforced by a second barrier layer to be used as sheets, bags or a lining of corrugated fiberboard containers (Imakura et al. 1992). Compression molding was applied to get polyethylene films containing γ-cyclodextrin/quercetin or β-cyclodextrin/α-tocopherol complexes as antioxidants, which not only delay oxidation of the packaged food during storage but also protect the polymer from oxidative degradation during melt processing (Siró et al. 2006; Koontz et al. 2010). The role of cyclodextrin was to reduce the loss of antioxidants during processing and prolong their release during application as food packaging. Complexation also reduced the loss of volatile natural preservatives, carvacrol and allylisothiocyanate during extrusion and thermo-molding of poly(lactic acid) films (Raouche et al. 2011).

Further examples are listed in Table 11.1.

Table 11.1 Examples for cyclodextrin-containing packaging films produced by melt processing

Polymer	Active	Cyclodextrin	Effects	References
Ethylene vinyl alcohol copolymer	Citral and cinnamaldehyde	β-CD	Antimicrobial packaging e.g. for chilled beef	Chen et al. (2019)
Plastic film	Isothiocyanates	n.s.	Antifungal packaging	Matsui et al. (1994)
Poly(butylene succinate)	D-Limonene	β-CD	Antibacterial packaging, delayed crystallization of the polymer	Mallardo et al. (2016)
Polyethylene	D-Limonene	β-CD	Long-lasting flavor for orange juice packaging laminated container	Yamada and Tsunoda (1987)
Polyethylene	Hinokitiol	n.s.	Antibacterial and antifungal films with controlled release, cucumber, broccoli, cherries and raw chicken were preserved for more than 7 days	Imakura et al. (1992)
Polyethylene	α-Tocopherol	β-CD	Antioxidative packaging	Siró et al. (2006), Koontz et al. (2010)
Polyethylene, polypropylene	Bromocynnamic aldehyde	n.s.	Antifungal packaging	Fujimura (1986)
Polyethylene terephthalate	Cinnamaldehyde	β-CD	Antibacterial packaging for fresh pork	Zhou et al. (2020)
Polypropylene	Rapeseed oil	α- or β-CD	Good sliding ability, easy to remove before consumption	Houchin (1989)
Polystyrene	Catechins	n.s.	Antimicrobial activity and antifogging film	Sawada and Fukukita (2004)
Polyvinyl chloride	Imazalil	β-CD	Antimicrobial packaging	Balogh et al. (2008)

n.s. not specified

11.2.2 Film Forming by Solution Casting

A simple way of producing films containing cyclodextrin or inclusion complexes is (1) preparation of common solution or suspension/emulsion of film-forming polymer(s) and cyclodextrin or inclusion complexes, (2) coating this solution on a carrier substrate and (3) removal of solvent. The resulting cast layer can be stripped from the carrier and used as a standalone film or can be used as multi-layered laminated film. Solution casting can be applied for non-thermoplastic raw materials such as cellulose derivatives. It is easy to incorporate additives, even heat-sensitive components as this technology does not need high temperature. Film thickness is determined by the concentration and volume of the solution casted. Plasticizers such as glycerol can be applied to improve flexibility.

This method was used e.g. for cellulose sulfate films with native cyclodextrins casted from aqueous solution (Chen and Liu 2016), and cellulose acetate films with cyclodextrin acetate from acetone (Asslisi 2011). Hydroxyethyl cellulose dissolved in water and mixed with soy lecithin and glycerol as plasticizers and resveratrol complexed by hydroxypropyl γ-cyclodextrin was layered on a previously prepared cellulose acetate film then dried to result in antimicrobial bilayer films (Silva et al. 2016). Chitosan/cyclodextrin films were prepared and filled with carvacrol to use for packaging of chicken fillets (Higueras et al. 2014). Oxidized nanocellulose as additive improved the mechanical properties of films casted from aqueous solution of clove oil ternary complex with chitosan and β-cyclodextrin citrate (Adel et al. 2019).

Poly(vinyl alcohol)-starch aqueous dispersion was mixed with lemongrass oil in complexed form and spread on glass surface then dried to obtain films with even distribution of the complex, good transparency, water permeability, antioxidant and antimicrobial properties (Chen et al. 2020a). Polystyrene films containing α-cyclodextrin/ethylene complex were casted from a solution in acetone to control ripening of the packaged fruits (Capozzi et al. 2017).

11.2.3 CD-Containing Coating

The packaging material (plastic or paper) can be covered by a thin film containing cyclodextrin or cyclodextrin-complexed preservatives to give a functionality to the packaging, such as barrier, odor absorption, antimicrobial, antioxidant properties. Cyclodextrin or complexes are obviously incorporated into the innermost layer of packaging material and the bioactive compounds are released from this food-contact layer on the effect of humidity.

There are several coating methods to build a thin film on a surface such as dip coating when the item is immersed and withdrawn from a solution of the film-forming components, and slot-die coating, blade coating, bar coating when the solution is placed directly on the item and the excess is removed in a second step determining the thickness (Aegerter and Mennig 2004).

A film layer containing isothiocyanate and/or terpene complex was coated to the inner surface of packaging material in contact with the food to achieve high antibacterial performance (Doi and Yashiki 1994). A polyethylene film containing cyclodextrin/limonene complex was laminated on a polyester film to obtain container for orange juice (Yamada and Tsunoda 1987).

Recent examples showed that the shelf life of tomato and mandarin was extended by spraying an acrylic solution of β-cyclodextrin-complexed carvacrol, oregano essential oil and cinnamon to get a lacquer coating on a paperboard surface (Buendía-Moreno et al. 2020; López-Gómez et al. 2020).

11.2.4 Grafting Cyclodextrin on Polymer Films

Grafting of cyclodextrin on the packaging material results in homogeneous distribution of cyclodextrin on the surface (Fig. 11.2c). This technique usually needs activation of cyclodextrin, to make it able to react with the surface, by introducing various functional groups. In some cases also the packaging material should be activated that is modified in a way to be a proper reaction partner for grafting.

- *Monochlorotriazinyl β-cyclodextrin* can be permanently fixed on natural and synthetic fibrous materials such as textiles, cellulose fibers and powders (Reuscher and Hirsenkorn 1996; Beermann et al. 2002; Rehmann et al. 2003; Furuta et al. 2006). More than 70μmol/g (11%) monochlorotriazinyl β-cyclodextrin content was achieved (de Bergamasco et al. 2007). It was also grafted on poly(vinyl alcohol) (PVA) films and then loaded with antimicrobial agents such as ferulic acid and allantoin to be used as a coating on polypropylene or polyethylene terephthalate for packaging of food products (Grigoriu et al. 2013). Coffee bags and tea bags were prepared by modifying tissue paper with monochlorotriazinyl β-cyclodextrin to avoid oxidative degradation and evaporation of flavor components during storage especially in multi-pack of bags where the last bag is used only after a shorter or longer storing (Wintersgill 2004)
- Grafting with *di- or polycarboxylic acid*, such as citric acid able to react both with cyclodextrin and cellulose is also a rather simple method: (1) the paper sheet or fibers are impregnated with a solution of cyclodextrin, citric acid and sodium hypophosphite catalyst, (2) after squeezing the excess of this solution they are dried at high temperature (160 °C) (Lavoine et al. 2014). The grafted cyclodextrin slowed down the release of carvacrol to ensure long-term antibacterial effect for packaging. Citric acid was used also to couple hydroxypropyl β-cyclodextrin to poly(vinyl alcohol) by heat processing the films casted from their common solution (Birck et al. 2014).
- *Epoxy compounds* such as 1,4-butanediol diglycidyl ether as crosslinking agent are another grafting agents to bind cyclodextrin on cellulose surface as they can react with both cyclodextrin and cellulose (de Bergamasco et al. 2007).

- *Click chemistry* was used to connect cyclodextrin to cellulose fibers by reacting mono-6-propargylamino-6-deoxy-β-cyclodextrin and azidated cellulose fibers to obtain fibers ready to be loaded with antibacterial compounds (Sun et al. 2020).
- *Direct grafting* of β-cyclodextrin on the carboxyl groups of TEMPO-oxidized cellulose nanofibers resulted in sustained release of carvacrol and as a consequence, also in long-lasting antibacterial activity (Saini et al. 2016; de Castro et al. 2018).
- Some further reactions for grafting cyclodextrin on cellulose were reviewed by Cova et al. (2018) concentrating mainly on textile applications. These reactions are to be explored for food packaging as well.

11.2.5 Crosslinking Cyclodextrin on the Surface of Packaging Material

Various crosslinking agents able to react with cyclodextrins but not with the packaging material such as polyethylene, polypropylene, polyesters can be used to create a cyclodextrin polymer layer on the surface of the support (Fig. 11.2d). The complex forming ability of cyclodextrins is preserved in this layer, so the cavities can be filled with bioactive compounds to be released in a controlled manner. There are, however, only a few examples of this fixing method in food packaging most probably due to the toxicity of some crosslinking agents which have to be removed from the packaging material with extreme care to avoid their migration into the food.

Non-toxic citric acid was used to fix β-cyclodextrin on the surface of polyethylene terephthalate pads, then the cavities were loaded with cinnamaldehyde, a natural antimicrobial agent to obtain sustained release antimicrobial pads with long-term antibacterial effect against *Escherichia coli* and *Staphylococcus aureus* (Zhou et al. 2020). The pads combined with tray packaging were successfully applied for cold fresh pork.

Cyclodextrin was incorporated into chitosan by toluene diisocyanate crosslinking reaction than coated on low-density polyethylene film to ensure the controlled release of allyl isothiocyanate as natural antibacterial agent (Huang et al. 2018a).

Also polyisocyanate was used for crosslinking cyclodextrin on polyethylene terephthalate fibers to include vitamin E (tocopherol) as antioxidant (Hirogaki et al. 2010).

11.2.6 Layer by Layer Coating

Layer by layer coating based on self-assembly can be obtained using electrostatic attraction between a positively charged film-forming polymer such as chitosan and a negatively charged polymer such as alginate, pectin. Cyclodextrin or complexes can be incorporated in any of the layers. The ionic cyclodextrin derivatives, such as the negatively charged carboxyalkyl cyclodextrins or the positively charged amino cyclodextrins can improve the adherence of the layers. Another possibility is the utilization of self-assembling via host-guest interaction: one layer grafted with cyclodextrin will readily adhere to the other layer grafted with the guest molecule.

A multilayer packaging material has been recently suggested by researchers from Spain: poly(lactic acid) was internally blended by ZnO nanoparticles, then the positively charged chitosan layer was deposited which was covered by the negatively charged carboxyethyl β-cyclodextrin layer with carvacrol within the cavities (Andrade-Del Olmo et al. 2019). The aim of applying chitosan layer in this three-layer structure is to reduce the lipophilicity of poly(lactic acid) and this way reducing adhesion of bacteria to the surface. The carboxyethyl β-cyclodextrin layer has similar effect in addition to providing controlled release of the antibacterial carvacrol (Fig. 11.2e). The obtained biodegradable films combine bacteria adhesion resistance, inorganic bactericide effect, and antimicrobial/antioxidant active release.

Alginate layer with citric acid to protect color and chitosan with oregano essential oil/β-cyclodextrin complex as antibacterial layer are alternately casted on the fresh-cut purple yam through a layer-by-layer assembly method to prolong the shelf life of the roots (Huang et al. 2018b). Similarly, alginate and L-menthol-beta-cyclodextrin-graft-chitosan layers were alternately applied to get an edible coating with self-healing property and sustained release of L-menthol (Yang et al. 2020). Applied on apples the fruits retained their firmness and suffered reduced weight loss compared to control for 30 days.

An example for building up multilayer films by supramolecular interactions when one polymer is modified with cyclodextrin, the other one with a guest was published by Xuan et al. (2019): alternating layers of poly(ethylenimine)-β-cyclodextrin and poly(acrylic acid)-adamantanamine were applied on fruits (Fig. 11.2f). Adamantanamine, a hydrophobic guest compound is included in the hydrophobic cavity of cyclodextrin via self-assembling and forms inclusion complex to keep together the two polymer layers. The biodegradable coatings with self-healing property were aimed for fresh-keeping banana by inhibiting and degrading ethylene.

11.2.7 Cyclodextrin-Enabled Active Packaging Material by Electrospinning

Electrospinning is a relatively simple and cost-effective method to produce thin fibrous mats made of various natural and synthetic polymers. This technology has the advantage of high fiber yield and relatively large specific surface area (fiber diameter in the range of tens of nanometers to several micrometers). As the process is performed at room temperature it allows nano-encapsulation of volatile and heat-sensitive active compounds especially if they are molecularly encapsulated with cyclodextrin (Table 11.2). The electrospun mats containing bioactive compounds

Table 11.2 Electrospun nanofibers with cyclodextrin enabled active compounds for food packaging

Active	Cyclodextrin	Polymer	Effects	References
Allyl isothiocyanate	β-CD	Soy protein/ poly(ethylene oxide, poly(lactic acid)	Humidity-triggered release	Vega-Lugo and Lim (2009)
Cinnamon essential oil	β-CD	Poly(lactic acid)	Prolonged shelf life of pork	Wen et al. (2016a, b)
		Poly(ethylene oxide)	Controlled release antibacterial	Lin et al. (2017)
		Poly(vinyl alcohol)	Delayed decay of mushroom	Pan et al. (2019)
		Chitosan/poly(vinyl alcohol)	Antifungal packaging against *Botrytis grey* mould of pomegranate	Munhuweyi et al. (2018)
Cinnamon essential oil + lysozyme	β-CD	Poly(vinyl alcohol)	Enhanced antimicrobial effect against bacteria and fungi	Munhuweyi et al. (2018)
Eucalyptus essential oil	β-CD	Zein	Antimicrobial effect	Dias Antunes et al. (2017)
Eugenol	γ-CD	Pullalan	Long lasting antioxidant	Celebioglu and Uyar (2021)
	β- and γ-CD	Poly(vinyl alcohol)	Enhanced durability and thermal stability	Kayaci et al. (2013a)
Ferulic acid	Hydroxypropyl β-CD	Gliadin (cereal protein)	Enhanced photostability and faster release into food stimulants	Sharif et al. (2018)
	γ-CD	Poly(vinyl alcohol)	Enhanced thermostability and sustained release and long-lasting antibacterial effect	Narayanan et al. (2020)
Gallic acid	Hydroxypropyl β-CD	Poly(lactic acid)	Slow release antioxidant	Aytac et al. (2016)
Geraniol	γ-CD	Poly(vinyl alcohol)	Enhanced stability against evaporation	Kayaci et al. (2014)
R-(+)-Limonene	β-CD	Pullalan	Controlled antimicrobial release	Fuenmayor et al. (2013)
Litsea cubeba essential oil	β-CD	Plant polysaccharide	Enhanced thermal stability, sustained release and long-lasting antibacterial effect	Cui et al. (2019)
1-Methyl cyclopropene	α-CD	Polystyrene	Ripening control	Neoh et al. (2017)

(continued)

Table 11.2 (continued)

Active	Cyclodextrin	Polymer	Effects	References
Oregano essential oil	α- and γ-CD	Poly(3-hydroxybutyrate-co-3-hydroxyvalerate)	Prolonged antimicrobial and antioxidant effect	Figuera-Lopez et al. (2020)
	β-CD	Chitosan/poly(vinyl alcohol)	Antifungal packaging against *Botrytis grey* mold of pomegranate	Munhuweyi et al. (2018)
Quercetin	β-CD	Zein	Enhanced solubility and stability	Wang et al. (2021)
	γ-CD	Zein	Sustained release bacteriostatic effect	Aytac et al. (2017a)
Tee tree oil	β-CD	Poly(ethylene oxide)	Prolonged shelf life of beef	Cui et al. (2018)
Thymol	γ-CD	Zein	Inhibiting the growth of bacteria on meat samples	Aytac et al. (2017b)
Thyme essential oil	β-CD	Gelatin	Low bacterial count without impact on color, texture and sensory properties of chicken	Lin et al. (2018)
α-Tocopherol	γ-CD	Poly(vinyl alcohol)	Enhanced oxidative stability of beef meat	Aytac et al. (2017c)
Triclosan	β- and γ-CD	Poly(vinyl alcohol)	Enhanced antibacterial activity	Kayaci et al. (2013b)

can be placed as a sachet inside the package. These mats can be also subjected to annealing, a thermal post-treatment to produce transparent films with low porosity and high mechanical stability (Figuera-Lopez et al. 2020). Another way of forming films from nanofibers is chemical crosslinking like crosslinking poly(vinyl alcohol) nanofibers with glutardialdehyde (Pan et al. 2019). Cyclodextrin can be also cross-linked on the surface of fibers: poly(butylene succinate-co-terephthalate) electrospun mats were immersed in aqueous solution containing cyclodextrin, citric acid and the catalyst and heated or all the components were mixed in the solution before electrospinning (Wei et al. 2016, 2018). The latter method ensures more even distribution of cyclodextrin within the membrane. With increasing the cyclodextrin polymer content in the range of 0–28% in the composite membrane the hydrophilicity, specific surface area and sorption capacity were increased making such membranes useful in active packaging designed for the removal of unwanted components.

Electrospun nanofibers can be used for sustained release of antimicrobial agents from the packaging film to the food (Kriegel et al. 2008). The trigger of the release is humidity: while the bioactive component is practically not released at 55% relative humidity even at high temperature it is released at high humidity (90%) (Mascheroni et al. 2013). The inclusion complex of γ-cyclodextrin and ferulic acid was encapsulated into poly(vinyl alcohol) (PVA) to form electrospun nanofibers (Narayanan et al. 2020). Both solubility and stability of phenolic acids can be improved by cyclodextrin complexation (Anselmi et al. 2006; Zhang et al. 2009; Wang et al. 2011).

Oregano essential oil contains thymol and carvacrol, both with antioxidant and antimicrobial properties useful in active packaging of food. When complexed with α- and γ-cyclodextrin and incorporated into poly(3-hydroxybutyrate-co-3-hydroxyvalerate) electrospun nanofibers then annealed to get transparent films for active packaging both the antimicrobial and antioxidant activity of the bioactive films were successfully maintained for up to 15 days (Figuera-Lopez et al. 2020).

A special off-odor removing active packaging material was developed by Shin et al. (2018) by electrospinning low-density polyethylene with triacetyl β-cyclodextrin. The hydrophobic fiber mat effectively removed sulfur odor compounds such as dimethyl disulfide, dimethyl sulfide and carbon disulfide.

11.2.8 CD-Modified Nanoparticles for Packaging

Application of cyclodextrin-containing nanoparticles is an emerging strategy in packaging. Instead of using simple binary complexes of the antimicrobial, antioxidant agent or other bioactive compound multicomponent nanoparticles are incorporated into the conventional packaging materials. Some examples are listed here:

- Chitosan nanoparticles with embedded vitamin C and E (a mixture of tocopherols and tocotrienols) formulated by sulfobutyl ether β-cyclodextrin were created by ionic gelation technique to be applied in food packaging (Aresta et al. 2013).
- Nanoparticles obtained by coating thyme essential oil/β-cyclodextrin with positively charged ε-polylysine were embedded into gelatin nanofibers prepared by electrospinning. The resulting nanofibers can be used as antimicrobial packaging able to reduce contamination by *Campylobacter jejuni*, a food borne pathogenic bacterium on poultry surface (Lin et al. 2018).
- Cinnamon oil encapsulated in nanosponges made by crosslinking of cyclodextrin with carbonyldiimidazole effectively killed foodborne bacteria such as *Esherichia coli*, *Yersinia enterolitica*, etc. (Simionato et al. 2019).
- The surface of TiO_2 nanoparticles was modified by α- and β-cyclodextrins and the cavities were loaded by benzoic acid and sorbic acid preservatives to be incorporated into packaging material for inhibiting microbial growth (Goñi-Ciaurriz et al. 2020).

11.2.9 Printing Ink, Tape, Label

Cyclodextrins not included in the packaging material but attached from outside with printing ink, tape or label is somewhat different approach. The antimicrobial, antioxidant, ripening-controlling, etc. agents as bioactive compounds are included in the cyclodextrin cavity.

Freshness-preserving printing ink containing also hinokitiol, a natural antimicrobial agent in cyclodextrin complexed form in addition to the pigment was simply printed on the packaging material by Naito et al. (1994).

Isothiocyanate ester included in cyclodextrin was incorporated either into the plastic layer (e.g. polypropylene) or into the adhesive layer of a tape to gradually release the volatile antibacterial agent when applied to food-packaging material (Komatsu et al. 1994).

Polystyrene films containing α-cyclodextrin/ethylene complex were casted and used as labels attached to the packaging to trigger the ripening of the packaged fruits on demand (Capozzi et al. 2016).

11.3 Applications of Cyclodextrin-Incorporated Packaging in Food Production

Active packaging aims at delaying the food spoilage by incorporating bioactive compounds, such as antimicrobial agents, antioxidants, ripening-control compounds into the packaging. These bioactive compounds are released from the packaging into the headspace between the packaging and the food, and adsorbed on the food surface from where migrated into the inner part of the food. After a while—usually in a few days—a partition equilibrium between packaging and food is reached (Siró et al. 2006; Higueras et al. 2014). When the bioactive compound is in cyclodextrin-complexed form, the release from the complex is the first step, followed by diffusion through the packaging layer, then partitioning into the food surface, then migrating into the food. The release rate is determined by the composition of the food (moisture, fat and acid content), temperature, concentration of the bioactive compound, its affinity toward cyclodextrin, its diffusion coefficient within the packaging material, its partition coefficient between the packaging material and the packaged food and between the packaging material and cyclodextrin (Wessling et al. 1999; Poverenov et al. 2013). The release is often a longer process compared to the uncomplexed alternative.

When "empty" cyclodextrin is incorporated into the packaging material to enhance the barrier function of packaging the interaction of polymer with cyclodextrin should be also considered in addition to the interaction of the volatile or migrating compound(s)—to be trapped—with both cyclodextrin and the polymer. Cyclodextrin can form inclusion complexes with polymers, such as polyethylene terephthalate, changing the polymer morphology (Tonelli 2003). The non-inclusion-complex-forming polymers, such as chitosan can also decrease the binding of other guests (Mura et al. 2007).

Designing a food packaging material, the direction of the release of the active components should be carefully planned, too. For example, in the case of a fungicide film used for packaging of hard cheese, most of the volatile fungicide must be released inside the package. An insect repellent film must release its

Fig. 11.3 Main applications of cyclodextrins in food packaging: films, cardboards, pads, edible coating, bottles, milk carton packages with antimicrobial and antioxidant properties, aroma-barrier and ripening-control effect

active component, contrary, outward. Other packaging must not release the active component at all. The feasible solution for the oriented release is the two-layered packaging composites, consisting from a hardly penetrable support and a carrier layer, which contains the component to be released (Szente and Fenyvesi 2018).

The release of bioactive compounds from the surfaces is triggered by humidity. They do not start to be deliberated from the complex until contact with moisture from packaged food.

The food packaging material can be formed into sheets, films, bags, containers, boxes, bottles and cups for beverages, etc. (Fig. 11.3).

In the followings several examples will be shown on cyclodextrin-enabled active food packaging with added functionality.

11.3.1 Antimicrobial Packaging

Antimicrobial packaging aims at retarding the growth rate of spoilage microorganisms without using preservatives added to the food. The antimicrobial agent is integrated into the packaging system in order to prevent microbial growth on food products and to enhance the shelf life (Mlalila et al. 2018). Some of these microorganisms, such as *Escherichia coli* and *Salmonella typhimurium*, are important pathogens, causing foodborne illnesses. Some yeasts and molds are also pathogenic. Antimicrobial agents in the food, however, even when applied at low concentrations have potential undesired health effects. The increasing demand for preservative-free food products forced the development of antimicrobial packaging which acts as a depo of these bioactive compounds deliberated gradually. So, the antimicrobial agents get into contact to the surface of food first where they have a low concentration just enough to exert the desired preserving (antibacterial, antifungal) effect. This way no adverse health consequences are expected. To prevent the losses during film processing and to maintain a humidity triggered, sustained release, the preservatives, especially the volatile ones are used in cyclodextrin-complexed form. For instance, plant extracts including essential oils are considered natural antibiotics. They are highly volatile, thermally unstable and easily oxidized. All these properties can be improved by complexation with cyclodextrins (Szente et al. 1988; Fenyvesi and Szente 2016; Kfoury et al. 2015a, b; Sebaaly et al. 2018).

The concentration of antimicrobial agent within the packaging has to be set to achieve proper antimicrobial effect during the storage. It should be taken into account, however, that these components migrate from the surface of the food into the inner parts and may compromise the sensory properties. For instance, carvacrol, a component of oregano oil frequently used as natural preservative, readily migrates into cold-fresh meat and might cause an off-taste (Higueras et al. 2014).

One of the earliest patent applications on antimicrobial food packaging with cyclodextrins claims hinokitiol, a natural monoterpenoid with antifungal properties encapsulated by β-cyclodextrin and laminated on the outer surface of the synthetic resin films to obtain wrapping film for packaging fruits and vegetables (Imakura et al. 1992).

Allyl isothiocyanate, an organosulfur compound in mustard, wasabi, radish and horseradish was used in cyclodextrin-complexed form

- in L-poly(lactide) and L-poly(lactide)-poly(caprolactone) co-polymer films to control the mold growth on packaged cheese (Plackett et al. 2006);
- in egg container as antibacterial lining (Saeki and Shirakawa 1996);
- in antimicrobial water absorbing sheets applied to retain freshness of perishable food (Hashizume and Saeki 1995);
- applied as nanoemulsified mustard combined with gelatin and hydroxypropyl β-cyclodextrin to extend the shelf life of turkey breast (Milani et al. 2020).

Microbiological studies proved the reduced count of microorganisms, yeasts and molds on the surface of food packaged with cyclodextrin-complexed antimicrobial agents (Table 11.3).

Similarly, to antimicrobial packaging, also packaging materials with insect repellent properties have been developed (Atsuta et al. 1989; Nakahara et al. 1995). For instance, cinnamon oil complexed with β-cyclodextrin was incorporated into starch-based insect-proof adhesives applied on cardboxes for cornflake cereal in order to prevent indianmeal moth, *Plodia interpunctella* larvae infestation (Kim et al. 2020).

Table 11.3 Some examples for studies on cyclodextrin-containing packaging with proved antimicrobial effect

Pathogen microbes	Antimicrobial agent	Cyclodextrin	Packaging film	Food	References
Botryris cinerea	Allyl isothiocyanate, carvacrol	β-CD	Poly(lactic acid)	Potato dextrose agar as model food	Raouche et al. (2011)
Escherichia coli	Tea tree oil	β-CD	Poly(ethylene oxide)	Beef	Cui et al. (2018)
E. coli, *Staphylococcus aureus*	Cinnamaldehyde	β-CD	Polyethylene terephthalate	Fresh pork	Zhou et al. (2020)
E. coli, *Listeria monocytogenes*	Cinnamaldehyde	β-CD	Poly(lactic acid)	Fruits	Zhang et al. (2020)
E. coli, *S. aureus*	Cinnamon essential oil	β-CD	Poly(vinyl alcohol)	Mushroom	Pan et al. (2019)
E. coli, *Salmonella typhimurium*, *S. aureus*, *L. monocytogenes*	Carvacrol, cinnamaldehyde and eugenol	β-CD	Chitosan	n.s.	Sun et al. (2014)
Penicillium expansum, *E. coli*, *L. monocytogenes*	Allyl isothiocyanate	β-CD	Paper	Fresh-cut onions	Piercey et al. (2012)
Yeast and fungi	Carvacrol	Hydroxypropyl β-CD	Chitosan	Chicken fillet	Higueras et al. (2014)
E. coli, *S. aureus*, *Bacillus subtilis*, *Aspergillus niger*	Mustard essential oil	β-CD	Cellulose sulfate	Microbial sensitive food products	Chen and Liu (2016)
Aerobic bacteria	Allyl isothiocyanate	β-CD	Chitosan coated on polyethylene	Fresh beef	Huang et al. (2018a)
Penicillium species (*P. commune* and *P. roqueforti*)	Allyl isothiocyanate	β-CD	L-poly(lactide) and copolymer with poly(caprolactone)	Cheese	Plackett et al. (2006)
Campylobacter jejuni, *E. coli*, *Arcobacter butzleri*	Resveratrol	Hydroxypropyl γ-CD	Hydroxyethylcellulose and cellulose acetate bilayer film	n.s.	Silva et al. (2016)
Aerobic bacteria, yeasts and molds	Cinnamaldehyde	β-CD	Chitosan and pectin	Fresh-cut melon	Moreira et al. (2014)
Shewanella putrefaciens	Lemongrass oil	β-CD	Poly(vinyl alcohol)-starch	Sea food	Chen et al. (2020a)
Aspergillus niger and S. aureus	Imazalil	β-CD	Poly(vinyl chloride)	Banana	Balogh et al. (2008)

n.s. not specified

Fat-containing food products containing unsaturated fatty acids are especially sensitive to oxidation leading to undesirable off-flavors, rancidity and potentially toxic oxidation products. The free radicals formed via oxidation are scavenged by antioxidants. Some antioxidants also delay the browning of fruit juices as well as the decline in vitamin C content (Pan et al. 2020).

Sophie Fourmentins's group developed a headspace gas chromatography method for simultaneous determination of the formation constants of the cyclodextrin complexes with the individual components of essential oils based on the volatility-reducing effect of complexation (Kfoury et al. 2015a). Studying nine different essential oils, β-cyclodextrin was found to have 3–4 times higher retention effect on the volatile components compared to α- and γ-cyclodextrin. The antioxidant effect characterized by radical scavenging was also improved.

Complexation improved the radical-scavenging effect of various essential oils, especially those with high phenolic content (Kfoury et al. 2015a). This improvement is only partly the consequence of inclusion complex formation; cyclodextrins themselves were shown to have antioxidant effect in preventing fruit juices from enzymatic browning, so they are recognized as "secondary antioxidants" (Núñez-Delicado et al. 1997; López-Nicolás et al. 2007, 2013).

The antioxidant activities of randomly methylated β-cyclodextrin-encapsulated thyme oil, lemon balm oil, lavender oil, peppermint oil and their active components were enhanced compared to their uncomplexed forms (Das et al. 2018).

Some further examples are listed in Table 11.4.

11.3.3 Edible Films

Edible films and coatings are thin protective layers on food products, which can be consumed together with the food without peeling them from the product (Hassan et al. 2018). These films are prepared from digestible natural compounds such as polysaccharides, proteins, lipids and composites—all of them are environmental friendly and sustainable being renewable and biodegradable. The technique is usually spraying the solution of film forming components on the food items and let the solvent evaporate.

Edible coatings such as waxing fruits and vegetables have been used in China since the twelfth century and from the beginning of the last century they have got additional functions such as preservation from spoiling microorganisms.

The idea of producing cyclodextrin-enabled edible film coating on food items to improve the shelf life goes back to 1983 when Nippon Oils Co. applied for a patent on spraying a mixture of cyclodextrin, glucose, maltose and other oligosaccharides on food products to form transparent edible films (Hiroshi 1981). Cyclodextrin was used as a carrier for deterioration inhibitor. Canadian researchers disclosed in 1992 a polysaccharide film with dinitrosyl ferrochemochrome encapsulated by β-cyclodextrin to be used in nitrite-free meat processing for improving the color (Rubin et al. 1992). After this, long time passed till recently the topic gained

Table 11.4 Some examples for packaging with proved antioxidant effect

Antioxidant	Cyclodextrin	Packaging film	Food	References
Butylated hydroxytoluene	β-CD	Polyethylene	n.s.	Agrawal et al. (2010)
Catechin	Sulfobutyl ether β-CD	Chitosan	Cold fresh meat	Jin et al. (2019)
Clove essential oil	β-CD	Alginate	Fresh food	Maestrello et al. (2017)
Curcumin	β-CD	Fish gelatin	Apple juice	Wu et al. (2018)
Eugenol	β-CD	Pullulan	n.s.	Celebioglu and Uyar (2021)
Gallic acid	β-CD	Chitosan	95% ethanol as food simulant	Aytac et al. (2016)
		Chitosan, gelatin	Fatty food	Rezaee et al. (2018)
Lemon grass oil	β-CD	Poly(vinyl alcohol), starch	Sea food	Chen et al. (2020a)
Mango peel extract	β-CD	Poly(vinyl alcohol), gelatin	Minced chicken	Kanatt and Chawla (2017)
Quercetin	β-CD	Polyethylene	n.s.	Koontz et al. (2010)
Rosemary extract	β-CD	Chitosan, carboxymethyl cellulose	n.s.	Kuang et al. (2016)
Tea polyphenols	β-CD	n.s.	Rice	Zhang and Wu (2017)
		Polyethylene	Orange juice	Pan et al. (2020)
	Sulfobutyl ether β-CD	Gelatin	Sun flower oil	Liu et al. (2015)
Tea polyphenols-Zn complex	β-CD	Gelatin	n.s.	Yang et al. (2015)
α-Tocopherol	β-CD	Polyethylene	95% ethanol as food simulant	Siró et al. (2006)
Carvacrol	β-CD	Alginate	Mushroom	Cheng et al. (2019)

n.s. not specified

11.3.2 Antioxidant Packaging

Antioxidants incorporated into the packaging may have dual function: they protect the polymer from oxidative degradation in the course of melt processing and also protect the packaged food from oxidation (Siró et al. 2006; Koontz et al. 2010). Natural antioxidants include ascorbic acid, tocopherols, and polyphenols found in essential oils. Synthetic antioxidants approved for food application are butylated hydroxyanisole, butylated hydroxytoluene and propyl gallate among others.

attention again as the demand for minimally processed food increased. Table 11.5 gives some examples.

11.3.4 Controlled Release of Gaseous Preservatives

The process of using gases to naturally preserve foods is called modified atmosphere packaging. Carbon dioxide is most commonly used to prevent the growth of bacteria and mold and delay oxidation induced spoilage of food. Carbon dioxide complexed with α-cyclodextrin reduced the mold and yeast growth on cottage cheese curd when simply mixed to curd prior to packaging this way realizing modified atmosphere packaging as the gas is slowly released from the complex on the effect of moisture within the curd (Ho et al. 2018).

Chlorine dioxide (ClO_2) is an effective antimicrobial capable of inhibiting viruses, bacteria, fungi and spores. It is a green-yellow gas under ambient conditions able to control microorganisms residing in the hard-to-reach areas of food products. It is safe for foods, its main application is disinfection of drinking water. Its complex with α-cyclodextrin is stable and non-explosive, releasing the preservative gas in an extended period (Chen et al. 2020b).

Both gas complexes could be embedded into packaging materials, too.

11.3.5 Ripening Control

A volatile phytohormone antagonist, 1-methyl cyclopropene is used widely as a post-harvest treatment to delay the ripening process of fruits and vegetables and inhibit color change and senescence during transport and storage. It is especially useful for tropical fruits like papaya ripening very fast and getting chilling injuries when stored in cold.

This highly reactive gas becomes a stable powder when complexed with α-cyclodextrin (Daly and Kourelis 2001; Konstansek 2003), while β-cyclodextrin cavity is too large for proper fit of this small molecule. The solution of the complex can be directly sprayed on the plant material or on the packaging material where the gas 1-methyl cyclopropene is slowly released. A new delivery system for cyclopropenes encapsulated within packaging materials was prepared by heat-sealing α-cyclodextrin/1-methyl cyclopropene complex powder into poly(vinyl alcohol) film, which released cyclopropenes at high humidity (Konstansek 2003). The complex was incorporated into several other common packaging films obtained either by heat-pressing or solution-casting method. The release rate was slow especially during the first 12 h. Elevated temperatures resulted in higher and faster release. Pressing films containing the complex above 100 °C reduced the amount of 1-methyl cyclopropene remaining in the film. Incorporation into low-density polyethylene films resulted in a higher and faster release than from polystyrene, poly(vinyl

Table 11.5 Examples for cyclodextrin-containing edible films for food packaging

Bioactive compound	Polymer	Cyclodextrin	Food	References
Antibiotic	Glucomannan, chitosan, guar gum and carragenan films	β-CD	Fruits, vegetables and hot-sterilized food	Xu (2006)
Antioxidant	Collagen	β-CD	Food	Zhao et al. (2017)
t-Cinnamaldehyde	Chitosan and pectin film	β-CD	Fresh-cut papaya and cantaloupe	Brasil et al. (2012), Martiñon et al. (2014)
	Chitosan	β-CD	Fresh-cut melon	Moreira et al. (2014)
Catechin + ε-polylysine	Gelatin + chitosan	Sulfobutyl ether β-CD	Cold fresh meat	Jin et al. (2019)
Chinese herb extract	Starch + cellulose	n.s.	Lollipop	Mi (2016a)
	Lactalalbumin + cellulose	n.s.	Chewing gum	Mi (2016b)
Curcumin	Gelatin		Carp fillet	Sun et al. (2019)
Essential oils	Cellulose sodium sulfate films	n.s.	Cookies, cakes, fruits and the like	Chen et al. (2014)
Lemon grass essential oil	Alginate	β-CD	Food	Bronzal et al. (2018)
L-Menthol	Alginate + chitosan	β-CD	Apple	Yang et al. (2020)
Natamycin	Pullalan and carboxymethyl cellulose films	β-CD	Fruits and vegetables	Li et al. (2012)
Oregano essential oil	Chitosan	β-CD	Fresh-cut purple yam	Huang et al. (2018b)
Perilla extract	Natural resin (gum lac)	Hydroxypropyl β-CD	Fruits and vegetables	Hu (2017)
Perilla essential oil	Konjac glucomannan + starch	n.s.	Food	Yuan et al. (2019)
Preservatives	Chitosan films	β-CD	Kiwi fruit or slices	Bao (2014)
Thyme camphor	Gelatin	β-CD	Food	Li et al. (2014)

n.s. not specified

chloride), and polypropylene packaging materials (Hotchkiss et al. 2007). Humidity-triggered release of the ripening inhibitor was reported from paper-based packaging (Wu et al. 2013), polyolefins (Wood et al. 2011) and recently from electrospun polystyrene nanofibers functionalized with α-cyclodextrin/1-methyl cyclopropene complex (Neoh et al. 2017). The technology is marketed for application as active packaging for in-transit treatments to enhance the shelf life of fresh fruits, vegetables and flowers (Agrofresh 2020; Cellresin 2020). There are preservative sachets and cards on the market ensuring extended shelf life to a variety of fruits and vegetables via controlled release of 1-methylcyclopropene upon absorbing water vapor (Wang et al. 2020).

In modern supply chains fruits are often harvested unripe to avoid the loss during transport and storage and the ripening process is initiated just before the delivery to the consumer. Ethylene gas is a phytohormone catalyzing the ripening process. Inclusion of ethylene into α-cyclodextrin cavity results in a controlled release formulation providing low rate of ethylene release in an extended period thus controlling the rate of ripening and avoiding overripening which happens often when ethylene gas is supplied from cylinder (Bazzano et al. 2016). The inclusion complex was incorporated into acrylic films produced by UV curing at low temperature not to lose gas during film preparation.

11.3.6 Enhanced Barrier Function, Odor Control, Aroma-Preserving Packaging

Barrier coating is a sorptive layer between packaging and the food to hinder the migration of harmful components such as plasticizers, residual additives, dyes, adhesives, and the like from packaging into food. There is an opposite requirement, too: to inhibit the release of flavor and aroma components, that is the packaging functions as a barrier for migration of volatile food components through the packaging into the environment. Both inward and outward migration rates are decreased by the presence of "empty" cyclodextrin which entraps atmospheric pollutants, such as cigarette smoke, exhaust gases and vapors penetrating from outside, the aroma components penetrating from inside as well as the harmful components migrating within the packaging, such as additives, stabilizers, plasticizers, fillers, colorants (Wood 2000; Wood and Beaverson 1999, 2003).

Cyclodextrin incorporated into the coating layer acts as a trap for such harmful components as bisphenol A, phthalates, mineral oils and benzophenons present in paints, adhesives, varnishes, printing ink on cellulosic packaging (SoLaPack project 2020). Cyclodextrin hindered the release of plasticizers in PVC films (Fenyvesi et al. 2007). Oil resistant packaging material was produced by incorporating hydrophilic α-cyclodextrin within the hydrophobic acrylic or polyurethane resins in order to reduce the migration of oil through the packaging (Monforton et al. 2007).

The smell of garlic, ripened cheese, etc. is reduced using a cyclodextrin-containing packaging because cyclodextrin can complex the odorous volatiles. The odorous antimicrobial agents, such as isothiocyanates, components of mustard oil show reduced volatility when they are trapped within the packaging in the form of cyclodextrin complexes (Chen and Liu 2016). These complexes incorporated into sandwich boxes have been used in Japan for long (Hashimoto 2002).

A fiber mat prepared by electrospinning low-density polyethylene with triacetyl β-cyclodextrin removed sulfur odor compounds such as dimethyl disulfide, dimethyl sulfide and carbon disulfide (Shin et al. 2018). A large range of differing malodors and permeants were trapped by the packaging containing cyclodextrin (Wood and Beaverson 2012).

Other food aromas (vanilla, coffee aroma, etc.) remain within the packaging due to complex formation. Edible films with aroma complexes applied on the surface of fresh cut fruits and vegetables extended the shelf life of food by hindering the escape of aroma (see above). Cellulose acetate films modified with cyclodextrins and used for packaging of ground, roasted coffee preserved the aroma and hindered oxidation better than the unmodified cellophane shown by the organoleptic properties and head space gas chromatograms (Asslisi 2011).

Laminated container with cyclodextrin/limonene complex in the packaging preserved 80% of D-limonene content in orange juice compared to 35% in a control container not containing the complex (Yamada and Tsunoda 1987). Inclusion complexes with flavor volatiles, such as D-limonene, α-pinene, and 2-methoxy-3-methylpyrazine, have been incorporated into low-density polyethylene powder by dry mixing and then thermally pressed into films providing extended food shelf life (Koontz and Marcy 2007).

11.3.7 Removal of Undesired Components from Food by Packaging

Cyclodextrin included into packaging material may act as a scavenger. Not only volatile unwanted components, such as odorous sulfur compounds can be captured by cyclodextrin within the packages (Shin et al. 2018), but also cholesterol content can be reduced. Cholesterol was removed from milk via immersing poly(vinyl alcohol) or ethylene-vinyl alcohol copolymer films containing β-cyclodextrin (López-de-Dicastillo et al. 2010, 2011a, b). The same films successfully removed aldehydes such as hexanal from packaged fried peanuts. Hexanal can be considered as indicator of lipid oxidation and product shelf life for nuts. Hexanal, a product of fatty acid peroxides degradation, also contributes to flavor deterioration of cooked meat during storage (Shahidi and Pegg 1994).

It was observed earlier that incorporating α-cyclodextrin into nylon film the sorption of aldehydes is enhanced by 2–10 times depending on the type of aldehyde (Suloff et al. 2003). The aldehyde-scavenging films showed selective scalping preferring smaller molecular weight aldehydes to larger ones.

11.4 Conclusion and Future Perspectives

The role of cyclodextrin in active packaging of food products is odor absorption, stabilizing and controlling the release of bioactive components which can be antibacterial, antifungal preservatives, antioxidants, flavoring compounds. The films are effective in preserving the quality of food, mostly fresh-cut vegetables and fruits, maintain the nutritional value, firmness and color and reduce the weight loss during storage. Depending on the functionality of the packaging, either empty cyclodextrins or cyclodextrin complexes can be incorporated into the packaging material using various technologies, such as embedding in thermoplasts during hot-melt processing and extrusion, casting films from common solutions, grafting or crosslinking cyclodextrin on appropriate surfaces. Recently electrospinning and layer-by-layer coating techniques have been developed for fixing cyclodextrin in the packaging.

Concerning the future perspectives, the environmental aspects are of growing importance therefore sustainable raw materials will gradually replace the non-renewable packaging. The European directive on the reduction of the impact of certain plastics on the environment aims to ban among other single-use plastics also food containers made of expanded polystyrene (Directive EU 2019/904 2019). Similar measures have been taken by other countries, e.g. Canada banned plastic bags and food ware (Canada 2020). Governments all over the world have taken actions to ban or charge the lightweight bags made usually of low-density polyethylene. General use of cellulose and protein-based packaging is expected. Novel forms of cellulose such as nanocellulose, including bacterial, nanocrystalline and nanofibrillated cellulose are to be applied in biodegradable and environmentally friendly bionanocomposites with various polymers, like proteins and starch-based matrices (Younas et al. 2018; Thomas et al. 2020). Plant proteins, such as corn zein, wheat gluten, soy protein, sunflower protein are emerging supplements to plastics. There is a still unexplored source of sustainable packaging material in the seas and oceans: the aquatic plants. Packaging films made of plant proteins need reinforcement by incorporating nanofillers, such as nanoclays, carbon nanotubes, graphene aerogel, etc. to improve the mechanical, water-barrier and gas-barrier properties.

The new materials require development in the production technology of packaging, too. Three dimensional (3D) printing is a technology of future which makes possible even printing food together with its packaging (Liu et al. 2017). 3D printing might help optimize package volume and shape. 3D printable cyclodextrin-containing polymers, such as poly(urethane-acrylate), multiacrylated cyclodextrins have already been developed for various applications (Cosola et al. 2020; Kim et al. 2020).

Due to stabilizing effect especially on volatile bioactive compounds and moisture triggered release properties, cyclodextrin and its complexes will surely have their role in the emerging packaging technologies, too.

References

Adel AM, Ibrahim AA, El-Shafei AM et al (2019) Inclusion complex of clove oil with chitosan/β-cyclodextrin citrate/oxidized nanocellulose biocomposite for active food packaging. Food Packag Shelf Life 20:100307. https://doi.org/10.1016/j.fpsl.2019.100307

Aegerter M, Mennig M (eds) (2004) Sol-gel technologies for glass producers and users. Springer, Heidelberg

Agrawal M, Kim YT, Tonelli A et al (2010) Cyclodextrin inclusion complex formation with butylated hydroxytoluene and its application in polyethylene film. J Appl Polym Sci 118:1184–1190. https://doi.org/10.1002/app.32543

Agrofresh (2020) Ethylene management: the core technology behind Smarter Freshness. https://www.agrofresh.com/about-us/ethylene-management/. Accessed 19 Oct 2020

Andrade-Del Olmo J, Pérez-Álvarez L, Hernáez E et al (2019) Antibacterial multilayer of chitosan and (2-carboxyethyl)-β-cyclodextrin onto polylactic acid (PLLA). Food Hydrocoll 88:228–236. https://doi.org/10.1016/j.foodhyd.2018.10.014

Anselmi C, Centini M, Ricci M et al (2006) Analytical characterization of a ferulic acid/γ--cyclodextrin inclusion complex. J Pharm Biomed Anal 40:875–881. https://doi.org/10.1016/j.jpba.2005.08.019

Aresta A, Calvano CD, Trapani A et al (2013) Development and analytical characterization of vitamin(s)-loaded chitosan nanoparticles for potential food packaging applications. J Nanopart Res 15:1592. https://doi.org/10.1007/s11051-013-1592-7

Asslisi C (2011) Active packaging of ground coffee using cyclodextrin-containing cellophane films. Dissertation, Corvinus University Budapest and CycloLab

Astray G, Gonzalez-Barreiro C, Mejuto JC et al (2009) A review on the use of cyclodextrins in foods. Food Hydrocoll 23:1631–1640. https://doi.org/10.1016/j.foodhyd.2009.01.001

Astray G, Mejuto JC, Simal-Gandara J (2020) Latest developments in the application of cyclodextrin host-guest complexes in beverage technology processes. Food Hydrocoll 106:105882. https://doi.org/10.1016/j.foodhyd.2020.105882

Atsuta K, Nakamura Y, Okabe T et al (1989) Tacky tape or sheet for preventing mold, microorganism and insect of vegetable, fruit or the like. JP Pat Appl JPH01153601, 15 Jun 1989

Ayala-Zavala JF, González-Aguilar GA (2010) Optimizing the use of garlic oil as antimicrobial agent on fresh-cut tomato through a controlled release system. J Food Sci 75:M398–M405. https://doi.org/10.1111/j.1750-3841.2010.01723.x

Aytac Z, Kusku SI, Durgun E et al (2016) Encapsulation of gallic acid/cyclodextrin inclusion complex in electrospun polylactic acid nanofibers: release behavior and antioxidant activity of gallic acid. Mater Sci Eng C 63:231–239. https://doi.org/10.1016/j.msec.2016.02.063

Aytac Z, Ipek S, Durgun E et al (2017a) Antioxidant electrospun zein nanofibrous web encapsulating quercetin/cyclodextrin inclusion complex. J Mater Sci 53:1527–1539. https://doi.org/10.1007/s10853-017-1580-x

Aytac Z, Ipek S, Durgun E et al (2017b) Antibacterial electrospun zein nanofibrous web encapsulating thymol/cyclodextrin-inclusion complex for food packaging. Food Chem 233:117–124. https://doi.org/10.1016/j.foodchem.2017.04.095

Aytac Z, Keskin NOS, Tekinay T et al (2017c) Antioxidant α-tocopherol/γ-cyclodextrin-inclusion complex encapsulated poly(lactic acid) electrospun nanofibrous web for food packaging. J Appl Polym Sci 134:44858. https://doi.org/10.1002/app.44858

Balogh K, Fenyvesi É, Makk J et al (2008) Cyclodextrin complexes of natural antimicrobial compounds in active packaging. In: Proceedings of fourteenth international cyclodextrins symposium, Kyoto, Japan, May 8–11, 2008, pp 320–323

Bao WQ (2014) Kiwi fruit preservative and preparation method thereof. Chin Pat Appl CN103704327, 09 Apr 2014

Baptista RC, Horita CN, Sant'Ana AS (2019) Natural products with preservative properties for enhancing the microbiological safety and extending the shelf-life of seafood: a review. Food Res Int 127:108762. https://doi.org/10.1016/j.foodres.2019.108762

Bazzano M, Barolo C, Buscaino R et al (2016) Controlled atmosphere in food packaging using ethylene–α-cyclodextrin inclusion complexes dispersed in photocured acrylic films. Ind Eng Chem Res 55:579–585. https://doi.org/10.1021/acs.iecr.5b03980

Beermann K, Buschmann H-J, Knittel D et al (2002) Methods of determining cyclodextrins on textile materials. Textilveredlung 37:17–22

Birck C, Degoutin S, Tabary N (2014) New crosslinked cast films based on poly(vinyl alcohol): preparation and physico-chemical properties. Exp Polym Lett 8:941–952. https://doi.org/10.3144/expresspolymlett.2014.95

Brasil IM, Gomes C, Puerta-Gomez A (2012) Polysaccharide-based multilayered antimicrobial edible coating enhances quality of fresh-cut papaya. LWT Food Sci Technol 47(1):39–45. https://doi.org/10.1016/j.lwt.2012.01.005

Bronzal LA, Martizano JO, Morata AS et al (2018) Alginate packaging film based on lemongrass (Cymbopogon citratus) essential oil/beta-cyclodextrin inclusion and the preparation method thereof. Phil Pat Appl PH12018050545

Buendía-Moreno L, Sánchez-Martínez MJ, Antolinos V et al (2020) Active cardboard box with a coating including essential oils entrapped within cyclodextrins and/or hallosyte nanotubes. A case study for fresh tomato storage. Food Contr 107:106763. https://doi.org/10.1016/j.foodcont.2019.106763

Canada (2020) Canada one-step closer to zero plastic waste by 2030. https://www.canada.ca/en/environment-climate-change/news/2020/10/canada-one-step-closer-to-zero-plastic-waste-by-2030.html. Accessed 22 Oct 2020

Capozzi LC, Bazzano M, Cavallero MC et al. (2016) Polymeric supports for controlled release of ethylene for food industry. Int Polym Proc 31:570–576. https://doi.org/10.3139/217.3233

Capozzi LC, Bazzano M, Sangermano M et al (2017) Inclusion complexes dispersed in polystyrene-based labels for fruit ripening on demand. Int J Food Sci Technol 53:389–394. https://doi.org/10.1111/ijfs.13596

Celebioglu A, Uyar T (2021) Electrohydrodynamic encapsulation of eugenol-cyclodextrin complexes in pullulan nanofibers. Food Hydrocoll 111:106264. https://doi.org/10.1016/j.foodhyd.2020.106264

Cellresin (2020) Cellresin Technologies. http://cellresin.com/. Accessed 29 Sept 2020

Chen G, Liu B (2016) Cellulose sulfate based film with slow-release antimicrobial properties prepared by incorporation of mustard essential oil and β-cyclodextrin. Food Hydrocoll 55:100–107. https://doi.org/10.1016/j.foodhyd.2015.11.009

Chen G, Liu B, Chen H et al (2014) Preparation method of long-acting antibacterial edible film with sustained release effect. Chin Pat Appl CN103910911, 07 Sep 2014

Chen H, Li L, Ma Y et al (2019) Development of active packaging film containing bioactive components encapsulated in β-cyclodextrin and its application. Food Hydrocoll 90:360–366. https://doi.org/10.1016/j.foodhyd.2018.12.043

Chen Z, Zong L, Chen C et al (2020a) Development and characterization of PVA-starch active films incorporated with β-cyclodextrin inclusion complex embedding lemongrass (Cymbopogon citratus) oil. Food Packag Shelf Life 26:100565. https://doi.org/10.1016/j.fpsl.2020.100565

Chen M, Chen X, Yam K (2020b) Encapsulation complex of chlorine dioxide in α-cyclodextrin: structure characterization and release property. Food Contr 107:106783. https://doi.org/10.1016/j.foodcont.2019.106783

Cheng M, Wang J, Zhang R et al (2019) Characterization and application of the microencapsulated carvacrol/sodium alginate films as food packaging materials. Int J Biol Macromol 141:259–267. https://doi.org/10.1016/j.ijbiomac.2019.08.215

Commission of European Communities No. 450/2009 (2009) On active and intelligent materials and articles intended to come into contact with food. https://eur-lex.europa.eu/legal-content/EN/ALL/?uri=CELEX%3A32009R0450. Accessed 1 Oct 2020

Cova TF, Murtinho D, Pais AACC et al (2018) Combining cellulose and cyclodextrins: fascinating designs for materials and pharmaceutics. Front Chem 6:271. https://doi.org/10.3389/fchem.2018.00271

Cosola A, Conti R, Grützmacher H *et al.* (2020) Multiacrylated cyclodextrin: A bio-derived photocurable macromer for VAT 3D printing. Macromol Mater Eng 2000350. https://doi.org/10.1002/mame.202000350

Crini G (2014) Review: A history of cyclodextrins. Chem Rev 114:10940–10975. https://doi.org/10.1021/cr500081p

Crini G, Fourmentin S, Fenyvesi É et al (2018) Cyclodextrins, from molecules to applications. Environ Chem Lett 16:1361–1375. https://doi.org/10.1007/s10311-018-0763-2

Cui H, Bai M, Lin L (2018) Plasma-treated poly(ethylene oxide) nanofibers containing tea tree oil/beta-cyclodextrin inclusion complex for antibacterial packaging. Carbohydr Polym 179:360–369. https://doi.org/10.1016/j.carbpol.2017.10.011

Cui H, Zhang C, Li C et al (2019) Preparation and antibacterial activity of *Litsea cubeba* essential oil/dandelion polysaccharide nanofiber. Ind Crop Prod 140:111739. https://doi.org/10.1016/j.indcrop.2019.111739

da Silva AM (2018) Room at the top as well as at the bottom: structure of functional food inclusion compounds. In: Aurora P, Dhingra N (eds) Cyclodextrin: a versatile ingredient. Intechopen, Vienna, pp 553–573. https://doi.org/10.5772/intechopen.74162

Daly J, Kourelis B (2001) Synthesis methods, complexes and delivery methods for the safe and convenient storage, transport and application of compounds for inhibiting the ethylene response in plants. US Pat 6,313,068, 06 Nov 2001

Das S, Gazdag Z, Szente L et al (2018) Antioxidant and antimicrobial properties of randomly methylated β cyclodextrin - captured essential oils. Food Chem 278:305–313. https://doi.org/10.1016/j.foodchem.2018.11.047

de Bergamasco RC, Zanin GM et al (2007) Grafting of cyclodextrins onto filter paper. J Incl Phenom Macrocycl Chem 57:75–78. https://doi.org/10.1007/s10847-006-9217-8

de Castro DO, Tabary N, Martel B et al (2018) Controlled release of carvacrol and curcumin: bio-based food packaging by synergism action of TEMPO-oxidized cellulose nanocrystals and cyclodextrin. Cellulose 25:1249–1263. https://doi.org/10.1007/s10570-017-1646-6

Dias Antunes M, da Silva Dannenberg G, Fiorentini ÂM et al (2017) Antimicrobial electrospun ultrafine fibers from zein containing eucalyptus essential oil/cyclodextrin inclusion complex. Int J Biol Macromol 104:874–882. https://doi.org/10.1016/j.ijbiomac.2017.06.095

Directive EU 2019/904 (2019) Directive (EU) 2019/904—on the reduction of the impact of certain plastic products on the environment. http://publications.europa.eu/resource/cellar/767969e4-e663-11e9-9c4e-01aa75ed71a1.0001.02/DOC_1. Accessed 21 Oct 2020

Doi T, Yashiki I (1994) Antibacterial film and manufacture thereof. Jap Pat Appl JPH06191562, 12 Jul 1994

Edo H, Yoshida N (1978) Fungicide. Jap Pat Appl JPS5587701, 02 Jul 1980

Espacenet (2020) https://worldwide.espacenet.com/. Accessed 06 Oct 2020

European Bioplastics (2020) Bioplastics, facts and figures. https://docs.european-bioplastics.org/publications/EUBP_Facts_and_figures.pdf. Accessed 24 Sep 2020

European Commission Regulation (2009) No. 450/2009 on active and intelligent materials and articles intended to come into contact with food. https://eur-lex.europa.eu/legal-content/EN/ALL/?uri=CELEX%3A32009R0450. Accessed 02 Oct 2020

Fenyvesi E, Szente L (2016) Nanoencapsulation of flavors and aromas by cyclodextrins. In: Grumezescu AM (ed) Encapsulations. Elsevier Inc, Amsterdam, pp 769–792. https://doi.org/10.1016/b978-0-12-804307-3.00018-1

Fenyvesi É, Balogh K, Siró I et al (2007) Permeability and release properties of cyclodextrin-containing poly(vinyl chloride) and polyethylene films. J Incl Phenom Macrocycl Chem 57:371–374. https://doi.org/10.1007/s10847-006-9256-1

Fenyvesi É, Vikmon M, Szente L (2016) Cyclodextrins in food technology and human nutrition: benefits and limitations. Crit Rev Food Sci Nutr 56:1981–2004. https://doi.org/10.1080/10408398.2013.809513

Figuera-Lopez KJ, Enescu D, Torres-Giner S et al (2020) Development of electrospun active films of poly(3-hydroxybutyrate-co-3-hydroxyvalerate) by the incorporation of cyclodextrin inclu-

sion complexes containing oregano essential oil. Food Hydrocoll 108, 106013. https://doi.org/10.1016/j.foodhyd.2020.106013
Fuenmayor CA, Mascheroni E, Cosio MS et al (2013) Encapsulation of R-(+)-limonene in edible electrospun nanofibers. Chem Eng Trans 32:1771–1776. https://doi.org/10.3303/TEC1332296
Fujimura Y (1986) Packaging material. Jap Pat Appl JPS6136361, 21 Feb 1986
Fukazawa R, Sato M (1986) Preservation of foods. Jap Pat Appl JP 86108359 (Chem Abstr 105:170928)
Fukazawa R, Sato M (1987) Freshness retention of mushroom. Jap Pat Appl JPS62236440, 16 Oct 1987
Furuta T, Kusuya Y, Neoh T et al (2006) Inclusion and release of hinokitiol into/from MCT-beta-CD fixed on Japanese washi paper. J Incl Phenom Macrocycl Chem 56:107–111
Geyer R, Jambeck JR, Law KL (2017) Production, use, and fate of all plastics ever made. Sci Adv 3:e1700782. https://doi.org/10.1126/sciadv.1700782
Goñi-Ciaurriz L, González-Gaitano G, Vélaz I (2020) Cyclodextrin-grafted nanoparticles as food preservative carriers. Int J Pharm 588:119664. https://doi.org/10.1016/j.ijpharm.2020.119664
Grigoriu A-M, Luca C, Horoba E, Dunca S (2013) Antimicrobial polymer films functionalized with cyclodextrins. Rev Chim 64:606–611
Hara K, Mikuni K, Hara K et al. (2002) Antimicrobial or insect-repellent water-soluble cyclodextrin film and method for producing the same. Jap Pat Appl JP2002029901, 29 Jan 2002
Hashimoto H (2002) Present status of industrial application of cyclodextrins in Japan. In: Abstract book of eleventh international cyclodextrin symposium, Reykjavik, Iceland, 5–8 May 2002
Hassan B, Chatha SAS, Hussain AI et al (2018) Recent advances on polysaccharides, lipids and protein based edible films and coatings: a review. Int J Macromol 109:1095–1107. https://doi.org/10.1016/j.ijbiomac.2017.11.097
Hashizume M, Saeki T (1995) Bactericidal water-absorbing sheet for keeping freshness of food. Jap Pat Appl JP 07123963, 16 May 1995
Higueras L, López-Carballo G, Hernández-Muñoz P (2014) Antimicrobial packaging of chicken fillets based on the release of carvacrol from chitosan/cyclodextrin films. Int J Food Microbiol 188:53–59. https://doi.org/10.1016/j.ijfoodmicro.2014.07.018
Hirogaki K, Kamitani J, Kimizu M et al (2010) Fixation of cyclodextrin onto polyester fiber 2: functional processing of fiber by fixing of cyclodextrin inclusion complex with isocyanate usable in water. J Fiber Sci Technol 66:141–146. https://doi.org/10.2115/fiber.66.141
Hiroshi K (1981) Edible film. Jap Pat Appl JPS5618555, 21 Feb 1981
Ho TM, Howes T, Bhandari BR (2018) An innovative approach to extend the shelf life of cottage cheese curds using food grade CO2-α-cyclodextrin complex powder: a preliminary study. J Food Process Preserv 42:e13514. https://doi.org/10.1111/jfpp.13514
Holobradi B, Kali G, Mészáros B (1986) Method for aromatizing papers particularly filter-papers and producing aromatized foodstuffs particularly filter teas. Hung Pat Appl HUT39343, 29 Sep 1986
Hotchkiss JH, Watkins CB, Sanchez DG (2007) Release of 1-methylcyclopropene from heat-pressed polymer films. J Food Sci 72:330–334. https://doi.org/10.1111/j.1750-3841.2007.00391.x
Houchin N (1989) Pull resin film sandwiched between laver and rice and rice ball therefrom. Jap Pat Appl JPH01257435, 13 Oct 1989
Hu X (2017) Air-permeable and edible fruit/vegetable preservation film-forming agent. Chin Pat Appl CN107494719, 22 Dec 2017
Huang Y, Zeng X, Zhu Q et al (2018a) Development of an active packaging with molecularly imprinted polymers for beef preservation. Packag Technol Sci 31:213–220. https://doi.org/10.1002/pts.2368
Huang C, Yin C, Liu Y et al (2018b) Preparation technology for establishing compound edible preservative film on surface of fresh-cut purple yam. Chin Pat Appl CN108713587
Imakura H, Yamada H, Fukazawa R (1992) Packaging film, material for keeping freshness of food and freshness-keeping method. Jap Pat Appl JPH04325069, 13 Oct 1992

Jajasena DD, Jo C (2014) Potential application of essential oils as natural antioxidants in meat and meat products: a review. Food Rev Int 30:71–90. https://doi.org/10.1080/87559129.2013.853776

Jamróz E, Kopel P (2020) Polysaccharide and protein films with antimicrobial/antioxidant activity in the food industry: a review. Polymers 12:1289. https://doi.org/10.3390/polym12061289

Jin G, Yue L, He L et al (2019) Antiseptic, anti-oxidation multi-effect active nano fresh-keeping film for cold fresh meat and preparation method thereof. Chin Pat Appl CN110272628, 24 Sep 2019

Joo M, Auras R, Almenar E (2011) Preparation and characterization of blends made of poly(l-lactic acid) and β-cyclodextrin: improvement of the blend properties by using a masterbatch. Carbohydr Polym 86:1022–1030. https://doi.org/10.1016/j.carbpol.2011.05.058

Kanatt SR, Chawla SP (2017) Shelf life extension of chicken packed in active film developed with mango peel extract. J Food Saf 38:e12385. https://doi.org/10.1111/jfs.12385

Kayaci F, Ertas Y, Uyar T (2013a) Enhanced thermal stability of eugenol by cyclodextrin inclusion complex encapsulated in electrospun polymeric nanofibers. J Agric Food Chem 61:8156–8165. https://doi.org/10.1021/jf402923c

Kayaci F, Umu OCO, Tekinay T et al (2013b) Antibacterial electrospun poly(lactic acid) (PLA) nanofibrous webs incorporating triclosan/cyclodextrin inclusion complexes. J Agric Food Chem 61:3901–3908. https://doi.org/10.1021/jf400440b

Kayaci F, Sen HS, Durgun E et al (2014) Functional electrospun polymeric nanofibers incorporating geraniol–cyclodextrin inclusion complexes: high thermal stability and enhanced durability of geraniol. Food Res Int 62:424–431. https://doi.org/10.1016/j.foodres.2014.03.033

Kfoury M, Auezova L, Greige-Gerges H et al (2015a) Promising applications of cyclodextrins in food: improvement of essential oils retention, controlled release and antiradical activity. Carbohydr Polym 131:264–272. https://doi.org/10.1016/j.carbpol.2015.06.014

Kfoury M, Auezova L, Ruellan S et al (2015b) Complexation of estragole as pure compound and as main component of basil and tarragon essential oils with cyclodextrins. Carbohydr Polym 118:156–164

Kim J, Yoon CS, Lee S et al (2020) Development of insect-proof starch adhesive containing encapsulated cinnamon oil for paper box adhesion to inhibit *Plodia interpunctella* larvae infestation. Food Sci 85:3363–3371. https://doi.org/10.1111/1750-3841.15425

Komatsu Y, Otani M, Shirakawa K (1994) Antibacterial pressure-sensitive tape. Jap Pat Appl JPH06158002, 07 Jun 1994

Konstansek CE (2003) Delivery systems for cyclopropenes. US patent 6,548,448, 26 Dec 2002

Koontz JL, Marcy JE (2007) Controlled release of active ingredients from polymer food packaging by molecular encapsulation with cyclodextrins. Polym Prepr (Am Chem Soc, Div Polymr Chem) 48(2):742

Koontz JL, Marcy JE, O'Keefe SF et al (2009) Cyclodextrin inclusion complex formation and solid-state characterization of the natural antioxidants α-tocopherol and quercetin. J Agric Food Chem 57:1162–1171. https://doi.org/10.1021/jf802823q

Koontz JL, Marcy JE, O'Keefe SF et al (2010) Polymer processing and characterization of LLDPE films loaded with α-tocopherol, quercetin, and their cyclodextrin inclusion complexes. J Appl Polym Sci 117:2299–2309. https://doi.org/10.1002/app.32044

Kriegel C, Arrechi A, Kit K et al (2008) Fabrication, functionalization, and application of electrospun biopolymer nanofibers. Crit Rev Food Sci Nutr 48(8):775–797. https://doi.org/10.1080/10408390802241325

Kuang Y, Wang H, Zhang S et al (2016) Preparation method of oil-soluble water-soluble type rosemary antioxidant. Chin Pat Appl CN105495272, 20 Apr 2016

Lavoine N, Givord C, Tabary C et al (2014) Elaboration of a new antibacterial bio-nano-material for food-packaging by synergistic action of cyclodextrin and microfibrillated cellulose. Innov Food Sci Emerg Technol 26:330–340. https://doi.org/10.1016/j.ifset.2014.06.006

Li B, Ma S, An C (2012) Vegetable and fruit coating preservative and preparation method thereof. Chin Pat Appl CN102316619, 11 Jan 2012

Li M, Wu Q, Liu Z (2014) Active antibacterial edible package with controlled release function and preparation method thereof. Chin Pat CN103865276, 18 Jun 2014

Lin L, Dai Y, Cui H (2017) Antibacterial poly(ethylene oxide) electrospun nanofibers containing cinnamon essential oil/beta-cyclodextrin proteoliposomes. Carbohydr Polym 178:131–140. https://doi.org/10.1016/j.carbpol.2017.09.043

Lin L, Zhu Y, Cui H (2018) Electrospun thyme essential oil/gelatin nanofibers for active packaging against *Campylobacter jejuni* in chicken. LWT 97:711–718. https://doi.org/10.1016/j.lwt.2018.08.015

Liu F, Antoniou J, Li Y et al (2015) Preparation of gelatin films incorporated with tea polyphenol nanoparticles for enhancing controlled-release antioxidant properties. J Agric Food Chem 63(15):3987–3995. https://doi.org/10.1021/acs.jafc.5b00003

Liu Z, Zhang M, Bhandari B et al (2017) 3D printing: printing precision and application in food sector. Trends Food Sci Technol 69:83–94. https://doi.org/10.1016/j.tifs.2017.08.018

López-de-Dicastillo C, Gallur M, Catala R et al (2010) Immobilization of β-cyclodextrin in ethylene-vinyl alcohol copolymer for active food packaging applications. J Membr Sci 353:184–191. https://doi.org/10.1016/j.memsci.2010.02.049

López-de-Dicastillo C, Jordá M, Catalá R et al (2011a) Development of active polyvinyl alcohol/β-cyclodextrin composites to scavenge undesirable food components. J Agric Food Chem 59:11026–11033. https://doi.org/10.1021/jf200749f

López-de-Dicastillo C, Catalá R, Gavara R (2011b) Food applications of active packaging EVOH films containing cyclodextrins for the preferential scavenging of undesirable compounds. J Food Eng 104:380–386. https://doi.org/10.1016/j.jfoodeng.2010.12.033

López-Gómez A, Ros-Chumillas M, Buendía-Moreno L et al (2020) Active cardboard box with smart internal lining based on encapsulated essential oils for enhancing the shelf life of fresh mandarins. Foods 9:590. https://doi.org/10.3390/foods9050590

López-Nicolás JM, Pérez-López AJ, Carbonell-Barrachina Á (2007) Use of natural and modified cyclodextrins as inhibiting agents of peach juice enzymatic browning. J Agric Food Chem 55:5312–5319. https://doi.org/10.1021/jf070499h

López-Nicolás JM, Rodríguez-Bonilla P, García-Carmona F (2013) Cyclodextrins and antioxidants. Crit Rev Food Sci Nutr 54:251–276. https://doi.org/10.1080/10408398.2011.582544

Maestrello C, Tonon L, Madrona G et al (2017) Production and characterization of biodegradable films incorporated with clove essential oil/β-cyclodextrin. Chem Eng Trans 57:1393–1398. https://doi.org/10.3303/CET1757233

Mallardo S, De Vito V, Malinconico M et al (2016) Poly(butylene succinate)-based composites containing β-cyclodextrin/d-limonene inclusion complex. Eur Polym J 79:82–96. https://doi.org/10.1016/j.eurpolymj.2016.04.024

Marques CS, Carvalho SG, Bertoli LD et al (2019) β-Cyclodextrin inclusion complexes with essential oils: obtention, characterization, antimicrobial activity and potential application for food preservative sachets. Food Res Int 119:499–509. https://doi.org/10.1016/j.foodres.2019.01.016

Marsin M, Muhamad II, Anis SNS et al (2020) Essential oils as insect repellent agents in food packaging: a review. Eur Food Res Technol 246:1519–1532. https://doi.org/10.1007/s00217-020-03511-1

Martina K, Binello A, Lawson D et al (2013) Recent applications of cyclodextrins as food additives and in food processing. Curr Nutr Food Sci 9:167–179. https://doi.org/10.2174/1573401311309030001

Martiñon ME, Moreira RG, Castell-Perez ME, Gomes C (2014) Development of a multilayered antimicrobial edible coating for shelf-life extension of fresh-cut cantaloupe (*Cucumis melo L.*) stored at 4 °C. LWT Food Sci Technol 56:341–350. https://doi.org/10.1016/j.lwt.2013.11.043

Mascheroni E, Fuenmayor CA, Cosio MS et al (2013) Encapsulation of volatiles in nanofibrous polysaccharide membranes for humidity-triggered release. Carbohydr Polym 98:17–25. https://doi.org/10.1016/j.carbpol.2013.04.068

Matencio A, Navarro-Orcajada S, García-Carmona F et al (2020) Applications of cyclodextrins in food science. A review. Trends Food Sci Technol 104:132–143. https://doi.org/10.1016/j.tifs.2020.08.009

Matsui T, Ootani M, Shirakawa K (1994) Antifungal food container. Jap Pat Appl JPH0670728, 15 Mar 1994

Mi W (2016a) Edible lollipop packaging film. Chin Pat Appl CN105254926, 20 Jan 2016

Mi W (2016b) Edible chewing gum packaging film. Chin Pat Appl CN105400213, 16 Mar 2016

Milani MA, Dana MG, Ghanbarzadeh B et al (2020) Effect of novel bioactive coating enriched with nanoemulsion of mustard essential oil on the quality of Turkey meat. J Food Nutr Res 59:71–80

Mlalila N, Hilonga A, Swai H et al (2018) Antimicrobial packaging based on starch, poly(3-hydroxybutyrate) and poly(lactic-coglycolide) materials and application challenges. Trends Food Sci Technol 74:1–11. https://doi.org/10.1016/j.tifs.2018.01.015

Monforton RJ, Popa MA, Plank DW et al (2007) Oil-resistant packaging. CA Pat Appl CA2578580, 17 Aug 2007

Moreira SP, de Carvalho WM, Alexandrino AC et al (2014) Freshness retention of minimally processed melon using different packages and multilayered edible coating containing microencapsulated essential oil. Int J Food Sci Technol 49:2192–2203. https://doi.org/10.1111/ijfs.12535

Munhuweyi K, Caleb OJ, van Reenen AJ et al (2018) Physical and antifungal properties of β-cyclodextrin microcapsules and nanofibre films containing cinnamon and oregano essential oils. LWT Food Sci Technol 87:413–422. https://doi.org/10.1016/j.lwt.2017.09.012

Mura P, Corti G, Maestrelli F et al (2007) The influence of chitosan on cyclodextrin complexing and solubilizing abilities towards drugs. J Incl Phenom Macrocycl Chem 59:307–313. https://doi.org/10.1007/s10847-007-9329-9

Naito S, Saga K, Imai T (1994) Freshness-preserving packaging material for fruit and vegetable and freshness preservation using the same. Jap Pat Appl JPH0616990, 25 Jan 1994

Nakahara M, Nakahara T, Sasaki Y et al (1995) Antibacterial insecticidal thermoplastic composition and its molded articles. Jap Pat Appl JPH07207165, 8 Aug 1995

Narayanan V, Mani MK, Thambusamy S (2020) Electrospinning preparation and spectral characterizations of the inclusion complex of ferulic acid and γ-cyclodextrin with encapsulation into polyvinyl alcohol electrospun nanofibers. J Mol Struct 1221:128767. https://doi.org/10.1016/j.molstruc.2020.128767

Neoh TL, Ariyanto HD, Menéndez Galvan P et al (2017) Controlled release of 1-methylcyclopropene from its functionalised electrospun fibres under constant and linearly ramped humidity. Food Addit Contam A 34:1690–1702. https://doi.org/10.1080/19440049.2017.1325520

Núñez-Delicado E, Sánchez-Ferrer A, García-Carmona F (1997) Cyclodextrins as secondary antioxidants: synergism with ascorbic acid. J Agric Food Chem 45:2830–2835. https://doi.org/10.1021/jf9609800

Oswell NJ, Thippareddi H, Pegg RB (2018) Practical use of natural antioxidants in meat products in the U.S.: a review. Meat Sci 145:469–479. https://doi.org/10.1016/j.meatsci.2018.07.020

Pan J, Ai F, Shao P et al (2019) Development of polyvinyl alcohol/β-cyclodextrin antimicrobial nanofibers for fresh mushroom packaging. Food Chem 300:125249. https://doi.org/10.1016/j.foodchem.2019.125249

Pan L, Zhang M, Lu L et al (2020) A polyethylene base moisture activating oxygen scavenging film co-extruded with tea polyphenols-β-cyclodextrin inclusion complex. Materials 13:3857. https://doi.org/10.3390/ma13173857

Piercey MJ, Mazzanti G, Budge SM et al (2012) Antimicrobial activity of cyclodextrin entrapped allyl isothiocyanate in a model system and packaged fresh-cut onions. Food Microbiol 30:213–218. https://doi.org/10.1016/j.fm.2011.10.015

Plackett DV, Holm VK, Johansen P et al (2006) Characterization of L-polylactide and L-polylactide–polycaprolactone co-polymer films for use in cheese-packaging applications. Packag Technol Sci 19:1–24. https://doi.org/10.1002/pts.704

Plackett D, Ghanbari-Siahkali A, Szente L (2007) Behavior of α- and β-cyclodextrin-encapsulated allyl isothiocyanate as slow-release additives in polylactide-co-polycaprolactone films. J Appl Polym Sci 105:2850–2857. https://doi.org/10.1002/app.26344

Poverenov E, Granit R, Gabai S (2013) Encapsulation and controlled release of antifungal propionic acid utilizing biodegradable active films based on natural polymers. Eur Food Res Technol 237:19–26. https://doi.org/10.1007/s00217-013-2011-0

Raouche S, Mauricio-Iglesias M, Peyron S et al (2011) Combined effect of high pressure treatment and anti-microbial bio-sourced materials on microorganisms' growth in model food during storage. Innov Food Sci Emerg Technol 12:426–434. https://doi.org/10.1016/j.ifset.2011.06.012

Rehmann L, Yoshii H, Furuta T (2003) Characteristics of modified beta-cyclodextrin bound to cellulose powder. Starch/Staerke 55:313–318. https://doi.org/10.1002/star.200390066

Research and Market (2020) Insights into the worldwide active packaging industry to 2025 - key drivers and restraints. Globe Newswire. https://www.globenewswire.com/news-release/2020/05/25/2038060/0/en/Insights-into-the-Worldwide-Active-Packaging-Industry-to-2025-Key-Drivers-and-Restraints.html. Accessed 01 Oct 2020

Reuscher H, Hirsenkorn R (1996) BETA W7 MCT – new ways in surface modification. J Incl Phenom Mol Recognit Chem 25:191–196. https://doi.org/10.1007/BF01041567

Rezaee M, Askar G, EmamDjomeh Z et al (2018) Effect of organic additives on physiochemical properties and anti-oxidant release from chitosan-gelatin composite films to fatty food simulant. Int J Biol Macromol 114:844–850. https://doi.org/10.1016/j.ijbiomac.2018.03.122

Rubin LJ, Diosady L, O'Boyle A et al (1992) Nitrite-free meat curative and process for curing meat. PCT Int Appl WO9203223, 05 Mar 1992

Saeki T, Shirakawa K (1996) Antibacterial egg container. Jap Pat Appl JPH08310577, 26 Nov 1996

Saini S, Quinot D, Lavoine N et al (2016) β-Cyclodextrin-grafted TEMPO-oxidized cellulose nanofibers for sustained release of essential oil. J Mater Sci 52:3849–3861. https://doi.org/10.1007/s10853-016-0644-7

Sawada E, Fukukita T (2004) Antifogging resin sheet. Jap Pat Appl JP2004137426, 13 May 2004

Scopus (2020) https://www.scopus.com/search/. Accessed 05 Oct 2020

Sebaaly C, Charcosset C, Fourmentin S et al (2018) Potential applications of cyclodextrin inclusion complexes, liposomes, and drug-in-cyclodextrin-in-liposome in food industry and packaging. In: Gruzemescu AM, Holban AM (eds) Role of materials science in food bioengineering, pp 187–234. https://doi.org/10.1016/b978-0-12-811448-3.00006-1

Shahidi F, Pegg RB (1994) Hexanal as an indicator of meat flavor deterioration. J Food Lipids 1:177–186. https://doi.org/10.1111/j.1745-4522.1994.tb00245.x

Sharif N, Golmakani M-T, Niakousari M et al (2018) Active food packaging coatings based on hybrid electrospun gliadin nanofibers containing ferulic acid/hydroxypropyl-beta-cyclodextrin inclusion complexes. Nanomaterials 8:919. https://doi.org/10.3390/nano8110919

Shin J, Lee EJ, Ahn DU (2018) Electrospinning of tri-acetyl-β-cyclodextrin (TA-β-CD) functionalized low-density polyethylene to minimize sulfur odor volatile compounds. Food Packag Shelf Life 18:107–114. https://doi.org/10.1016/j.fpsl.2018.10.005

Shin J, Kathuria A, Lee YS (2019) Effect of hydrophilic and hydrophobic cyclodextrins on the release of encapsulated allyl isothiocyanate (AITC) and their potential application for plastic film extrusion. J Appl Polym Sci 136:48137. https://doi.org/10.1002/app.48137

Silva Â, Duarte A, Sousa S et al (2016) Characterization and antimicrobial activity of cellulose derivatives films incorporated with a resveratrol inclusion complex. LWT Food Sci Technol 73:481–489. https://doi.org/10.1016/j.lwt.2016.06.043

Simionato I, Domingues FC, Nerín C et al (2019) Encapsulation of cinnamon oil in cyclodextrin nanosponges and their potential use for antimicrobial food packaging. Food Chem Tox 132:110647. https://doi.org/10.1016/j.fct.2019.110647

Siró I (2012) Active and intelligent packaging of food. In: Bhat R, Alias AK, Paliyath G (eds) Progress in food preservation. Wiley, Hoboken, NJ, pp 23–48. https://doi.org/10.1002/9781119962045.ch2

Siró I, Fenyvesi E, Szente L et al (2006) Release of alpha-tocopherol from antioxidative low-density polyethylene film into fatty food simulant: influence of complexation in beta-cyclodextrin. Food Addit Contam 23:845–853. https://doi.org/10.1080/02652030600699064

SoLaPack (2020) Preventing migration of critical substances through the application of a sorption layer on packaging material of cellulose origin. http://cobro.org.pl/english/index.php/dzialalnosc-naukowa/projekty-badawcze/10-projekty/28-solapack. Accessed 26 Oct 2020

Suloff EC, Marcy JE, Blakistone BA et al (2003) Sorption behavior of selected aldehyde-scavenging agents in poly(ethylene terephthalate) blends. J Food Sci 68:2028–2033. https://doi.org/10.1111/j.1365-2621.2003.tb07013.x

Sun X, Sui S, Ference C et al (2014) Antimicrobial and mechanical properties of β-cyclodextrin inclusion with essential oils in chitosan films. J Agric Food Chem 62:8914–8918. https://doi.org/10.1021/jf5027873

Sun X, Guo X, Ji M et al (2019) Preservative effects of fish gelatin coating enriched with CUR/βCD emulsion on grass carp (*Ctenopharyngodon idellus*) fillets during storage at 4 °C. Food Chem 272:643–652. https://doi.org/10.1016/j.foodchem.2018.08.040

Sun L, Yang S, Qian X et al (2020) Cyclodextrin and cellulose combination product developed by click chemistry: fascinating design for inclusion of ciprofloxacin. Cellulose 27:5955–5970. https://doi.org/10.1007/s10570-020-03200-y

Szejtli J (2004) Cyclodextrins and molecular encapsulation. In: Nalwa HS (ed) Encyclopedia of nanoscience and nanotechnology. American Scientific Publishers, Los Angeles, pp 283–304

Szente L, Fenyvesi É (2018) Cyclodextrin-enabled polymer composites for packaging. Molecules 23:1556. https://doi.org/10.3390/molecules23071556

Szente L, Szejtli J (2004) Cyclodextrins as food ingredients. Trends Food Sci Technol 15:137–142. https://doi.org/10.1016/j.tifs.2003.09.019

Szente L, Harangi J, Szejtli J (1988) Long-term storage stability studies on flavor-β-cyclodextrin complexes. In: Huber O, Szejtli J (eds) Proc int symp cyclodextrins, 4th edn. Kluwer, Dordrecht, Netherland, pp 545–549

Thomas P, Duolikun T, Rumjit NP (2020) Comprehensive review on nanocellulose: recent developments, challenges and future prospects. J Mech Behav Biomed Mater 110:103884. https://doi.org/10.1016/j.jmbbm.2020.103884

Tian B, Xiao D, Hei T et al (2020) The application and prospects of cyclodextrin inclusion complexes and polymers in the food industry: a review. Polym Int 69:597–603. https://doi.org/10.1002/pi.5992

Tonelli AE (2003) Reorganization of the structures, morphologies, and conformations of polymers by coalescence from their crystalline inclusion compounds formed with cyclodextrins. Macromol Symp 203:71–88. https://doi.org/10.1002/masy.200351306

Vega-Lugo A-C, Lim L-T (2009) Controlled release of allyl isothiocyanate using soy protein and poly(lactic acid) electrospun fibers. Food Res Int 42:933–940. https://doi.org/10.1016/j.foodres.2009.05.005

Verstichel S, De Wilde B, Fenyvesi É et al (2004) Investigation of the aerobic biodegradability of several types of cyclodextrins in a laboratory-controlled composting test. J Polym Environ 12:47–55. https://doi.org/10.1023/B:JOOE.0000010050.52967.94

Wang J, Cao Y, Sun B et al (2011) Characterisation of inclusion complex of trans-ferulic acid and hydroxypropyl-β-cyclodextrin. Food Chem 124:069e1075. https://doi.org/10.1016/j.foodchem.2010.07.080

Wang Q, Yan H, Mou W (2020) Fruit, vegetable and flower freshness protection package. Chin Pat Appl CN210338844, 17 Apr 2020

Wang Z, Zou W, Liu L et al (2021) Characterization and bacteriostatic effects of β-cyclodextrin/quercetin inclusion compound nanofilms prepared by electrospinning. Food Chem 338:127980. https://doi.org/10.1016/j.foodchem.2020.127980

Wei Z, Liu Y, Hu H et al (2016) Biodegradable poly(butylene succinate-co-terephthalate) nanofibrous membranes functionalized with cyclodextrin polymer for effective methylene blue adsorption. RSC Adv 6:108240–108246. https://doi.org/10.1039/C6RA22941G

Wei Z, Pan Z, Li F et al (2018) Poly(butylene succinate-: co-terephthalate) nanofibrous membrane composited with cyclodextrin polymer for superhydrophilic property. RSC Adv 8:1378–1384. https://doi.org/10.1039/C7RA12068K

Wen P, Zhu D-H, Feng K et al (2016a) Fabrication of electrospun polylactic acid nanofilm incorporating cinnamon essential oil/β-cyclodextrin inclusion complex for antimicrobial packaging. Food Chem 196:996–1004. https://doi.org/10.1016/j.foodchem.2015.10.043

Wen P, Zhu D-H, Wu H et al (2016b) Encapsulation of cinnamon essential oil in electrospun nanofibrous film for active food packaging. Food Contr 59:66–376. https://doi.org/10.1016/j.foodcont.2015.06.005

Wessling C, Nielsen T, Leufven A et al (1999) Retention of alpha-tocopherol in low-density polyethylene (LDPE) and polypropylene (PP) in contact with foodstuffs and food simulating liquids. J Sci Food Agric 79:1635–1641. https://doi.org/10.1002/(SICI)1097-0010(199909)79:12<1635::AID-JSFA413>3.0.CO;2-%23

Wintersgill S (2004) Treated fibres for making a beverage infusion package with improved freshness and a beverage infusion package made therefrom. PCT Int Appl WO 2004067841, 12 Aug 2004

Wood WE (2000) Improved aroma barrier properties in food packaging with cyclodextrins. In: Proceedings of the10th international cyclodextrin symposium, Ann Arbor, MI, USA, 21–24 May 2000, pp 338–347

Wood WE, Beaverson NJ (1998) Rigid polymeric beverage bottles with improved resistance to permeant elution. US Pat US5837339, 17 Nov 1998

Wood WE, Beaverson NJ (1999) Packaging system comprising cellulosic web with a permeant barrier or contaminant trap. US Pat US5985772, 16 Nov 1999

Wood WE, Beaverson NJ (2003) Sealing element for vessel or container closures having improved barrier properties. US Pat US2003207056, 6 Nov 2003

Wood WE, Beaverson NJ (2012) Malodor absorbent polymer and fiber. Eur Pat Appl EP2414450, 8 Feb 2012

Wood WE, Beaverson NJ, Kuduk WJ (2011) Maturation or ripening inhibitor release from polymer, fiber, film, sheet or packaging. PCT Int Appl WO2011081877, 07 Jul 2011

Wu B, Zhang Y, Wu Z et al (2013) Preparation method of cyclopropene preservative paper. Chin Pat Appl CN102964615, 13 Mar 2013

Wu J, Sun X, Guo X et al (2018) Physicochemical, antioxidant, in vitro release, and heat sealing properties of fish gelatin films incorporated with β-cyclodextrin/curcumin complexes for apple juice preservation. Food Bioprocess Technol 11:447–461. https://doi.org/10.1007/s11947-017-2021-1

Xu PJ (2006) Edible water resistant, high temperature resistant composite gel antibiotic film and its preparing method. Chin Pat Appl CN1781978, 07 Jun 2006

Xuan H, Zhu Y, Ren J et al (2019) Intrinsic self-healing and biocompatibility of carbon nitride coatings via inhibiting or degrading ethylene for fruit preservation. J Mater Sci 54:9282–9290. https://doi.org/10.1007/s10853-019-03383-1

Yamada K, Tsunoda H (1987) Food packaging materials containing cyclodextrin-flavoring agent inclusion compounds. Jap Pat Appl JP 87263047 (Chem Abstr 108:203557)

Yang L, Qu Y, Xu Y et al (2015) Method for preparing antibacterial and antioxidant composite film by using scale. Chin Pat Appl CN104974531, 14 Oct 2015

Yang Y, Ren J, Luoi C et al (2020) Fabrication of l-menthol contained edible self-healing coating based on guest-host interaction. Colloids Surf A Physicochem Eng Asp 597:124743. https://doi.org/10.1016/j.colsurfa.2020.124743

Yoshitomi T, Tanaka Y, Yoshida R et al (1976) Antiseptic and anticorrosive packaging films. Jap Pat Appl JP 51140964, 4 Dec 1976 (Chem Abstr 86:91993)

Younas M, Noreen A, Sharif A et al (2018) A review on versatile applications of blends and composites of CNC with natural and synthetic polymers with mathematical modeling. Int J Biol Macromol 124:591–626. https://doi.org/10.1016/j.ijbiomac.2018.11.064

Yuan C, Zou Y, Cui B et al (2019) Preparation method and application of antibacterial active food packaging film. Chin Pat Appl CN110078958, 02 Aug 2019

Zhang C, Wu C (2017) Preservative of rice. Chin Pat Appl CN107087680, 25 Aug 2017

Zhang M, Li J, Zhang L et al (2009) Preparation and spectral investigation of inclusion complex of caffeic acid with hydroxypropyl-beta-cyclodextrin. Spectrochim Acta A Mol Biomol Spectr 7:1891–1895. https://doi.org/10.1016/j.saa.2008.07.014

Zhang L, Huang C, Xu Y et al (2020) Synthesis and characterization of antibacterial polylactic acid film incorporated with cinnamaldehyde inclusions for fruit packaging. Int J Biol Macromol 164:4547–4555. https://doi.org/10.1016/j.ijbiomac.2020.09.065

Zhao X, Song W, Li W et al (2017) Collagen membranes crosslinked by β-cyclodextrin polyrotaxane monoaldehyde with good biocompatibilities and repair capabilities for cornea repair. RSC Adv 7:28865–28875. https://doi.org/10.1039/c7ra03994h

Zhou Z, Liu Y, Liu Z, Fan L (2020) Sustained-release antibacterial pads based on nonwovens polyethylene terephthalate modified by β-cyclodextrin embedded with cinnamaldehyde for cold fresh pork preservation. Food Packag Shelf Life 26:100554. https://doi.org/10.1016/j.fpsl.2020.100554

Chapter 12
Encapsulation for Masking Off-Flavor and Off-Tasting in Food Production

Lajos Szente, Tamás Sohajda, and Éva Fenyvesi

12.1 Sensing Off-Flavor and Off-Tasting in Food Products

Aroma is defined as an odor, sensed through the nose and retronasally, and also through the back of the mouth where the nasal and mouth cavities are interlinked. Taste is the sense experienced by the tongue and describes sensations of saltiness, sweetness, sourness, bitterness or umami. Flavor is defined as a combination of both aroma and taste.

Taste is the perception when a substance in the mouth binds to taste receptors located on taste buds in the oral cavity, mostly on the tongue. All the basic tastes can be detected over the full surface of the tongue, but the tip is the most sensitive to the sweet taste, the posterolateral and mediolateral edges to salt and sour, respectively, while bitter taste is perceived the most on the back of the tongue. These are the four basic tastes. According to Japanese people the taste of monosodium glutamate, "*umami*" is different from the four basic tastes. Astringency has also been debated as a taste, but since gustatory as well as non-gustatory epithelia react to an astringent stimulus, it is not considered a taste. In addition to the four (five) basic tastes there are many additional taste sensations, such as pungency (burning sensation caused by chili peppers and ginger), coolness (fresh or minty sensation), starchy, fat and various metallic tastes (Purves et al. 2001). Recent research revealed also the receptors of fat taste on taste bud cells which can sense fatty acids independently of other tastes (Keast and Costanzo 2015). Metallicness is generally considered rather an off-flavor when present in food or drink.

The tongue is covered with thousands of small bumps called papillae, where hundreds of taste buds are located. Yet taste receptors are not only found on the tongue; scientists recently discovered the existence of taste receptors that reside in

L. Szente · T. Sohajda · É. Fenyvesi (✉)
Cyclodextrin Research & Development Laboratory Ltd., Budapest, Hungary
e-mail: szente@cyclolab.hu; sohajda@cyclolab.hu; fenyvesi.e@cyclolab.hu

T. M. Ho et al. (eds.), *Functionality of Cyclodextrins in Encapsulation for Food Applications*, https://doi.org/10.1007/978-3-030-80056-7_12

the gastrointestinal tract, internal organs, and central nervous system. The taste receptors within the gut control the process of digestion, take part in regulating caloric intake and insulin secretion (Deportere 2014).

Despite the different locations, taste and aroma perception do not seem to be completely independent; they combine in the brain. Aroma is the most important part of flavor. Aromas are perceived on the olfactory epithelium in the nasal cavity. These compounds can reach the epithelium either orthonasally (by sniffing) or retronasally (from the back of the mouth). Humans have approximately 400 individual receptors which are responsible for the thousands of aromas we can distinguish (Trimmer et al. 2019).

Off-flavor means unwanted and unpleasant taste and odor resulting from natural spoilage of food. The unpleasant odors and tastes arising from contamination of a food by some chemical(s) are called taint (Pegg and Shahidi 2012).

Off-flavor of water has a consequence for the quality of beverages produced by using it. Water quality is critically important in food and beverage production. Thus, chlorinated water—otherwise safe for drinking—should be avoided in winery due to the formation of haloanisoles responsible for cork taint (Tarasov et al. 2017).

Soy proteins and other vegetable proteins have characteristic off-flavors, both taste and odor, and this is the main reason why their applications in food are not emerging as expected based on their nutritional values (Damodaran and Arora 2013).

In many cases, traces of a single component can change the flavor of a food or beverage making it unsuitable for consumption. For instance, geosmin gives earthy off-flavor to cocoa and beer, 2,4,6-trichloroanisole gives mouldy flavor to wine, and methional gives the smell of potato to milk (Collins et al. 2012; Cadwallader and Singh 2009).

12.2 Characterization of Off-Flavor and Off-Tasting

Traditionally, taste and smell attributes are measured by human panels. Odor measurement requires the human nose as tester. Well-trained human assessors are needed to get valid test results with high repeatability. Human nose is very sensitive and can be trained, but the uncertainty of the result is high due to person-to-person differences and also the personal fitness/fatigue.

There is large interindividual variation in taste perception, too. Supertasters are individuals who show superior taste response across multiple qualities especially to bitterness.

Exact measurement of taste and smell is not easy because of complexity of the mixtures of components determining the taste and perception. For instance, in the case of coffee, over 800 different compounds have been identified as playing a role in determining the coffee aroma (Grosch 1998), out of which 35 are considered the most potent aroma elements (Holscher et al. 1990). The other difficulty is the various threshold concentrations of the various components the panelists can distinguish.

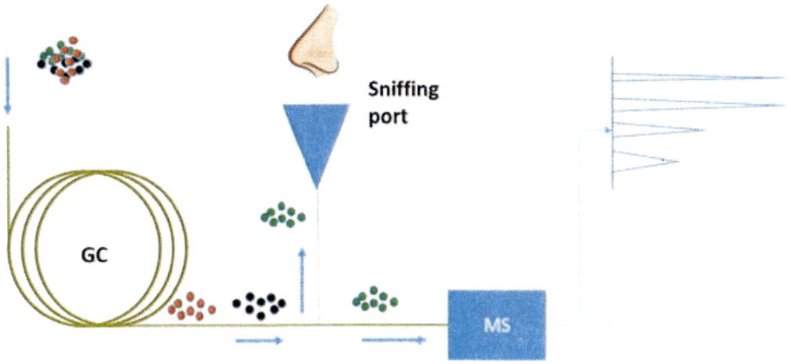

Fig. 12.1 Scheme of gas chromatography olfactometry

A combination of instrumental measurement with human perception is the gas chromatography-olfactometry (Fig. 12.1) (van Ruth 2001). Head-space solid phase microextraction or purge and trap extracts are analysed by gas chromatography-mass spectrometry parallel to the sensory evaluation of the extracts (Odor Observatory 2020). High-resolution gas chromatography-mass spectrometry provides information on the molecular composition of an odor sample, while gas chromatography sniffing helps to identify compound(s) responsible for a specific odor in a sample. Odor measurement requires well-trained human nose as tester to get valid test results with high repeatability.

Electronic-noses and -tongues mimic the human perception. Electronic nose was developed to measure smell by mimicking human olfactory system while eliminating operator fatigue. The sample from a headspace is delivered to the array of detectors (sensors) where the electronic properties are changed upon contact with volatile compounds.

Various sensors have been developed, such as quartz crystal microbalance, surface acoustic wave detectors, mass spectrometers, polymer composite sensors, metal-oxide-semiconductors, etc. (Wilson and Baietto 2009). The bio-electronic nose contains biosensor with olfactory receptors-proteins cloned from biological organisms. The output from individual sensors are collected and integrated to get a distinct digital response pattern (electronic fingerprint). Based on the obtained aroma signature pattern the flavor can be identified—without separating the individual components—using a reference library aroma patterns.

The electronic tongue is an instrument measuring tastes. The multichannel sensory system detects the components responsible for taste. Similarly to human receptors, each sensor has a spectrum of reactions different from the other. The information given by the sensors are combined to obtain a unique fingerprint (Vlasov and Legin 1998). Most of the detection thresholds of sensors are similar to or even better than those of human receptors.

Gas chromatography coupled with ion-mobility spectrometry detector is a recent technique applied for food flavor analysis. This is a very fast method to detect various substances within a couple of minutes, and to analyse the evolution of volatile organic compounds profiles over time (Wang et al. 2020). It can be used for evaluation of food freshness, changes in aroma during processing and storage and recognising off-flavor components.

12.3 Causes for Off-Flavor and Off-Tasting

Off-flavor of foods comes from undesirable components already present in the raw materials or formed during food processing and storage as a result of chemical or biochemical reactions. Off-flavor of milk, for instance, might be transmitted from the feed of cow, derived from microbial processes, heat treatment, lipase-catalysed hydrolysis, oxidation and other processes (Bassette et al. 1986; Deeth 2011). Off-flavor can be often detected in drinking water and sometimes in beer and wine due to non-optimal fermentation.

The off-flavor of water derives from geosmin and 2-methyl iso-borneol (Fig. 12.2). These compounds are produced by various bacteria, mainly by blue-green algae (cyanobacteria) and are responsible for the earthy odour of water (Lin et al. 2019). They can be accumulated in fatty tissues of freshwater fishes and associated with the muddy smell of carp and catfish (Lovell 1983; Lindholm-Lehto and Vielma 2019). The human nose is extremely sensitive to geosmin.

Certain yeasts such as Brettanomyces spp. in wines can lead to the accumulation of 4-ethylphenol and 4-ethylguaiacol which cause off-odor and spoil the wine (Fig. 12.3).

The food products, especially the fresh foods are deteriorated due to physiological aging, biochemical changes, high respiration rate and high ethylene production. These factors contribute considerably to development of off-flavors in addition to discoloration, reduced firmness, acidification, and microbial spoilage (Pasha et al. 2014).

Plant- or dairy-based proteins are used to increase the protein content of many foods. Yet these can have "beany" or metallic off-flavors that reduce consumer acceptance. Similarly, negative off-flavors can come from functional ingredients, such as texturants or stabilizers (e.g., hydrolysed proteins), or nutraceuticals.

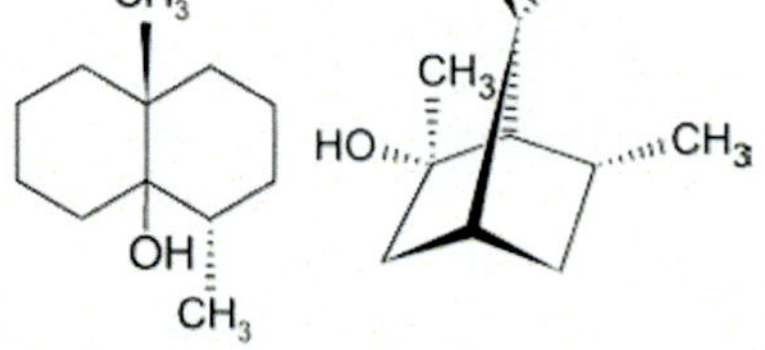

Fig. 12.2 Chemical structure of geosmin and 2-methyl isoborneol, typical odorants of water

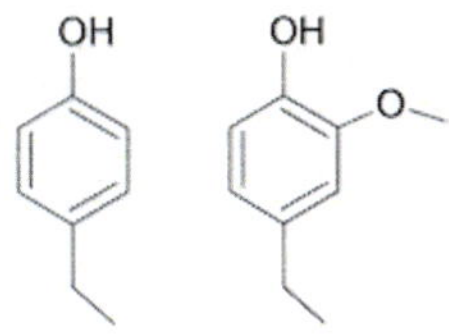

Fig. 12.3 Chemical structure of 4-ethylphenol and 4-ethylguaiacol, typical off-flavors of wine

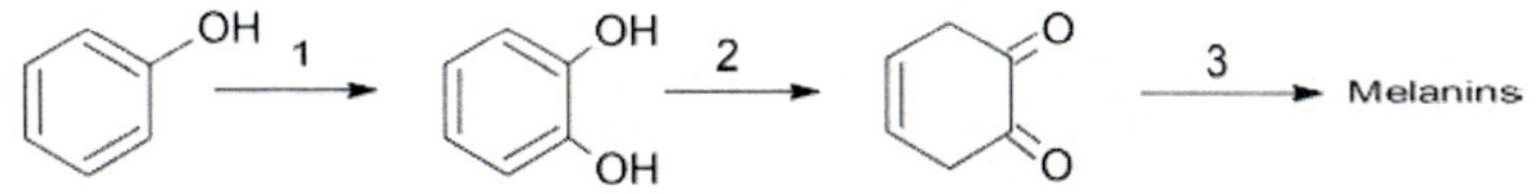

Fig. 12.4 Enzymatic browning of phenols in plants catalysed by polyphenol oxidases (1) Hydroxylation of monophenols to diphenols, (2) Oxidation of o-diphenols to o-diquinones, (3) Reaction with proteins and polymerization resulting in brown pigments, melanins

12.3.1 Enzymatic Browning

Food browning is the consequence of oxidation. In this process affecting both the taste and flavor of food, phenols are oxidized into quinones which then react with amino and sulfhydryl groups of proteins and are polymerized forming the brown polymers, melanins. The first reaction is catalysed by polyphenol oxidases present in the food. These copper-containing enzymes have catechol oxidase activity and often also tyrosinase activity, thus they catalyse oxidation of o-diphenols to their corresponding o-quinones and hydroxylation of monophenols to o-diphenols, respectively (Fig. 12.4) (Sullivan 2014; Jiang et al. 2016). For instance, chlorogenic acid, the major phenolic compound of plant-derived foods, is also oxidized to a highly reactive o-quinone intermediate which then can interact with NH_2 groups of lysine, SCH_3 groups of methionines and indole rings of tryptophan in nucleophilic addition and in polymerization reactions. These transformations destroy essential amino acids, reduce digestibility and nutritional quality, and may also result in the formation of off-flavored compounds (Sapers 1993).

The other enzymes in connection with browning are peroxidases performing single electron oxidation on a wide variety of compounds in the presence of hydrogen peroxide. These reactions also result in products with brown flavor. Hydrogen peroxide might be generated during oxidation of phenols in reactions catalysed by polyphenol oxidases. Mainly polyphenol oxidases but also peroxidases are responsible for quality loss due to oxidation of phenolics (Tomás-Barberán and Espín 2001).

The phenolic compounds participating in browning may be classified into four groups: simple phenols, phenolic acids, cinnamic acid derivatives, and flavonoids (Mesquita and Queiroz 2013).

Enzymatic browning can be observed in fruits (apricots, pears, bananas, grapes, etc.), vegetables (potatoes, mushrooms, lettuce, etc.) and also in seafood (shrimps, spiny lobsters and crabs), thus in all foods rich in oxidizable phenols. Food industry aims at preventing enzymatic browning in order to avoid adverse consumer response to the altered organoleptic and visual properties of products. The simplest way for controlling enzymatic browning is exclusion of oxygen by using modified atmosphere packaging. Thermal treatment such as blanching in hot water or microwave heating and high-pressure processing might help, too. Reducing agents, such as ascorbic acid, citric acid are commonly used browning-inhibitors. Sulphites became restricted because of their negative effects on human health. Natural antioxidants, such as curcumin, resveratrol, coumarin and quercetin proved to be inhibitors of polyphenol oxidase (López-Nicolás et al. 2013). Nowadays the antioxidant properties of essential oils have been also explored for this purpose (Zhang et al. 2019). The application of CDs is an innovative approach as well (see Sect. 12.4.2).

12.3.2 Rancidification

When fats and oils are exposed to air, light or moisture hydrolytic and oxidative transformations, called rancidification, occur. These processes result in products of undesirable taste and odor. In addition to hydrolytic and oxidative rancidity enzymatic rancidity can contribute to off-flavor.

In the presence of water, triglycerides are hydrolysed first to di- then monoglycerides and finally free fatty acids as reactions in hydrolytic rancidity. Especially the short-chain fatty acids, such as butyric acid have unacceptable odor (Pegg and Shahidi 2012). In milk, the reaction is catalysed by the indigenous milk lipoprotein lipase or bacterial lipases produced by *Pseudomonas* (Deeth 2011).

The oxidative rancidity involves free-radical chain reactions. This autooxidation of lipids catalysed by heat, light, ionizing radiation and metal ions results in hydroperoxides which further degrade to various low molecular-weight alkanes, alkenes, aldehydes, alcohols, furanes, etc. with characteristic odour. For instance, C_3-C_{12} aldehydes, such as propanal, 1-penten-3-one, hexanal, (E,E)-2,4-heptadienal and (E,Z)-2,6-nonadienal, 2,4-decadienal, produced by oxidation of unsaturated fatty acids give the typical rancid flavor to cooked meat during storage (Pegg and Shahidi 2012). Oxidation of long-chain fatty acids, such as docosahexaenoic acid and eicosapentaenoic acid, both abundant in fish oil, results in fishy smell and metallic off-flavor (Miyashita et al. 2018).

Hydroperoxides are formed also by photooxidation of lipids, but these are different from those produced by autooxidation and also result in different volatile breakdown products. To avoid this process, it is advised to keep fatty food in dark.

It is not possible to prevent the decomposition of hydroperoxides once they are formed; so it is important to prevent or retard their formation as long as possible. This may be achieved to some degree by incorporating synthetic or natural antioxidants into foods to extend shelf life and reduce off-flavor development.

To avoid rancidification food industry applies antioxidants, first of all natural antioxidants, such as vitamin E, rosemary and carotenoids like β-carotene and lycopene. Rosemary extract is one of the favorite antioxidant additives used for meat products (Pegg and Shahidi 2012). Citric and ascorbic acid, as well as nitrites and nitrates are used for the same purpose. CDs can protect fatty acids from oxidation in this way preventing the reactions leading to rancid flavor (see Sect. 12.4.3).

12.3.3 Off-Flavor of High-Protein Products

Beany, earthy, rancid, green, grassy and muddy are the main beany-related flavors characteristic to food from beans and peas. Beany flavors in raw pulse flours, protein isolates and concentrates are mainly derived from hydrolytic and oxidative degradation of lipids and amino acids especially after thermal processing (MacLeod et al. 1988; Xu et al. 2019).

Soy is rich in proteins and phytochemicals, but the human consumption is low due to the beany and grassy odor and bitter and astringent taste (Damodaran and Arora 2013). The volatile components released and perceived as off-flavor are by-products of oxidative degradation reactions of non-volatile precursors during processing and storage of soy protein isolates.

Soy flour, protein concentrate and isolate contain 50%, 70% and 90% proteins, respectively. Soy protein is produced from dehulled, defatted soybean by removing the soluble non-protein compounds such as carbohydrates and salts. The approximately 20% lipids are extracted but the residual lipids such as free unsaturated fatty acids and phospholipids bound to proteins are subjected to *oxidative degradation* during storage generating small molecular alcohols, aldehydes, ketones and esters of characteristic smell (Damodaran and Arora 2013). Pyrazines, furans, furanones, aldehydes and ketones were detected in the headspace of thermally treated pulse products with beany flavor (Heiniö et al. 2003).

Lipohydroperoxides are generated from lipids under high moisture conditions due to the activity of lipoxygenase enzyme (Rackis et al. 1979). The decomposition of the lipohydroperoxides gives the compounds, such as n-hexanal, 3-cis-hexenal, n-pentylfuran, 2(1-pentenyl)furan and ethyl vinyl ketone responsible for the grassy-beany flavor.

Not only the polyunsaturated fatty acids such as oleic acid, linoleic acid and linolenic acid are precursors of off-flavor in legume protein isolates but also some phytochemicals (isoflavones, saponins and phenolic acids), mostly present in protein-bound state, contribute to the bitter taste (Damodaran and Arora 2013).

Soy milk is a stable emulsion of oil, water and protein produced by soaking and grinding soybeans, boiling the mixture and filtering out remaining particulates. Looking like dairy milk soy milk is used in the vegan diet and in case of lactose-intolerance but the beany flavor is not attractive for most of consumers. As the soy milk market including beverages and non-dairy desserts is dynamically growing and expected to reach USD 11.08 billion by 2025 according to a report of Grand View Research, Inc. (2020) the improvement of flavor is a crucial point. Flavoring usually with vanilla or chocolate is a common solution to suppress beany flavor. Extraction of lipids, enzyme treatments, heat treatments and breeding soybeans with low beany flavor might result in more acceptable flavor and aroma (Chiba et al.

1979). These methods, however, may result in loss of protein functionality (Damodaran and Arora 2013), so gentler treatments, like cyclodextrin treatment, are needed.

12.3.4 *Bitter Taste*

Bitterness is usually not an off-taste, but indigenous in many foods, and is one of the most unpleasant tastes. The perception of bitterness evolved in animals to avoid potentially toxic or harmful substances, while plants evolved bitter-tasting compounds to prevent consumption. Bitter compounds are structurally different, they interact with one of the 25 known different bitter receptors in the mouth (Chandrashekar et al. 2006).

The aversion to bitterness is important for detecting toxic compounds in food but many beneficial nutrients such as polyphenols from green tea, cocoa, or caffeine also taste bitter. Citrus peels, grapefruit, and mustard greens are bitter. The bitter taste is originated often from phytonutrients, such as polyphenols, flavonoids, isoflavones and terpenes having positive health effect, increase appetite, promote healthy digestion and boost microbiome. Some studies indicate that some bitter components might modulate eating behavior and energy intake (Bray and Carek 2020). Still most people try to avoid bitter food, so reducing bitterness is important in various foods.

Chlorogenic acid is a bitter component of coffee beans which is decomposed to caffeic acid and D-(−)-quinic acid upon roasting (Fig. 12.5). Lactones and phenylindanes are also formed during roasting especially at high temperature (Farah et al. 2005). Both are bitter but while lactones are pleasant the phenylindanes give harsh bitter taste.

12.4 Cyclodextrin Encapsulation for Masking Off-Flavor and Off-Tasting

In theory, there are three distinct techniques to mask or block a flavor: encapsulation, peripheral interaction (receptor inhibition in the mouth), and central cognitive interaction (suppression in the central nervous system) (Coupland and Hayes 2014). In this chapter an overview on masking or blocking off-taste and off-flavor using encapsulation by CDs will be given.

CDs, the cyclic oligosaccharides produced from starch by enzymatic fermentation are odorless and have a very slight sweet taste (Szejtli 2004; Szejtli and Szente 2005; Linde et al. 2011; Crini 2014; Crini et al. 2018). The natural CDs consisting of 6, 7 or 8 glucose units, the so called α-, β- and γ-CDs, have been approved in food products: α- and γ-CD without limitations, while an acceptable daily intake (ADI)

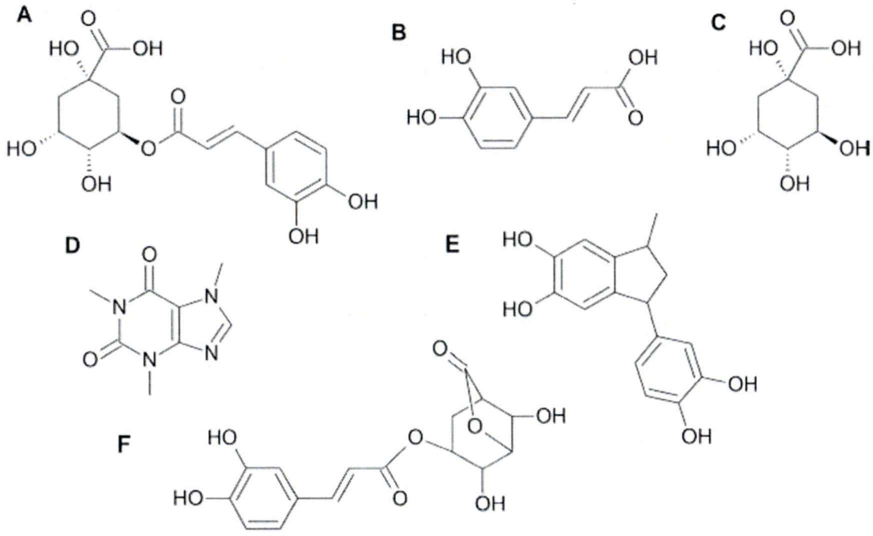

Fig. 12.5 Some characteristic bitter components of roasted coffee: chlorogenic acid (**a**), caffeic acid (**b**), quinic acid (**c**), caffeine (**d**), phenylindane (**e**) and a chlorogenic acid lactone (4-caffeinyl-1,5-quinidine) (**f**)

of 5 mg/kg has been established for β-CD (EFSA 2016). The regulatory situation in different countries has been recently reviewed by da Silva (2018). Several reviews on various application of CDs in food industry mention also taste and odor masking in addition to improving stability of flavors, vitamins, colorants, polyphenols, unsaturated fatty acids, etc., removing cholesterol, application in packaging and so forth (Szente and Szejtli 2004; Astray et al. 2009; Martina et al. 2013; Fenyvesi et al. 2016; Astray et al. 2020; Tian et al. 2020; Matencio et al. 2020; see also the previous chapters in this book). These applications as many others are based on the special feature of CDs: their ability to include apolar molecules or their moieties into their slightly apolar cavities forming host-guest inclusion complexes hold together mainly by hydrophobic interactions (Szejtli 2004).

It has been known for long that CDs can stabilize flavors (Szente et al. 1988; Decock et al. 2006; Astray et al. 2010; Fenyvesi et al. 2016; Kfoury et al. 2016; Fenyvesi and Szente 2016). These host molecules improve the thermal stability of included guests by reducing volatility of the flavor components and therefore hindering the change in aroma during storage. For instance, molecular encapsulation of coffee flavor resulted in enhanced stability against light and oxidation as well as in negligible loss of components due to diminished volatility and as a consequence the sensory properties were maintained during long-term storage (Szente and Szejtli 1986).

Armstrong's group measured the association constants of 33 malodorous compounds and odor precursors with native CDs using nuclear magnetic resonance spectroscopy titration or capillary electrophoresis (Lantz et al. 2006). These malodorous compounds are formed in food stored improperly. For instance, the odorous butyric and valeric acid demonstrate high affinity for complexation with both α- and β-CD, while isobutyric acid with β-CD only. The γ-CD was ineffective in this respect. These data encourage researchers to use mainly β-CD for masking the off-odor provided by these compounds.

12.4.1 Removal of Undesirable Components by Complexation

The simplest way of reducing off-taste and off-flavor is the removal of the responsible components from food or beverages either by precipitation or by sorption.

The concentration of malodorous pollutants, geosmin and 2-methyl borneol (see their structure in Fig. 12.2) can be diminished in water by β-CD (Perly et al. 1998; Baudin et al. 2000). These stinking compounds cannot be completely removed by the traditional techniques such as ozonation, adsorption on activated carbon, biodegradation and filtration (Nerenberg et al. 2000), but >99% removal rate was observed using β-CD-based nanoporous polyurethanes crosslinked by hexamethylene diisocyanate which adsorbs these components (Mamba et al. 2007; Mhlanga et al. 2007). In this experiment 0.5 g sorbent was used in 500 mL water containing ~100 ng/L geosmin in flow-through set-up. In another experiment, an epichlorohydrin-crosslinked β-CD polymer adsorbed 77% of geosmin from a water with 20μg/L geosmin concentration using lower adsorbent/water ratio (10 mg sorbents/500 mL water) in batchwise operation (Rashmawi et al. 2008).

CD and ascorbic acid incorporated into filter of coffee or tea bag are used for reducing trihalomethane content of beverages (Suzuki and Masuno 1994).

CD was applied as a competitive binder of the volatile flavor components bound to soy protein, e.g. 2-nonanone was removed almost completely (94%) by using β-CD in the acceptor phase of dialysis experiments (Arora and Damodaran 2010). Applying together with phospholipase A_2 α-CD performed better (>95%) than β-CD, while γ-CD was ineffective in removal of hexanal, a typical off-flavor-causing compound from the protein precipitated after the treatment (Zhu and Damodaran 2018).

CDs are able to complex polyunsaturated fatty acids, precursors of beany flavor in soy protein products (Szejtli and Bánky-Előd 1975; Rajnavölgyi et al. 2014; Malanga et al. 2016; Fenyvesi et al. 2016) but β-CD was not efficient in removing protein-bound phospholipids consisting of two fatty acid chains most probably due to the steric hindrance of binding both fatty acid chains. Applying together with enzyme phospholipase A_2, which hydrolysed the phospholipid into lysophospholipid and fatty acid, >70% phospholipid was removed (Arora and Damodaran 2011; Damodaran and Arora 2011). The removal rate was further improved up to 99% by combining the technology with sonication in order to detach phospholipids from proteins. The residual β-CD and its complex with phospholipid/fatty acid were removed by ultrafiltration.

When β-CD was added to enzymatic hydrolysate of soy protein for deodorization, the complex was filtered out to get soybean peptides of improved odor and taste (Zhou 2018). Also α-CD had similar effect: reduced both bitterness and the characteristic soy odor (Linde et al. 2009). Similar outcome of β-CD treatment was observed on whey protein hydrolysate (Yang et al. 2012a). CDs can include the main amino acids with bitter taste with affinity for complexation in the following order: tryptophan > tyrosine > phenylalanine > proline > histidine > isoleucine (Linde et al. 2010). These bitter amino acids can be removed from protein hydrolysates by sorption on CD polymers (Szente et al. 1981).

The bitter components in orange and grapefruit juices form complexes with β-CD, so adding it to citrus juices the bitter taste was reduced (Masaru and Konno 1982). Both naringin and limonin (Fig. 12.6) responsible for the bitter taste could be removed by adsorption on CD polymers obtained by crosslinking with epichlorohydrin (Shaw and Wilson 1983, 1985; Shaw et al. 1984). In addition to these bitter components also naringenin-7-O-rutinoside, coumarins, and flavonoids were removed in some extent, but total acidity and ascorbic acid content remained unchanged. A further improvement of debittering process was achieved by using fluid bed column: the same polymer was regenerated 21 times without apparent loss of capacity and removing 30–60% and 30–67% of naringin and limonin, respectively, resulting in juices with significantly improved flavor (Wagner et al. 1988; Wilson et al. 1989). Thai tangerine juice was debittered similarly and 50–60% of limonin removal was achieved (Mongkolkul et al. 2006). CD polymers crosslinked with various crosslinking agents under nonconventional conditions (ultrasound, microwave) were compared to find >95% naringin removal by the polymers crosslinked with epichlorohydrin and hexamethylene diisocyanate (Binello et al. 2008). These polymers could be easily regenerated by a counter-current ethanol wash (Wagner et al. 1988) or by aqueous sodium carbonate solution (Ujházi and Szejtli 1989). Molecular imprinted polymer with high affinity to bind naringin was prepared by crosslinking CD in the presence of naringin (Chen et al. 2011).

A more intensive debittering process was achieved by applying β-CD together with naringinase enzyme (Zhang et al. 2020).

Volatile phenols contributing to off-odor can be removed from wine by adsorption on CD polymer crosslinked with hexamethylene diisocyanate (Dang et al.

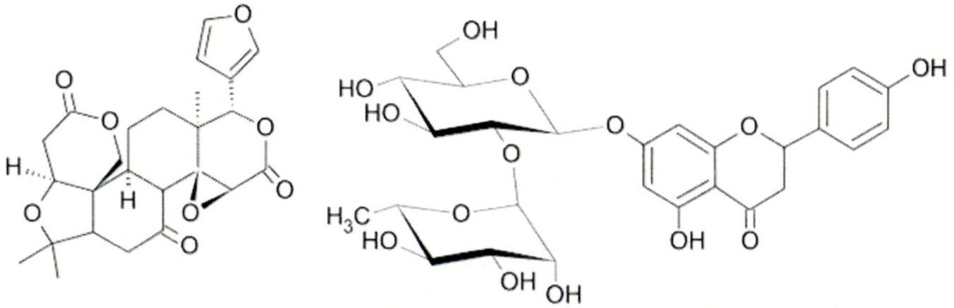

Fig. 12.6 Chemical formulae of limonin and naringin, the main bitter components of citrus juices

2020). The polymer adsorbed four volatile phenols (guaiacol, 4-methylguaiacol, 4-ethylguaiacol and 4-ethylphenol) with high efficiency. It was not studied if other important components of wine were also adsorbed changing the taste or not, but the polymer was suggested for solid-phase microextraction in analysis of volatiles in wine.

CDs incorporated into food packaging material may scavenge the off-flavor components formed during storage. For instance, interior can coating containing CDs binds aldehydes and ketones (Bobo 1993); packaging films or bottles with CD can capture off-flavor components (Wood and Beaverson 1996, 1998).

12.4.2 Inhibition of Enzymatic Browning

Browning with off-flavor development of fresh-cut fruits and vegetables as well as their juices is a challenge for food industry which aims to market these minimally processed foods similar in sensory characteristics to the fresh products. Early studies showed that browning of apple juice could be effectively reduced by combining β-CD with ascorbic acid or phosphates and β-CD was more effective than α- and γ-CD (Sapers et al. 1989; Gacche et al. 2003). Later studies showed that α-CD was superior to β-CD in peach juice (López-Nicolás et al. 2007a). The limited effect of β-CD in this case was attributed to its limited aqueous solubility. Therefore its water-soluble derivatives such as maltosyl, hydroxyethyl and random methyl β-CD were also applied (Hicks et al. 1996; Casado-Vela et al. 2006); then thiol-β-CD gave good results (Manta et al. 2013). The mechanism involves the complexation of polyphenol oxidase substrates, such as chlorogenic acid, caffeic acid and catechol and this inhibits their proper landing on the active site of the enzyme (Irwin et al. 1994; Ohnishi and Matsubara 1996; Rodrigues et al. 2002). Only the free substrates can be oxidized. In case of thiol-β-CD the interaction of thiol group with copper ions in the active site of enzyme may also contribute to this inhibition effect (Manta et al. 2013). It cannot be excluded, however, especially at higher concentrations, that CD also interacts some hydrophobic amino acids of the enzyme this way reducing its activity (Alvarez-Parrilla et al. 2007).

The affinity for complex formation characterized by association constants is different for different phenolic compounds and for different CDs. For instance, pyrocatechol is complexed in a highest extent by hydroxypropyl β-CD, followed by β-CD and α-CD, while the affinity of the phenolic compounds follows the order of (+)-catechin > (−)-epicatechin > chlorogenic acid > caffeic acid > pyrocatechol (Billaud et al. 1995) (Fig. 12.7). The higher the association constant the higher is the inhibition of oxidation by polyphenol oxidase. This strong correlation between the association constants and inhibition was valid only for single phenolic solutions. In solutions containing two or more phenolic compounds, however, various effects were observed: depending on the chemical nature of phenols present β-CD reduced in a high or a small extent the enzyme activity or in some cases it even enhanced it slightly. This explains why the inhibition of enzyme browning by β-CD was found to depend on apple variety, and also on the type of fruit. The effect is not predictable, depends on the composition of phenolics in the fruit juice.

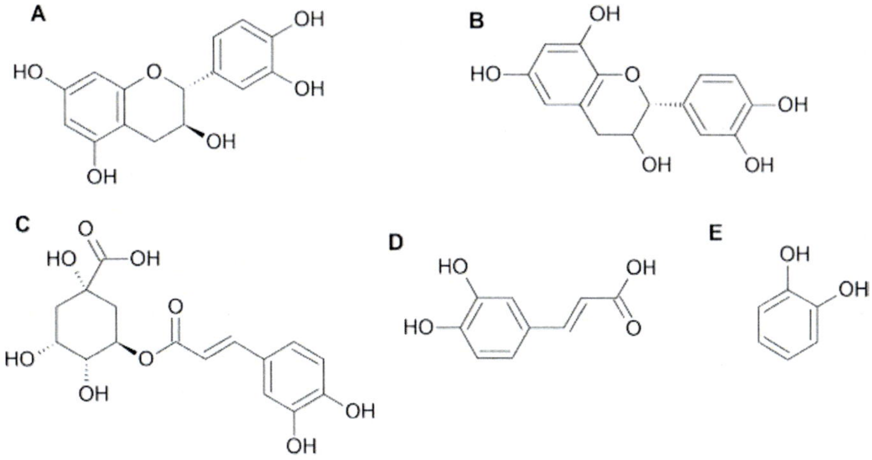

Fig. 12.7 Chemical structure of typical phenolic compounds responsible for enzymatic browning: catechin (**a**), (−)-epicatechin (**b**), chlorogenic acid (**c**), caffeic acid (**d**) and pyrocatechol (**e**)

CD was suggested to have intrinsic antioxidant effect: maltosyl β-CD was found to act as a "secondary antioxidant", reducing browning of apple, peach and pear juice and enhancing the naturally occurring antioxidant capacity of a food (López-Nicolás and García-Carmona 2007; López-Nicolás et al. 2007a, b). An opposite effect was observed for banana juice: CDs promoted browning (López-Nicolás et al. 2007c). This unexpected phenomenon was explained by the presence of different hydrophobic and hydrophilic (non-complexing) phenols. Treating banana pulp with β-CD, hydroxypropyl or maltosyl β-CD, no delay in browning was observed because dopamine, the natural substrate in banana is water soluble and does not readily form inclusion complex (Sojo et al. 1999). There were a few unsuccessful attempts for using CD as anti-browning agents, for instance, β-CD was not effective in controlling the browning of fresh-cut persimmon and eggplants (Ghidelli et al. 2013, 2014). The substrate specificity of inhibition by β-CD was observed for potato polyphenol oxidase, too (Jiang and Penner 2019). While polyphenol oxidase from apple or endive cannot attack the complexed phenols, mushroom tyrosinase was able to act also on the complex (Fayad et al. 1997).

High pressure processing of fruit and vegetable juices ensures quality and safety, but the increasing enzymatic browning should be suppressed via controlling oxidation of chlorogenic acid and (−)-epicatechin. This can be achieved by complexation with β-CD. Indeed, a β-CD concentration of 15 mM successfully prevented browning of apple juice treated at 500 MPa (Martínez-Hernández et al. 2019).

Eluting the juices through a column of water-insoluble β-CD polymer resulted in a huge improvement: while the control samples browned within a few minutes to a few hours, the treated samples did not show any browning for days (Hicks et al. 1996). The effect was explained by filtering out cell debris and particulate material and with this also the particulate-bound polyphenol oxidase. Some further examples are listed in Table 12.1.

Table 12.1 Inhibition of enzyme browning of fruits and vegetables and their juices

Fruit	CD	Effect of treatment	Reference
Asparagus juice	β-CD	Flavor protection	Li et al. (2010)
Avocado pulp	β-CD	Delayed browning	Fuentes Campo et al. (2019)
Banana pulp	Maltosyl β-CD	Dopamine, the main substrate in banana is not complexed	Sojo et al. (1999)
Celery juice	β-CD	Protecting flavor and vitamins	Yang (2019)
Fruit jam	β-CD	Stable flavor	Ma (2014)
Kiwi fruit juice	β-CD	Stabilized flavor and aroma	Zhao et al. (2015)
Orange/grape juice	n.s.	Maintaining the flavor and aroma	Sumioka (1991)
Pomegranate juice	β-CD	Long flavor conservation	Wang (2012)
Purple sweet potato juice	n.s.	Flavor protection	Tao et al. (2011)
Quince fruit pulp	Hydroxypropyl β-CD	Sequestration of tert-butyl catechol	Orenes-Piñero et al. (2006)
Sweet potato juice	Maltosyl β-CD	Reduced off-odour	Tamaki et al. (2007)
Tomato juice	Random methylated β-CD	Substrate-dependent effect	Casado-Vela et al. (2006)

n.s. not specified

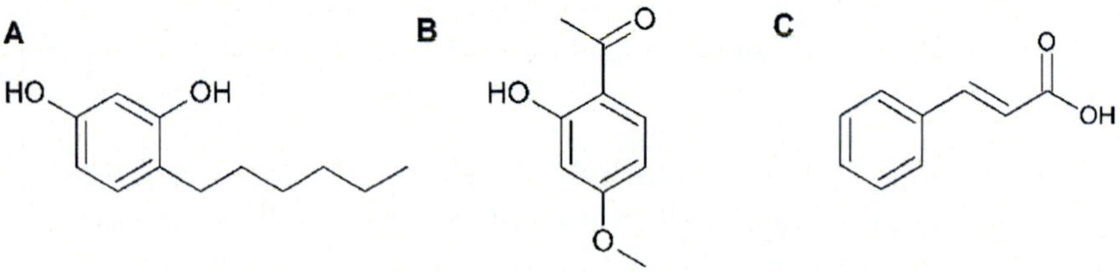

Fig. 12.8 Polyphenol oxidase inhibitors: hexylresorcinol (**a**), paeonol (**b**) and cinnamic acid (**c**)

Several antioxidants are inhibitors of polyphenol oxidases (Fig. 12.8). Their complex formation can either improve or reduce this inhibition. For instance, the inclusion complex of hexylresorcinol showed synergic effect in inhibition of apple polyphenol oxidase (Alvarez-Parrilla et al. 2007), while only a weak inhibitory effect was observed in peaches (de la Rosa et al. 2010). β-CD enhanced the aqueous solubility and thermal stability of paeonol without compromising the inhibition efficiency (Fuentes Campo et al. 2019). On the contrary, cinnamic acid, another natural inhibitor of polyphenol oxidase, when applied together with β-CD, cannot exert its inhibiting effect due to complex formation (Sojo et al. 1999). Similarly to cinnamic acid, also the effect of sodium dodecyl sulfate, an activator of the enzyme, is suppressed by complexation (Laveda et al. 2000).

Edible coatings containing antioxidants complexed by CDs can protect fresh-cut surfaces of fruits and vegetables from oxidation (see the details in Chap. 11 in this book).

12.4.3 Stabilization Against Oxidation of Fats and Oils

There are different strategies to avoid off-flavor caused by hydrolysis or oxidation of lipids (1) preventing lipid oxidation by antioxidants, (2) encapsulating lipids or (3) masking the forming low molecular oxidation products (Böttcher et al. 2015).

Plant oils with a high content of polyunsaturated fatty acid-containing triglycerides are stabilized against autoxidation by addition of γ-CD, which forms inclusion complexes with the oils. For example, evening primrose oil complexed with γ-CD was found to be stable (Regiert et al. 1996; Wimmer et al. 1997; Schmied et al. 2000). γ-CD was not only effective for preservation of plant oils but also improved oxidative stability of fish oil rich in ω-3 fatty acids such as eicosapentaenoic acid and docosahexaenoic acid (Schmied et al. 2000; Lee et al. 2013). γ-CD partly covered the smell and aftertaste of fish oil (Kobayashi et al. 2007). This effect was further improved by sodium caseinate addition serving as a stabilizer of the γ-CD/triglyceride inclusion complex (Lee et al. 2013) or by encapsulation in whey protein (Na et al. 2011). On the other hand, β-CD was found to block oxidation of seal blubber oil and with this to hinder the development of unpleasant taste and odor (Shahidi and Wanasundara 1995). Also α-CD protected docosahexaenoic acid from oxidation when included in fish meal paste (Yoshii et al. 1996). Perilla oil or linseed oil rich in ω-3 fatty acids were stabilized by β-CD in beverages containing lactic acid bacteria (Yoo et al. 2010). Similarly other seed oils, such as grape, pomegranate, lemon, peony, hemp and celery seed oils could be protected from oxidation and the unpleasant smell did not occur (Kim et al. 2007a; Zhao and Du 2015; Wang et al. 2018a; Zheng et al. 2019; Xiao et al. 2019; Xu et al. 2020). Having complexed the free fatty acids by β-CD they were protected from oxidation even in pure oxygen (Szejtli and Bánky-Előd 1975).

The affinity for complex formation follows the order of fatty acids > mono- > di- > triglycerides (Okada et al. 1989). While α-CD is preferred by the saturated fatty acids, β-CD by the (poly)unsaturated ones (Szente and Fenyvesi 2017).

The goaty flavor in goat milk and yogurt produced from it can be reduced by adding α-CD which complexes the branched chain fatty acids, such as 4-methyl octanoic acid responsible for this unpleasant smell and taste (Bordignon Luiz and Fett 2000; Young et al. 2012). When powdered milk is diluted with CD the original taste can be restored after being dissolved in water (Terao et al. 2007).

The effect of CD complexation can be combined with that of antioxidants such as tocopherol, polyphenols, ascorbic acid, sesamol and gossypol in order to further improve the oxidative stability of oils and fats and avoid this way formation of off-flavors (Park et al. 2008; Sekhon-Loodu et al. 2013). It should be, however, taken into account that some of these antioxidants also form complexes and a competition for the CD cavities may occur resulting in reduced effect instead of synergism.

12.4.4 Odor-Masking by Stabilization Against Volatilization

The unpleasant smell of volatile compounds can be reduced by molecular encapsulation with CD (Szejtli 2004; Szente and Szejtli 2004). Typical examples are essential oils, such as garlic, fennel, lemon, onion, chamomile oils (Szente and Szejtli 1988; Cabral Marques 2010). For, instance, odourless garlic powder was developed (Kawashima 1986; Komaki 1996). Tanaka and Uesawa filed a patent application in 1991 on CD as food deodorizer.

Trimethylamine gives the typical fishy smell and volatile compounds responsible for beany flavor are formed upon oxidation and hydrolysis of lipids and proteins. Both can be reduced by CD complexation. For instance, the soy-protein-bound off-flavor compounds, such as 2-nonanone were complexed by β-CD to improve the flavor of soy protein products (Arora and Damodaran 2010). α-CD added to soymilk reduced the volatile components such as pentanal, hexanal, hexanol, heptanal, benzaldehyde, 1-octen-3-ol, 2-pentylfuran and nonanal associated with beany odor as detected by gas chromatography, however, trained panellists did not detect the improvement in flavor due to the low concentration of CD added which was not enough to reduce the concentration of these off-flavor molecules in the headspace (Suratman et al. 2004). In another study, both sensory evaluation and solid phase micro-extraction gas chromatography of soymilk showed that β-CD was effective odor scavenger of hexanal, 2-pentanone, 2-heptenal, and 2,4-nonadienal at both 0.1% w/v and 1.0% w/v CD concentration (Norton 2003). Combining β-CD complexation with heat treatment resulted in significant decrease of critical beany flavor compounds, such as hexanal, hexanol, 1-octen-3-ol and trans-2-octenal (Shi et al. 2017). Sensory evaluation indicated that the soymilk with the addition of 0.75% β-CD at 60 °C during heat treatment had the best flavor quality. Adding CD to the peanut protein isolate during spray drying is an effective approach in reducing beany odor due to complexation of volatiles with α-CD performing better than β- or γ-CD (Cui et al. 2020).

The off-odor of 4-ethylphenol and 4-ethylguaiacol in wine can be masked by adding β-CD (Botelho et al. 2011). Similarly the off-flavor caused by decanal, octanol, (E)-2-octenol, and (E)-2-decenal forming upon heat treatment of watermelon can be reduced by β-CD (Yang et al. 2020).

Also the characteristic smell of vitamin B1 can be decreased when added to the food in CD-complexed form (Moritaka and Yamamoto 1981).

Further examples are listed in Table 12.2.

Table 12.2 Some examples on applications of CD complexation to reduce off-odor of various food products via decreasing volatility

Food	CD	Effect of complexation	References
Abalone peptide	β-CD	Reduced fishy smell	Bao et al. (2018a)
Asparagus juice	β-CD	Unique flavor	Li et al. (2010)
Canned green tea	α-, β- or γ-CD	Elimination of retort odor	Takeo and Kinugasa (1989)
Crab sauce seasoning	β-CD	Removing fishy flavor	Chen et al. (2019)
Dried fish	β-CD	Removing fishy smell	Zhao et al. (2019)
Dried shrimp	β-CD	Reduced fishy taste and smell	Xun and Weng (2014)
Fermented food by using *Bacillus natto*	n.s.	Traditional malodorous healthy food in Japan	Sekiguchi et al. (1982), Ota (1996)
Fish rice noodles	β-CD	Reduced fishy smell	Liang (2020)
Mussel flesh	β-CD	Reduced off-flavor	Wang et al. (2011a, b)
Propolis	n.s.	Low in off-odor	Akutsu and Sakamoto (2000)
Radish juice	β-CD	Removal of odor	Liu et al. (2020)
Sea cucumber	β-CD	Reduced fishy smell	Bao et al. (2018b)
Seaweed extract	β-CD	Reduced fishy smell	Zhu et al. (2020)
Seaweed noodles	β-CD	Decreased fishy smell	Liu et al. (2017)
Soy bean milk	n.s.	Reduced grassy smell and bitter taste	Suda et al. (1999)
Safflower-yellow powdered colorant	n.s.	Natural food colorant without off-flavor	Murakawa et al. (1985)

n.s. not specified

12.4.5 *Masking Malodor and Off-Tasting*

A review of Szejtli and Szente (2005) on elimination of bitter taste of foods by CDs clearly shows that the mechanism is based on enwrapping the bad-tasting molecule by CD in this way hindering its interaction with taste buds in the oral cavity. There exists also an opposite trend: making soluble the bitter, astringent components otherwise insoluble can enhance the perception of bitter, astringent taste. For instance the astringency of spray-dried pomegranate juice is slightly increased when γ-CD is added (Watson et al. 2017).

The unpleasant off-taste (bitterness) of coffee or tea or other hot-water extracts can be reduced by adding small amount (0.1–1%) of β-CD which complexes the components formed upon overbrewing or being exposed to air for longer time (Hamilton and Heady 1970). Both chlorogenic acid and its decomposition products caffeic acid and D-(−)-quinic acid contribute to bitter taste of coffee and form stable inclusion complexes with β-CD (Aree 2019; Aree and Jongrungruangchok 2018). The bitter taste of catechin, giving the characteristic bitter taste to tea, can be masked by adding β-CD (Terao and Nakada 2006; Xu et al. 2011; Ho et al. 2017, 2019) or by combining β-CD complexation with sweeteners, such as sucrose or rebaudiose (Gaudette and Pickering 2012). Various tea drinks can be debittered by β-CD (Takatsu and Otsuka 2006; Kim et al. 2007b; Sakai and Yamashita 2008; Cheng 2017).

The bitter components of grapefruit were removed before spray drying by β-CD treatment to get instant beverage (Chen 2015). Citrus juices, such as lemon and ponkan juice were debittered by using β-CD (3–7 g/L) (Gao et al. 2013; Ma et al. 2011).

The bitter compounds in ginseng extract and ginkyo leaf extract were complexed by γ-CD to make the taste more acceptable (Lee et al. 2008; Endo and Hirata 2019).

Encapsulation of flavors with CD generates fewer off-flavors during frozen storage and during microwave cooking. For example, raspberry glaze formulations for microwaving contain β- and γ-CDs for flavor encapsulation (McBride et al. 2002). The pungent taste of piperine, a bioactive component in black pepper with anti-inflammatory effect can be modulated by complexation (Ezawa et al. 2017).

Propylene glycol is a food additive of bitter taste, which can be improved by adding CD (Imai and Yajima 1980).

Foods fortified with bioactive components having unpleasant taste or odor often contain CD to improve the sensorial properties, e.g. the offensive odor of soy saponins is reduced by γ-CD (Murakami and Minami 1992); unpleasant taste and odor of vitamins added to food or beverages are improved (Oishi et al. 2004); the bitter taste and characteristic odor of peptides are masked (Fuji et al. 2006; Ide et al. 2007); the sulfur smell and hot taste of α-lipoic acid is reduced by branched CDs (Furuse and Nagata 2007); astringency of polyphenols is decreased (Sakai and Yamashita 2008); even the pungent taste and powerful smell of natto, a traditional Japanese food of fermented soy beans, is reduced (Sekiguchi et al. 1982; Takada 1992).

Steviol glycosides as non-caloric sweeteners not only have low aqueous solubility but also astringent and metallic taste and a persistent after taste or lingering taste. Both can be improved by CD according to the patent of Coca Cola Co. (Upreti et al. 2011). Also the stability of aspartame, another preferred sweetener apt to hydrolysis, can be improved by complexation (Brewster et al. 1991; Prankerd et al. 1992; Sohajda et al. 2009).

Odour-free docosahexaenoic acd solid beverage and soft sweet is produced by spray drying this ω-3 fatty acid together with aromas and β-CD (Lin et al. 2012; Wang et al. 2018b).

Various herbs are extracted in the presence of CD to improve the taste of the health food or beverages, e.g. *Fomes japonicas* (Shinoda and Ikeda 1985), *Gymnema silvestre* (Hane and Kenmasa 1989; Yumoto et al. 1989; Terao et al. 2003), *Gynostemma pentaphyllum* (Takeshita 1987), *Fructus schisandrae* (Peng 2016), *Tribulus terrestris*, *Actium* (Ding et al. 2016), *rhizoma picrorhizae* (Dai and Wu 2019), and *Sophora viciifolia* (Fan et al. 2009).

The astringent taste and irritating smell of allylisothiocyanate in mustard is reduced by adding β-CD (Zhang et al. 2016). The taste of Japanese horseradish can be also improved similarly (Yamamoto et al. 1991; Hao 2015).

Some further examples are listed in Tables 12.3 and 12.4.

Table 12.3 Sea food with improved taste and smell

Food	CD	Effect	Reference
Abalone peptide	β-CD	No bitter taste and reduced fishy smell	Bao et al. (2018a)
Soft-shelled turtle products	β-CD	Decreased fishy smell	Xu and Qi (2001)
Dried and powdered eel	β-CD	No fishy or other peculiar smell	Cai et al. (2009)
Mussel flesh softened by protease	β-CD	Removing fishy smell	Wang et al. (2011b)
Dried shrimp treated with glutamine transaminase	β-CD	Masking fishy and bitter taste	Xun and Weng (2014)
Salmon frame protein hydrolysate	β-CD	Debittering	Singh et al. (2020)
Sea cucumber polysaccharide	β-CD	Removing fishy smell	Bao et al. (2018b)
Sweet corn seaweed vegetable granules	β-CD	Masking fishy smell	Zhu et al. (2020)

Table 12.4 Examples for CD-added foods with improved taste

Food	CD	Effect	References
Apple polyphenols in chewing tablets	γ-CD	Masked astringent taste	Azechi et al. (2010)
Beef granules	β-CD	Reduced meaty and fishy smell	Wang et al. (2015)
Bean sprout juice	β-CD	Removed offensive fishy smell	Hong (2004)
Bergamot liquor	β-CD	Removed bitter taste and astringency	Lu and Li (2005)
Broad bean hydrolysate wine	β-CD	Reduced bitter taste and good oxidative stability	Yang et al. (2012b)
Candy with menthol	n.s.	Improved cool and refreshing taste	Guo et al. (2018)
Carrot juice	Hydroxypropyl β-CD	Improved taste	Domoto et al. (2009)
Fried rabbit meat crisps	β-CD	Delicious flavor without fishy smell	Gan et al. (2010)
Food/beverages with ginkgo leaf extract	n.s.	Suppressed bitter taste	Yi (2004), Endo and Hirata (2019)
Ginseng-containing food	n.s.	Controlling bitter taste	Cho et al. (2010, 2019), Kim and Yoon (2009), Tamamoto et al. (2010)
Goose liver	β-CD	Debittering and removal of fishy smell	Lyu et al. (2017)
Green barley juice	β-CD	Fresh taste	Wang et al. (2016)
Hop extract/plant extracts	Branched CD	Debittering	Wada and Yamamoto (2012), Urata and Urata (2011)
Lemon jam	β-CD	Debittering	Li et al. (2018)
Lotus root extract	α- and β-CD	Reduced astringency	Wang et al. (2011a)
Fruit/vegetable juices	β-CD	Debittering	Jiang et al. (2017)
Mulberry leaf extract	β-CD	Reduced bitter taste	Liu et al. (2012)
Orange and pomelo peel	Hydroxypropyl β-CD	Reduced bitter taste	Wu et al. (2014)

(continued)

Table 12.4 (continued)

Food	CD	Effect	References
Pastry with artichoke	Ethoxy-β-CD	Improved taste and flavor	Liu (2018)
Pomegranate wine with mushroom	β-CD	Removed bitter taste of mushroom	Huang (2018)
Pomelo juice	β-CD	Reduced bitterness	Lin and Chen (1994)
Squab amino acid beverage	β-CD	Improved smell and taste	Yao et al. (2017)

12.5 Conclusion and Future Perspectives

In this chapter, the various causes of off-flavor and off-taste of food products were explored and the possibilities of CD treatment were represented based on approximately 260 references including also patents. Several examples were shown to illustrate how CDs can be used for the removal of undesirable components from food, inhibit enzymatic browning, stabilize fats and phenolic compounds against oxidation, reduce off-flavor by stabilizing the volatile compounds, and mask malodor and off-taste by complexation.

Soy and fish protein isolates and the easier digestible protein hydrolysates are of increasing importance in nutrition of the growing population in the world. The acceptance of such products is hindered by peculiar taste and flavor. Both bitterness and fishy/beany off-flavor can be reduced by CD treatment during processing paving the way for the CD application in the food industry of the future.

References

Akutsu K, Sakamoto T (2000) Water-extracted propolis extract and its production. Jap Pat Appl JP2000316498, 21 Nov 2000

Alvarez-Parrilla E, de la Rosa LA, Rodrigo-Garcia J et al (2007) Dual effect of beta-cyclodextrin (beta-CD) on the inhibition of apple polyphenol oxidase by 4-hexylresorcinol (HR) and methyl jasmonate (MJ). Food Chem 101:1346–1356. https://doi.org/10.1016/j.foodchem.2006.03.040

Aree T (2019) Understanding structures and thermodynamics of β-cyclodextrin encapsulation of chlorogenic, caffeic and quinic acids: implications for enriching antioxidant capacity and masking bitterness in coffee. Food Chem 293:550–560. https://doi.org/10.1016/j.foodchem.2019.04.084

Aree T, Jongrungruangchok S (2018) Structure–antioxidant activity relationship of β-cyclodextrin inclusion complexes with olive tyrosol, hydroxytyrosol and oleuropein: deep insights from X-ray analysis, DFT calculation and DPPH assay. Carbohydr Polym 199:661–669. https://doi.org/10.1016/j.carbpol.2018.07.019

Arora A, Damodaran S (2010) Competitive binding of off-flavor compounds with soy protein and βcyclodextrin in a ternary system: a model study. J Am Oil Chem Soc 87:673–679. https://doi.org/10.1007/s11746-009-1535-8

Arora A, Damodaran S (2011) Removal of soy protein-bound phospholipids by a combination of sonication, β-cyclodextrin, and phospholipase A2 treatments. Food Chem 127:1007–1013. https://doi.org/10.1016/j.foodchem.2011.01.073

Astray G, Gonzalez-Barreiro C, Mejuto JC et al (2009) A review on the use of cyclodextrins in foods. Food Hydrocoll 23:1631–1640. https://doi.org/10.1016/j.foodhyd.2009.01.001

Astray G, Mejuto JC, Morales J et al (2010) Factors controlling flavors binding constants to cyclodextrins and their applications in foods. Food Res Int 43:1212–1218

Astray G, Mejuto JC, Simal-Gandara J (2020) Latest developments in the application of cyclodextrin host-guest complexes in beverage technology processes. Food Hydrocoll 106:105882. https://doi.org/10.1016/j.foodhyd.2020.105882

Azechi Y, Miyagi T, Takeda K et al (2010) Granules or chewable tablets containing polyphenols and water-soluble cellulose derivatives to mask unpleasant taste and odour and manufacture thereof. Jap Pat Appl JP 2010154769, 26 Dec 2008

Bao W, Ma P, Yang D et al (2018a) Abalone peptide odour removal method. Chin Pat Appl CN108077940, 29 May 2018

Bao W, Ma P, Yang D et al (2018b) Odour-smell-removing technology of sea cucumber polypeptides. Chin Pat Appl CN108065412, 25 May 2018

Bassette R, Fung DYC, Mantha VR et al (1986) Off-flavors in milk. Crit Rev Food Sci Nutr 24:1–52. https://doi.org/10.1080/10408398609527431

Baudin C, Pean C, Perly B et al (2000) Inclusion of organic pollutants in cyclodextrins and derivatives. Int J Environ Anal Chem 77:233–242. https://doi.org/10.1080/03067310008032685

Billaud C, Regaudie E, Fayad N et al (1995) Effect of cyclodextrins on polyphenol oxidation catalyzed by apple polyphenol oxidase. In: Lee CY, Whitaker JR (eds) Enzymatic browning and its prevention, ACS symposium series, vol 600. American Chemical Society, Washington, pp 295–312. https://doi.org/10.1021/bk-1995-0600.ch023

Binello A, Robaldo B, Barge A et al (2008) Synthesis of cyclodextrin-based polymers and their use as debittering agents. J Appl Polym Sci 107:2549–2557. https://doi.org/10.1002/app.27249

Bobo WS (1993) Interior can coating compositions containing cyclodextrins. US Pat US5177129, 05 Jan 1993

Bordignon Luiz MT, Fett R (2000) Elimination of off-flavor from goat milk and milk products by using b-cyclodextrin. BR 9802169, 08 Feb 2000

Botelho G, Valiau C, Moreira da Silva A (2011) Effect of cyclodextrins on off-odours removal of red wine: an innovative approach. Ciencia e Tec Vitiv 26:63–68

Böttcher S, Steinhäuser U, Drusch S (2015) Off-flavor masking of secondary lipid oxidation products by pea dextrin. Food Chem 169:492–498. https://doi.org/10.1016/j.foodchem.2014.05.006

Bray SC, Carek PJ (2020) How bitter taste influences nutrition and health in primary care. J Fam Med Prim Care 9:3205–3208. https://www.jfmpc.com/text.asp?2020/9/7/3205/290786

Brewster ME, Loftsson T, Baldvinsdóttir J (1991) Stabilization of aspartame by cyclodextrins. Int J Pharm 75:R5–R8. https://doi.org/10.1016/0378-5173(91)90257-O

Cabral Marques HM (2010) A review on cyclodextrin encapsulation of essential oils and volatiles. Flav Fragr J 25:313–326. https://doi.org/10.1002/ffj.2019

Cadwallader K, Singh T (2009) Flavors and off-flavors in milk and dairy products. In: McSweeney P, Fox P (eds) Advanced dairy chemistry. Springer, New York, pp 631–690. https://doi.org/10.1007/978-0-387-84865-5_14

Cai H, Chen S, Yang Y et al (2009) Eel full nutrient powder and preparation method thereof. Chin Pat Appl CN101491341, 29 Jul 2009

Casado-Vela J, Selles R, Bue R (2006) Influence of developmental stage, cultivar, and hexapeptide and cyclodextrin inhibitors on polyphenol oxidase activity from tomato fruits. J Food Biochem 30:623–640. https://doi.org/10.1111/j.1745-4514.2006.00084.x

Chandrashekar J, Hoon MA, Ryba NJ et al (2006) The receptors and cells for mammalian taste. Nature 444:288–294. https://doi.org/10.1038/nature05401

Chen F (2015) Grapefruit, golden silk date and fruit vinegar aged wine and preparation method thereof. Chin Pat Appl CN104774732, 15 Jul 2015

Chen Z, Ma X, Chen R et al (2011) Method for preparing rod-like imprinted polymer (RMIP) used for high-efficiently separating naringin in water phase. Chin Pat Appl CN 102002169, 6 Apr 2011

Chen D, Liu T, Liang Z, Fan S (2019) Ghost crab sauce seasoning capable of remove fishy flavor and enhancing freshness and its preparation method. Chin Pat Appl CN110089725, 06 Aug 2019

Cheng J (2017) Processing method of distinctive yellow bud tea beverage. Chin Pat Appl CN107251966, 17 Oct 2017

Chiba H, Takahashi N, Sasaki R (1979) Enzymatic improvement of food flavor II. Removal of beany flavor from soybean products by aldehyde dehydrogenase. Agric Biol Chem 43:1883–1889. https://doi.org/10.1271/bbb1961.43.1883

Cho HH, Lee JW, Son DW et al (2010) Food compositions for comprising red ginseng and chongkukjang. Kor Pat Appl KR 101021925, 18 Mar 2011

Cho HH, Seo YM, Oh SS (2019) Composition for controlling bitter taste. Kor Pat Appl KR20190139625, 18 Dec 2019

Collins TS, Hjelmerald A, Ebeler SE (2012) Analysis of haloanisoles in corks and wines. In: Toth S, Mussinan C (eds) Recent advances in the analysis of food and flavors. American Chemical Society, Washington, pp 109–127. https://doi.org/10.1021/bk-2012-1098.ch008

Coupland JN, Hayes JE (2014) Physical approaches to masking bitter taste: lessons from food and pharmaceuticals. Pharm Res 31:2921–2939. https://doi.org/10.1007/s11095-014-1480-6

Crini G (2014) Review: A history of cyclodextrins. Chem Rev 114:10940–10975. https://doi.org/10.1021/cr500081p

Crini G, Fourmentin F, Fenyvesi É et al (2018) Cyclodextrins, from molecules to applications. Environ Chem Lett 16:1361–1375. https://doi.org/10.1007/s10311-018-0763-2

Cui L, Kimmel J, Zhou L et al (2020) Combining solid dispersion-based spray drying with cyclodextrin to improve the functionality and mitigate the beany odor of pea protein isolate. Carbohydr Polym 245:116546. https://doi.org/10.1016/j.carbpol.2020.116546

da Silva AM (2018) Room at the top as well as at the bottom: structure of functional food inclusion compounds. In: Aurora P, Dhingra N (eds) Cyclodextrin: a versatile ingredient. Intechopen, Vienna, pp 553–573. https://doi.org/10.5772/intechopen.74162

Dai B, Wu O (2019) Food retention-removing Jianer granule shielding bitter taste and preparation method and application thereof. Chin Pat Appl CN110123972, 16 Aug 2019

Damodaran S, Arora A (2011) Process for removing phospholipids and off-flavors from proteins and resulting protein product. US Pat US2011045128, 24 Feb 2011

Damodaran S, Arora A (2013) Off-flavor precursors in soy protein isolate and novel strategies for their removal. Annu Rev Food Sci Technol 4:327–346. https://doi.org/10.1146/annurev-food-030212-182650

Dang C, Jiranek V, Taylor DK et al (2020) Removal of volatile phenols from wine using cross-linked cyclodextrin polymers. Molecules 25:910. https://doi.org/10.3390/molecules25040910

de la Rosa LA, Mercado-Mercado G, Rodrigo-García J et al (2010) Peach polyphenol oxidase inhibition by β-cyclodextrin and 4-hexylresorcinol is substrate dependent. CyTA J Food 8:87–93. https://doi.org/10.1080/19476330903146013

Decock G, Fourmentin S, Surpateanu GG (2006) Experimental and theoretical study on the inclusion compounds of aroma components with β-cyclodextrins. Supramol Chem 18:477–482. https://doi.org/10.1080/10610270600665749

Deeth HC (2011) Milk lipids | lipolysis and hydrolytic rancidity. In: Fuquay JW, Fox PF, McSweeney PLH (eds) Encyclopedia of dairy sciences. Academic, San Diego, CA, pp 721–726. https://doi.org/10.1016/B978-0-12-374407-4.00343-5

Deportere I (2014) Taste receptors of the gut: emerging roles in health and disease. Gut 63:179–190. https://doi.org/10.1136/gutjnl-2013-305112

Ding P, Li C, Ding L (2016) Blood sugar-lowering burdock composite beverage and preparation method thereof. Chin Pat Appl CN105433160, 30 Mar 2016

Domoto T, Takei T, Makamura M (2009) Carrot extract-containing beverage. Jap Pat Appl JP2009219435, 01 Oct 2009-10-01

EFSA (2016) Re-evaluation of β-cyclodextrin (E 459) as a food additive. EFSA J 14:e04628. https://doi.org/10.2903/j.efsa.2016.4628

Endo S, Hirata A (2019) Gingko leaf extract-containing food/beverage composition. Jap Pat Appl JP2019180337, 24 Oct 2019

Ezawa T, Inoue Y, Murata I et al (2017) Characterization of the dissolution behavior of piperine/cyclodextrins inclusion complexes. AAPS PharmSciTech 19:923–933. https://doi.org/10.1208/s12249-017-0908-9

Fan J, Zhao T, Cao J et al (2009) *Sophora viciifolia hance* soft candy and preparation method thereof. Chin Pat Appl CN101366435, 18 Feb 2009

Farah A, de Paulis T, Trugo LC et al (2005) Effect of roasting on the formation of chlorogenic acid lactones in coffee. J Agric Food Chem 53:1505–1513. https://doi.org/10.1021/jf048701t

Fayad N, Marchal L, Billaud C (1997) Comparison of β-cyclodextrin effect on polyphenol oxidation catalyzed by purified polyphenol oxidase from different sources. J Agric Food Chem 45:2442–2446. https://doi.org/10.1021/jf9607932

Fenyvesi E, Szente L (2016) Nanoencapsulation of flavors and aromas by cyclodextrins. In: Grumezescu AM (ed) Encapsulations. Academic, Amsterdam, pp 769–792. https://doi.org/10.1016/b978-0-12-804307-3.00018-1

Fenyvesi É, Vikmon M, Szente L (2016) Cyclodextrins in food technology and human nutrition: benefits and limitations. Crit Rev Food Sci Nutr 56:1981–2004. https://doi.org/10.1080/10408398.2013.809513

Fuentes Campo A, Sancho MI, Melo G et al (2019) In vitro and in vivo inhibition of Hass avocado polyphenol oxidase enzymatic browning by paeonol, β-cyclodextrin, and paeonol:β-cyclodextrin inclusion complex. J Biosci Bioeng 127:703–709. https://doi.org/10.1016/j.jbiosc.2018.11.009

Fuji K, Yasuda A, Kawaraichi N, Tokomura T (2006) Peptide-containing beverage. Jap Pat Appl JP2006075064, 23 Mar 2006

Furuse N, Nagata K (2007) Lipoic acid-containing beverage and production method. Jap Pat Appl JP2007006771, 18 Jan 2007

Gacche RN, Zore GB, Ghole VS (2003) Kinetics of inhibition of polyphenol oxidase mediated browning in apple juice by β-cyclodextrin and l-ascorbate-2-triphosphate. J Enzyme Inhib Med Chem 18:1–5. https://doi.org/10.1080/1475636031000062874

Gan P, Sheng J, Zhong Z (2010) Fried rabbit meet crisps and the manufacture method thereof. Chin Pat Appl CN101731639, 16 Jun 2010

Gao J, He H, Zhu Y et al (2013) Method for debitterizing lemon juice. Chin Pat Appl CN103238905, 14 Aug 2013

Gaudette NJ, Pickering GJ (2012) Optimizing the orosensory properties of model functional beverages: the influence of novel sweeteners, odorants, bitter blockers, and their mixtures on (+)-catechin. J Food Sci 77:226–232. https://doi.org/10.1111/j.1750-3841.2012.02707.x

Ghidelli C, Rojas-Argudo C, Mateos M et al (2013) Effect of antioxidants in controlling enzymatic browning of minimally processed persimmon 'Rojo Brillante'. Postharv Biol Technol 86:487–493. https://doi.org/10.1016/j.postharvbio.2013.07.034

Ghidelli C, Mateos M, Rojas-Argudo et al (2014) Effect of antioxidants on enzymatic browning of eggplant extract and fresh-cut tissue. J Food Process Preserv 38:1501–1510. https://doi.org/10.1111/jfpp.12109

Grand View Research, Inc. (2020) https://www.grandviewresearch.com/press-release/global-soy-milk-market. Accessed 02 Dec 2020

Grosch W (1998) Flavor of coffee. A review. Die Nahrung 42:344–350. https://doi.org/10.1002/(sici)1521-3803(199812)42:06<344::aid-food344>3.0.co;2-v

Guo X, Liu R, Zheng J et al (2018) Press candy with super-long cool sense lasting time and preparation method thereof. Chin Pat Appl CN107751522, 06 Mar 2018

Hamilton RM, Heady RE (1970) Eliminating undesirable taste from coffee and tea extracts and products. US Pat 3528819, 15 Sept 1970

Hane HY, Kenmasa G (1989) Method for reducing bitter taste of extract of *Gymnema silvesta*. Jap Pat Appl JPS642552, 06 Jan 1989-01-06

Hao Y (2015) Horseradish seasoning sauce production process. Chin Pat Appl CN104719861, 24 Jun 2015

Heiniö RL, Katina K, Wilhelmson A et al (2003) Relationship between sensory perception and flavor-active volatile compounds of germinated, sourdough fermented and native rye following the extrusion process. LWT Food Sci Technol 36:533–545. https://doi.org/10.1016/S0023-6438(03)00057-4

Hicks KB, Haines RM, Tong CBS et al (1996) Inhibition of enzymatic browning in fresh fruit and vegetable juices by soluble and insoluble forms of β-cyclodextrin alone or in combination with phosphates. J Agric Food Chem 44:2591–2594. https://doi.org/10.1021/jf960181h

Ho S, Thoo YY, Young DJ et al (2017) Inclusion complexation of catechin by β-cyclodextrins: characterization and storage stability. LWT Food Sci Technol 86:555–565. https://doi.org/10.1016/j.lwt.2017.08.041

Ho S, Thoo YY, Young DJ et al (2019) Stability and recovery of cyclodextrin encapsulated catechin in various food matrices. Food Chem 275:594–599. https://doi.org/10.1016/j.foodchem.2018.09.117

Holscher W, Vitzthum OG, Steinhart H (1990) Identification and sensorial evaluation of aroma-impact-compounds in roasted Colombian coffee. Café, Cacao, Thé 34:205–212

Hong TK (2004) Bean sprouts juice and production of the same by heating or blanching bean sprouts in fry pan without oil within short time and then squeezing juice from blanched bean sprouts. Kor Pat Appl KR20040087268, 13 Oct 2004

Huang Z (2018) Brewing technology of pomegranate and *ganoderma lucidum* health-care wine. Chin Pat Appl CN108315181, 24 Jul 2018

Ide D, Shimizu T, Mori M (2007) Peptide-containing food and drink reduced in bitter taste and astringency, and method for producing the same. Jap Pat Appl JP2007097465, 19 Apr 2007

Imai Y, Yajima M (1980) Edible propylene glycol preparation. Jap Pat Appl JPS5571456, 29 May 1980

Irwin PL, Pfeffer PE, Doner LW et al (1994) Binding geometry, stoichiometry, and thermodynamics of cyclomalto-oligosaccharide (cyclodextrin) inclusion complex formation with chlorogenic acid, the major substrate of apple polyphenol oxidase. Carbohydr Res 256:13–27. https://doi.org/10.1016/0008-6215(94)84223-x

Jiang S, Penner MH (2019) The nature of β-cyclodextrin inhibition of potato polyphenol oxidase-catalyzed reactions. Food Chem 298:125004. https://doi.org/10.1016/j.foodchem.2019.125004

Jiang Y, Duan X, Qu H et al (2016) Browning: enzymatic browning. In: Caballero B, Finglas P, Toldra F (eds) Encyclopedia of food and health. Elsevier, Amsterdam, pp 508–514. https://doi.org/10.1016/B978-0-12-384947-2.00090-8

Jiang D, Wu J, Ma L et al (2017) Fruit and vegetable compound beverage and preparation method. Chin Pat Appl CN106923110, 07 Jul 2017

Kawashima Z (1986) Preparation of odourless garlic. Jap Pat Appl JPS6191128, 09 May 1986

Keast RS, Costanzo A (2015) Is fat the sixth taste primary? Evidence and implications. Flavor 4:5. https://doi.org/10.1186/2044-7248-4-5

Kfoury M, Hadaruga NG, Hadaruga D et al (2016) Cyclodextrins as encapsulation material for flavors and aroma. Chapter 4. In: Grumezescu AM (ed) Encapsulations. Academic, Amsterdam, pp 127–192. https://doi.org/10.1016/B978-0-12-804307-3.00004-1

Kim TY, Yoon BK (2009) Composition containing extract of ginseng of which bitterness is masked. Kor Pat Appl KR100910176, 30 Jul 2009

Kim JH, Shin JK, Park SH et al (2007a) Grape seed oil which improves oxidative stability and method thereof. Kor Pat Appl KR100773050, 02 Nov 2007

Kim SH, Kim YK, Baik JH (2007b) Instant powdered green tea with excellent preference. Kor Pat Appl KR20070068630, 02 Jul 2007

Kobayashi K, Hamazaki K, Fujioka S et al (2007) The effect of n-3 PUFA/gamma-cyclodextrin complex on serum lipids in healthy volunteers - a randomized, placebo-controlled, double-blind trial. Asia Pacific J Clin Nutr 16:429–434

Komaki H (1996) Production of odourless garlic. Jap Pat Appl JPH08336370, 24 Dec 1996

Lantz AW, Rodriguez MA, Wetterer SM et al (2006) Estimation of association constants between oral malodour components and various native and derivatized cyclodextrins. Anal Chim Acta 557:184–190. https://doi.org/10.1016/j.aca.2005.10.005

Laveda F, Nunez-Delicado E, Garcia-Carmona F (2000) Reversible sodium dodecyl sulfate activation of latent peach polyphenol oxidase by cyclodextrins. Arch Biochem Biophys 379:1–6. https://doi.org/10.1006/abbi.2000.1838

Lee SH, Yu HJ, Cho NS et al (2008) A method for preparing the inclusion complex of ginseng extract with gamma-cyclodextrin, and the composition comprising the same. PCT Int Appl WO2008127063, 23 Oct 2008

Lee CM, Kim MH, Na HS (2013) The effect of caseinate on inclusion complexes of gamma-cyclodextrin for oxidative stabilization of fish oil. Biotechnol Bioprocess Eng 18:507–513. https://doi.org/10.1007/s12257-012-0752-4

Li G, Li L, Yang Y et al (2010) Asparagus beverage and preparation method thereof. Chin Pat Appl CN101828742, 15 Sep 2010

Li X, Qiao S, Cheng Y (2018) Method for preparing lemon jam by fermentation of spent edible fungus substrate. Chin Pat Appl CN107822039, 23 Aug 2018

Liang G (2020) Corn-flavored fish rice noodles. Chin Pat Appl CN110731449, 31 Jan 2020

Lin C, Chen Y (1994) Shaddock's granular and fruit juice beverage and preparing process thereof. Chin Pat Appl CN1092625, 28 Sept 1994

Lin X, Zhang B, Chen J et al (2012) Nutritive soft sweet having high DHA content yet not fishy smell and manufacture method thereof. Chin Pat Appl CN102422960, 28 Oct 2012

Lin TF, Watson S, Dietrich AM et al (2019) Taste and odour in source and drinking water: causes, controls, and consequences. IWA Publishing, London, UK

Linde GA, Junior AL, Faria EV et al (2009) Taste modification of amino acids and protein hydrolysate by α-cyclodextrin. Food Res Int 42:814–818. https://doi.org/10.1016/j.foodres.2009.03.016

Linde GA, Junior AL, Faria EV et al (2010) The use of 2D NMR to study beta-cyclodextrin complexation and debittering of amino acids and peptides. Food Res Int 43:187–192. https://doi.org/10.1016/j.foodres.2009.09.025

Linde GA, Laverde A, Colauto NB (2011) Changes to taste perception in the food industry: use of cyclodextrins. In: Preedy V, Watson R, Martin C (eds) Handbook of behavior, food and nutrition. Springer, New York. https://doi.org/10.1007/978-0-387-92271-3_8

Lindholm-Lehto PC, Vielma J (2019) Controlling of geosmin and 2-methylisoborneol induced off-flavors in recirculating aquaculture system farmed fish-a review. Aquac Res 50:9–28. https://doi.org/10.1111/are.13881

Liu Z (2018) Composite health-care pastry containing globe artichoke and preparation method of pastry. Chin Pat Appl CN108617732, 09 Oct 2018

Liu S, Hua J, Luo X et al (2012) Method for producing mulberry leaf oral liquid. Chin Pat Appl CN102429146, 02 May 2012

Liu S, Sun Z, Liu J et al (2017) Production method of deodorized instant kelp noodles. Chin Pat Appl CN106616568, 10 May 2017

Liu Y, Xiao Y, Liu Y et al (2020) Technical method for deodorizing radish juice and obtained radish juice. Chin Pat Appl CN111387399, 10 Jul 2020

López-Nicolás JM, García-Carmona F (2007) Use of cyclodextrins as secondary antioxidants to improve the color of fresh pear juice. J Agric Food Chem 55:6330–6338. https://doi.org/10.1021/jf070819x

López-Nicolás JM, Pérez-López AJ, Carbonell-Barrachina Á et al (2007a) Use of natural and modified cyclodextrins as inhibiting agents of peach juice enzymatic browning. J Agric Food Chem 55:5312–5319. https://doi.org/10.1021/jf070499h

López-Nicolás JM, Núñez-Delicado E, Sánchez-Ferrer A (2007b) Kinetic model of apple juice enzymatic browning in the presence of cyclodextrins: the use of maltosyl-β-cyclodextrin as secondary antioxidant. Food Chem 101:1164–1171. https://doi.org/10.1016/j.foodchem.2006.03.018

López-Nicolás JM, Pérez-López AJ, Carbonell-Barrachina Á et al (2007c) Kinetic study of the activation of banana juice enzymatic browning by the addition of maltosyl-β-cyclodextrin. J Agric Food Chem 55:9655–9662. https://doi.org/10.1021/jf0713399

López-Nicolás JM, Rodríguez-Bonilla P, García-Carmona F (2013) Cyclodextrins and antioxidants. Crit Rev Food Sci Nutr 54:251–276. https://doi.org/10.1080/10408398.2011.582544

Lovell RT (1983) Off-flavors in pond-cultured channel catfish. Water Sci Technol 15:67–73. https://doi.org/10.2166/wst.1983.0131

Lu X, Li Y (2005) Method of preparing debitterized and astringency removed liquor from bergamot. Chin Pat Appl CN 1635095, 07 Jun 2005

Lyu Q, Li J, Lu P et al (2017) Degassing and deodorizing processing technology for goose livers. Chin Pat Appl CN106509592, 22 Mar 2017

Ma J (2014) Compound fruit jam and production technique thereof. Chin Pat Appl CN103478520, 01 Jan 2014

Ma C, Huang Q, Gao Y (2011) Preparation method of bitter removed ponkan juice. Chin Pat Appl CN102204705, 05 Oct 2011

MacLeod G, Ames J, Betz NL (1988) Soy flavor and its improvement. Crit Rev Food Sci Nutr 27:219–399. https://doi.org/10.1080/10408398809527487

Malanga M, Szemán J, Fenyvesi É (2016) "Back to the future": a new look at hydroxypropyl beta-cyclodextrins. J Pharm Sci 105:2921–2931

Mamba BB, Krause RW, Malefetse TJ et al (2007) Removal of geosmin and 2-methylisorboneol (2-MIB) in water from Zuikerbosch Treatment Plant (Rand Water) using β-cyclodextrin polyurethanes. Water SA 33:223–227. https://doi.org/10.4314/wsa.v33i2.49066

Manta C, Peralta-Altier G, Gioia L et al (2013) Synthesis of a thiol-β-cyclodextrin, a potential agent for controlling enzymatic browning in fruits and vegetables. J Agric Food Chem 61:11603–11609. https://doi.org/10.1021/jf403063s

Martina K, Binello A, Lawson D et al (2013) Recent applications of cyclodextrins as food additives and in food processing. Curr Nutr Food Sci 9:167–179. https://doi.org/10.2174/1573401311309030001

Martínez-Hernández GB, Álvarez-Hernández MH, Artés-Hernández F (2019) Browning control using cyclodextrins in high pressure-treated apple juice. Food Bioprocess Technol 12:694–703. https://doi.org/10.1007/s11947-019-2242-6

Masaru M, Konno A (1982) Citrus food containing cyclodextrin. US4332825, 1 Jun 1982

Matencio A, Navarro-Orcajada S, García-Carmona F et al (2020) Applications of cyclodextrins in food science. A review. Trends Food Sci Technol 104:132–143. https://doi.org/10.1016/j.tifs.2020.08.009

McBride C, Qi H, Hedges A (2002) Flavor stabilization in foods. PCT Int Appl WO 2002049455, 27 Jun 2002

Mesquita VLV, Queiroz C (2013) Enzymatic browning. In: Eskin MNA, Shahidi F (eds) Biochemistry of foods, pp 387–418. https://doi.org/10.1016/b978-0-08-091809-9.00010-8

Mhlanga SD, Mamba BB, Krause RW et al (2007) Removal of organic contaminants from water using nanosponge cyclodextrin polyurethanes. J Chem Technol Biotechnol 82:382–388. https://doi.org/10.1002/jctb.1681

Miyashita K, Uemura M, Hosokawa M (2018) Effective prevention of oxidative deterioration of fish oil: focus on flavor deterioration. Annu Rev Food Sci Technol 9:209–226. https://doi.org/10.1146/annurev-food-030117-012320

Mongkolkul P, Rodart P, Pipatthitikorn T et al (2006) Debittering of tangerine Citrus reticulata blanco juice by β-cyclodextrin polymer. J Incl Phenom Macrocycl Chem 56:167–170. https://doi.org/10.1007/s10847-006-9078-1

Moritaka S, Yamamoto H (1981) Vitamin B1-enriched rice and enriched cleaned barley. Jap Pat Appl JPS56131349, 14 Oct 1981

Murakami F, Minami M (1992) Soya saponin-containing food and soya saponin pharmaceutical preparation. Jap Pat Appl JPH04364130, 16 Dec 1992

Murakawa K, Muraoka H, Kiyama H (1985) Preparation of dyestuff of safflower green powder. Jap Pat Appl JPS60156761, 16 Aug 1985

Na HS, Kim JN, Kim JM et al (2011) Encapsulation of fish oil using cyclodextrin and whey protein concentrate. Biotechnol Bioprocess Eng 16:1077–1082. https://doi.org/10.1007/s12257-011-0099-2

Nerenberg R, Rittmann BE, Soucie WJ (2000) Ozone/biofiltration for removing MIB and geosmin. J Am Water Works Assoc 92:85–95. https://doi.org/10.1002/j.1551-8833.2000.tb09073.x
Norton JL (2003) Efficacy of odor scavengers in reducing odor compounds in water, milk, and soymilk. Thesis submitted to the Faculty of the Virginia Polytechnic Institute and State University. https://vtechworks.lib.vt.edu/handle/10919/45094. Accessed 14 Dec 2020
Odor Observatory (2020) https://odourobservatory.org/measuring-odour/gas-chromatography-olfactometry/. Accessed 24 Nov 2020
Ohnishi M, Matsubara T (1996) Complex formation between sugar ligands and catechol. Studies on the fluorescence titration with TNS and its effect on the polyphenol oxidase-catalyzed browning reaction. Starch/Staerke 48:223–238. https://doi.org/10.1002/star.19960480607
Oishi M, Sumiyoshi H, Yamamoto M (2004) Food or beverage and method for producing the same. Jap Pat Appl JP2004305116, 04 Nov 2004
Okada Y, Koizumi K, Ogata K, Ohfuji T (1989) Inclusion complexes of lipids with branched cyclodextrins. Chem Pharm Bull 37:3096–3099. https://doi.org/10.1248/cpb.37.3096
Orenes-Piñero E, Garcia-Carmona F, Sanchez-Ferrer A (2006) Latent polyphenol oxidase from quince fruit pulp (*Cydonia oblonga*): purification, activation and some properties. J Sci Food Agric 86:2172–2178. https://doi.org/10.1002/jsfa.2593
Ota T (1996) Fermented food. Jap Pat Appl JPH08149960, 11 Jun 1996
Park IK, Kim MH, Kim MK et al (2008) Microencapsulation processing method of omaga-3 fatty acid, method for making the food and food. Kor Pat Appl KR100810868, 06 Mar 2008
Pasha I, Saeed F, Sultan MT et al (2014) Recent developments in minimal processing: a tool to retain nutritional quality of food. Crit Rev Food Sci Nutr 54:340–351. https://doi.org/10.1080/10408398.2011.585254
Pegg RB, Shahidi F (2012) Off-flavors and rancidity in foods. In: Nollet LML (ed) Handbook of meat, poultry and seafood quality. Wiley, Hoboken, NJ, pp 127–139. https://doi.org/10.1002/9781118352434.ch9
Peng J (2016) Fructus schisandrae beverage with health function and preparation method of *fructus schisandrae* beverage. Chin Pat Appl CN105876580, 24 Aug 2016
Perly B, Baudin C, Gosselin P (1998) Water purification using cyclodextrin. PCT Int Appl WO9818722, 07 May 1998
Prankerd RJ, Stone HW, Sloan KB et al (1992) Degradation of aspartame in acidic aqueous media and its stabilization by complexation with cyclodextrins or modified cyclodextrins. Int J Pharm 88:189–199. https://doi.org/10.1016/0378-5173(92)90316-T
Purves D, Augustine GJ, Fitzpatrick D et al (eds) (2001) Neuroscience, 2nd edn. Oxford, Sinauer Associates. https://www.ncbi.nlm.nih.gov/books/NBK10833/
Rackis JJ, Sessa DJ, Honig DH (1979) Flavor problems of vegetable food proteins. J Am Oil Chem Soc 56:262–271. https://doi.org/10.1007/BF02671470
Rajnavölgyi É, Laczik R, Kun V et al (2014) Effects of RAMEA-complexed polyunsaturated fatty acids on the response of human dendritic cells to inflammatory signals. Beilstein J Org Chem 10:3152–3160. https://doi.org/10.3762/bjoc.10.332
Rashmawi K, Oskouie AK, Noll KE (2008) Development and evaluation of a copolymer for removal of taste and odor causing compounds from lake Michigan. Water J Environ Eng 134:531–535. https://doi.org/10.1061/(asce)0733-9372(2008)134:7(531)
Regiert M, Wimmer T, Moldenhauer JP (1996) Application of γ-cyclodextrin for the stabilization and/or dispersion of vegetable oils containing triglycerides of. J Incl Phenom Macrocycl Chem 25:213–216. https://doi.org/10.1007/BF01041571
Rodrigues E, Vaz S, Gil V et al (2002) Inclusion of polyphenol oxidase substrates in β-cyclodextrin: a 1H-NMR Study. J Incl Phenom 44:395–397. https://doi.org/10.1023/A:1023007600469
Sakai A, Yamashita M (2008) Polyphenol composition. Jap Pat Appl JP2008118933, 29 May 2008
Sapers GM (1993) Browning of foods: control by sulfites, antioxidants, and other means. Food Technol 47:75–84
Sapers GM, Hicks KB, Phillips JG et al (1989) Control of enzymatic browning in apple with ascorbic acid derivatives, polyphenol oxidase inhibitors, and complexing agents. J Food Sci 54:997–1002. https://doi.org/10.1111/j.1365-2621.1989.tb07931.x

Schmied G, Harrison M, Polchinski P (2000) Process for the stabilisation of acylglycerols comprising high amounts of ω-3 polyunsaturated fatty acids by means of gamma-cyclodextrin. EP1041136, 04 Oct 2000

Sekhon-Loodu S, Warnakulasuriya SN, Rupasinghe HP et al (2013) Antioxidant ability of fractionated apple peel phenolics to inhibit fish oil oxidation. Food Chem 140:189–196. https://doi.org/10.1016/j.foodchem.2013.02.040

Sekiguchi R, Yamazaki M, Suzuki K et al (1982) Natto. Jap Pat Appl JPS5729263, 17 Feb 1982

Shahidi F, Wanasundara UN (1995) Oxidative stability of encapsulated seal blubber oil. In: Flavor technology, ACS symp ser, vol 610. American Chemical Society, Washington, pp 139–151

Shaw PE, Wilson CW (1983) Debittering citrus juices with β-cyclodextrin polymer. J Food Sci 48:646–467. https://doi.org/10.1111/j.1365-2621.1983.tb10811.x

Shaw PE, Wilson CW (1985) Reduction of bitterness in grapefruit juice with β-cyclodextrin polymer in a continuous-flow process. J Food Sci 50:1205–1207. https://doi.org/10.1111/j.1365-2621.1985.tb13054.x

Shaw PE, Tatum JH, Wilson CW (1984) Improved flavor of navel orange and grapefruit juices by removal of bitter components with β-cyclodextrin polymer. J Agric Food Chem 32:832–836. https://doi.org/10.1021/jf00124a034

Shi X, Lv Y, Guo S (2017) Effects of heat treatment and β-cyclodextrin addition on soymilk flavor. Nongye Gongcheng Xuebao/Trans Chin Soc Agric Eng 33:293–300

Shinoda M, Ikeda G (1985) Method for eliminating bitter taste of *Fomes japonicas*. JPS60126226, 05 Jul 1985

Singh A, Idowu AT, Benjakul S et al (2020) Debittering of salmon (Salmo salar) frame protein hydrolysate using 2-butanol in combination with β-cyclodextrin: impact on some physicochemical characteristics and antioxidant activities. Food Chem 321:126686. https://doi.org/10.1016/j.foodchem.2020.126686

Sohajda T, Béni S, Varga E et al (2009) Characterization of aspartame–cyclodextrin complexation. J Pharm Biomed Anal 50:737–745. https://doi.org/10.1016/j.jpba.2009.06.010

Sojo MM, Nunez-Delicado E, Garcia-Carmona F (1999) Cyclodextrins as activator and inhibitor of latent banana pulp polyphenol oxidase. J Agric Food Chem 47:518–523. https://doi.org/10.1021/jf9808978

Suda I, Hajika M, Sato T et al (1999) Production of soybean milk using soybean lacking in lipoxygenase and not having grassy smell and bitter taste. Jap Pat Appl JPH11332496, 07 Dec 1999

Sullivan ML (2014) Beyond brown: polyphenol oxidases as enzymes of plant specialized metabolism. Front Plant Sci 5:783. https://doi.org/10.3389/fpls.2014.00783

Sumioka S (1991) Production of hand-kneaded noodle containing fruit juice. Jap Pat Appl JPH03266951, 27 Nov 1991

Suratman LLI, Jeon IJ, Schmidt KA (2004) Ability of cyclodextrins to entrap volatile beany flavor compounds in soymilk. J Food Sci 69:109–113. https://doi.org/10.1111/j.1365-2621.2004.tb15499.x

Suzuki Y, Masuno E (1994) Beverage filter and manufacture thereof. Jap Pat Appl JPH06211216, 02 Aug 1994

Szejtli J (2004) Cyclodextrins and molecular encapsulation. In: Nalwa HS (ed) Encyclopedia of nanoscience and nanotechnology, vol 2. American Scientific Publishers, Washington, pp 283–304

Szejtli J, Bánky-Elód E (1975) Inclusion complexes of unsaturated fatty acids with amylose and cyclodextrin. Staerke 27:368–376. https://doi.org/10.1002/star.19750271104

Szejtli J, Szente L (2005) Elimination of bitter, disgusting tastes of drugs and foods by cyclodextrins. Eur J Pharm Biopharm 61:115–125. https://doi.org/10.1016/j.ejpb.2005.05.006

Szente L, Fenyvesi É (2017) Cyclodextrin-lipid complexes: cavity size matters. Struct Chem 28:479–492. https://doi.org/10.1007/s11224-016-0884-9

Szente L, Szejtli J (1986) Molecular encapsulation of natural and synthetic coffee flavor with β-cyclodextrin. J Food Sci 51:1024–1027. https://doi.org/10.1021/bk-1988-0370.ch016

Szente L, Szejtli J (1988) Stabilization of flavors by cyclodextrins. In: Risch S, Reineccius GA (eds) Flavor encapsulation, ACS symposium series, vol 370. American Chemical Society, Washington, pp 148–157. https://doi.org/10.1021/bk-1988-0370.ch016

Szente L, Szejtli J (2004) Cyclodextrins as food ingredients. Trends Food Sci Technol 15:137–142. https://doi.org/10.1016/j.tifs.2003.09.019
Szente L, Szejtli J, Specht M et al (1981) A process for decreasing the phenylalanine content of protein hydrolysates and meat-aroma concentrates prepared therefrom. FR2472348, 03 Jul 1981
Szente L, Harangi J, Szejtli J (1988) Long-term storage stability studies on flavor-β-cyclodextrin complexes. In: Huber O, Szejtli J (eds) Proc int symp cyclodextrins, 4th edn. Kluwer, Dordrecht, Netherland, pp 545–549
Takada Y (1992) Functional powdery salt-free soybean fermented food having. Jap Pat Appl JPH0451860, 20 Feb 1992
Takatsu H, Otsuka K (2006) Producing process of packaged green tea beverage. Jap Pat Appl JP2006174748, 06 Jul 2006
Takeo C, Kinugasa H (1989) Elimination of retort odour in green tea beverage. Jap Pat Appl JPH01174328, 10 Jul 1989
Takeshita M (1987) Method for mitigating bitter taste of extract of *Gynostemma pentaphyllum*. Jap Pat Appl JPS62151166, 06 Jul 1987
Tamaki K, Tamaki T, Suzuki Y (2007) Deodorisation of off-odour during sweet potato juice production by employing physical and chemical deodorants. Food Chem 105:454–541. https://doi.org/10.1016/j.foodchem.2007.03.064
Tamamoto LC, Schmidt SJ, Lee SY (2010) Sensory properties of ginseng solutions modified by masking agents. J Food Sci 75:S341–S347. https://doi.org/10.1111/j.1750-3841.2010.01749.x
Tanaka T, Uesawa Y (1991) Food deodorizer and deodorization of food using the same. Jap Pat Appl JPH03219842, 27 Sep 1991
Tao L, Liu S, Zhu F et al (2011) Purple sweet potato and oat bud juice beverage, and preparation method thereof. Chin Pat Appl CN102204700, 05 Oct 2011
Tarasov A, Rauhut D, Jung R (2017) “Cork taint” responsible compounds. Determination of haloanisoles and halophenols in cork matrix: a review. Talanta 175:82–92. https://doi.org/10.1016/j.talanta.2017.07.029
Terao K, Nakada D (2006) Method for reducing bitter taste of catechin-containing composition. Jap Pat Appl JP2006115772, 11 May 2006
Terao K, Inoue Y, Kaneda T (2003) Composition containing *Gymnema sylvestre* extract and having reduced bitterness. Jap Pat Appl JP2003095986, 03Apr 2003
Terao K, Nakada D, Konishi M et al (2007) Method for producing powdered milk. Jap Pat Appl JP 2007275029, 25 Oct 2007
Tian B, Xiao D, Hei T et al (2020) The application and prospects of cyclodextrin inclusion complexes and polymers in the food industry: a review. Polym Int 69:597–603. https://doi.org/10.1002/pi.5992
Tomás-Barberán FA, Espín JC (2001) Phenolic compounds and related enzymes as determinants of quality in fruits and vegetables. J Sci Food Agric 81:853–876. https://doi.org/10.1002/jsfa.885
Trimmer C, Keller A, Murphy NR et al (2019) Genetic variation across the human olfactory receptor repertoire alters odor perception. Proc Natl Acad Sci 116:9475–9480. https://doi.org/10.1073/pnas.1804106115
Ujházi A, Szejtli J (1989) Removal of naringin from aqueous solution with cyclodextrin bead polymer. Gordian 3:43–45
Upreti M, Prakash I, Chen YL (2011) Solubility enhanced terpene glycoside(s). PCT Int Appl WO2011097620, 11 Aug 2011
Urata T, Urata Y (2011) Composition for oral ingestion. Jap Pat Appl JP2011162539, 25 Aug 2011
van Ruth SM (2001) Methods for gas chromatography-olfactometry: a review. Biomol Eng 17:121–128. https://doi.org/10.1016/S1389-0344(01)00070-3
Vlasov YU, Legin A (1998) Non-selective chemical sensors in analytical chemistry from “electric nose” to “electric tongue”. Fresenius J Anal Chem 361:255–260
Wada K, Yamamoto T (2012) Manufacture of inclusion compound of hop-derived component and cyclodextrin, and manufacture of food containing hop-derived component. Jap Pat Appl JP2012019695, 02 Feb 2012

Wagner CJ, Wilson CW, Shaw PE (1988) Reduction of grapefruit bitter components in a fluidized β-cyclodextrin polymer bed. J Food Sci 53:516–518. https://doi.org/10.1111/j.1365-2621.1988.tb07745.x

Wang J (2012) Flavor protection method of pomegranate juice. Chin Pat Appl CN102783684, 21 Nov 2012

Wang SH, Wang JC, Zhai B et al (2011a) The discussion about the function of cyclodextrin eliminate the astringency of the lotus root juice. J Qinghai Norm Univ (Nat Sci Ed). http://en.cnki.com.cn/Article_en/CJFDTOTAL-QHSD201101013.htm. Accessed 09 Dec 2020

Wang J, Shen X, Yue M et al (2011b) Method for preparing ready-to-eat mussel flesh products. Chin Pat Appl CN102232580, 09 Nov 2011

Wang Z, Zhang W, Fang Y et al (2015) Appetizing beef granules and preparation method thereof. Chin Pat Appl CN105077308, 25 Nov 2015

Wang J, Gu Z, Yang R et al (2016) Barley green flavor protection technology and product thereof. Chin Pat Appl CN105249140, 20 Jan 2016

Wang K, Dai L, Zhang B et al (2018a) Processing technique for preparing debitterized lemon seed oil by cold pressing. Chin Pat Appl CN107794114, 13 Mar 2018

Wang H, Fu K, Yuan Y, Zhou W, Qian F (2018b) Odour-free DHA solid beverage capable of promoting mentality and invigorating brain and preparation method thereof. Chin Pat Appl CN108041394, 18 May 2018

Wang S, Chen HC, Sun B (2020) Recent progress in food flavor analysis using gas chromatography–ion mobility spectrometry (GC–IMS). Food Chem 315:126158. https://doi.org/10.1016/j.foodchem.2019.126158

Watson MA, Lea JM, Bett-Garber KL (2017) Spray drying of pomegranate juice using maltodextrin/cyclodextrin blends as the wall material. Food Sci Nutr 5:820–826. https://doi.org/10.1002/fsn3.467

Wilson AD, Baietto M (2009) Applications and advances in electronic-nose technologies. Sensors (Basel) 9:5099–5148. https://doi.org/10.3390/s90705099

Wilson CW, Wagner CJ, Shaw PE (1989) Reduction of bitter components in grapefruit and navel orange juices with beta-cyclodextrin polymers or XAD resins in a fluidized bed process. J Agric Food Chem 37:14–18. https://doi.org/10.1021/jf00085a004

Wimmer T, Regiert M, Moldenhauer JP (1997) Stabilization and dispersion of plant oils containing polyunsaturated fatty acid residues by complexation with γ-cyclodextrin. Germ Pat Appl DE19612658, 02 Oct 1997

Wood WE, Beaverson NJ (1996) Barrier packaging films from a thermoplastic and a compatible cyclodextrin derivative. PCT Int Appl WO9600260, 04 Jan 1996

Wood WE, Beaverson NJ (1998) Rigid polymeric beverage bottles with improved resistance to permeant elution. US Pat US5837339, 17 Nov 1998

Wu F, Li H, Zhao L (2014) Food or feed additive produced by debitterizing orange peel, flavedo and shaddock peel, and preparation method of food or feed additive. Chin Pat Appl CN103783279, 14 May 2014

Xiao J, Cong R, Ma F et al (2019) Hemp seed oil microcapsule and preparation method and application thereof. Chin Pat Appl CN109549191, 02 Apr 2019

Xu K, Qi Q (2001) Whole soft-shelled turtle product without fishy smell and its production and storage method. Chin Pat Appl CN1294873, 16 May 2001

Xu Y, Yin J, Yuan H et al (2011) Processing method of high catechin beverage with low bitter taste. Chin Pat Appl CN102265953, 07 Dec 2011

Xu M, Jin Z, Lan Y et al (2019) HS-SPME-GC-MS/olfactometry combined with chemometrics to assess the impact of germination on flavor attributes of chickpea, lentil, and yellow pea flours. Food Chem 280:83–95. https://doi.org/10.1016/j.foodchem.2018.12.048

Xu W, Liang M, Su W (2020) Solid dispersion of celery seed oil and preparation method of solid dispersion. Chin Pat Appl CN111249322, 09 Jun 2020

Xun Y, Weng M (2014) Method for producing dried shrimp meat slice from low-value shrimp meat having odor. Chin Pat Appl CN103829284, 04 Jun 2014

Yamamoto M, Takeda F, Maruo T (1991) Processed Japanese horseradish product. Jap Pat Appl JPH0367559, 22 Mar 1991

Yang C (2019) Asparagus celery juice composite functional beverage and preparation method thereof. Chin Pat Appl CN110495539, 26 Nov 2019

Yang S, Mao XY, Li FF et al (2012a) The improving effect of spray-drying encapsulation process on the bitter taste and stability of whey protein hydrolysate. Eur Food Res Technol 235:91–97. https://doi.org/10.1007/s00217-012-1735-6

Yang X, Liu Y, Dang B et al (2012b) Preparation method of broad bean polypeptide wine. Chin Pat Appl CN102443516, 09 May 2012

Yang X, Yang F, Liu Y et al (2020) Off-flavor removal from thermal-treated watermelon juice by adsorbent treatment with β-cyclodextrin, xanthan gum, carboxymethyl cellulose sodium, and sugar/acid. LWT Food Sci Technol:109775. https://doi.org/10.1016/j.lwt.2020.109775

Yao H, Jia T, Chen X et al (2017) Squab amino acid oral liquid, preparation method and application thereof. Chin Pat Appl CN107095185, 29 Aug 2017

Yi, H. (2004) Method for removing bitter and astringent taste of ginkgo extract and its use. Chin Pat Appl CN1554276, 15 Dec 2004

Yoo IJ, Lee KI, Lee EY, Kwak JY (2010) Efficient method for masking off flavor of perilla oil added to lactic acid bacteria beverage. Kor Pat Appl KR20100091270, 19 Aug 2010

Yoshii H, Furuta T, Yasunishi A et al (1996) Oxidation stability of eicosapentaenoic and docosahexaenoic acid included in cyclodextrins. J Incl Phenom Macrocycl Chem 25:217–220. https://doi.org/10.1007/BF01041572

Young OA, Gupta RB, Sadooghy-Saraby S (2012) Effects of cyclodextrins on the flavor of goat milk and its yogurt. J Food Sci 77:122–127. https://doi.org/10.1111/j.1750-3841.2011.02557.x

Yumoto T, Gunji Y, Takahashi T et al (1989) Removal of bitterness and astringency of essence of leaf of *Gymnema Sylvestre*. Br Jap Pat Appl JPH01102028, 19 Apr 1989

Zhang M, Wang T, Liu Y et al (2016) Gentle type yellow mustard sauce easy to smear and preparation method thereof. Chin Pat Appl CN105231413, 13 Jan 2016

Zhang G, Gu L, Lu Z et al (2019) Browning control of fresh-cut Chinese yam by edible coatings enriched with an inclusion complex containing star anise essential oil. RSC Adv 9:5002–5008. https://doi.org/10.1039/C8RA08295B

Zhang M, Li L, Ye Y et al (2020) Debittering processing and moisture-absorption resisting method of lyophilized citrus paradise. Chin Pat Appl CN111067000, 28 Apr 2020

Zhao J, Du J (2015) Pomegranate seed oil microcapsule and preparation method thereof. Chin Pat Appl CN104622921, 20 May 2015

Zhao Z, Luo Y, Xie W et al (2015) Preparation method of selenium-enriched kiwifruit yogurt ice cream. Chin Pat Appl CN105028888, 11 Nov 2015

Zhao Z, Li B, Jin Y et al (2019) Barbecue-flavored tiny dried fish and preparation method thereof. Chin Pat Appl CN110013006, 16 Jul 2019

Zheng J, Tian H, Xiang H et al (2019) Solid peony seed oil and processing method thereof. Chin Pat Appl CN109805110, 28 May 2019

Zhou Y (2018) Preparation method of soybean active peptide. CN108220374, 29 Jun 2018

Zhu D, Damodaran S (2018) Removal of off-flavor-causing precursors in soy protein by concurrent treatment with phospholipase A 2 and cyclodextrins. Food Chem 264:319–325. https://doi.org/10.1016/j.foodchem.2018.05.045

Zhu J, Dong X, Qi H et al (2020) Sweet corn seaweed vegetable granules and preparation method thereof. Chin Pat Appl CN111357922, 03 Jul 2020

Chapter 13
Alpha-Cyclodextrin Functions as a Dietary Fiber

Keita Chikamoto and Keiji Terao

13.1 Introduction

Dietary fiber is recognized to be an essential nutrient for humans. It can be characterized according to molecular size, solubility, fermentability, and source, all of which modify its effects on health (Tungland and Meyer 2002; Anderson et al. 2009). Numerous clinical studies have shown that dietary fiber has health-promoting effects, such as assisting with glycemic control, restricting serum lipid concentration, and improving gut health and the gut environment (Aller et al. 2004; Holscher 2017). Dietary fiber is defined as plant material that cannot be digested by human enzymes, and includes lignin, cellulose, and polysaccharides. Furthermore, in recent years, indigestible oligosaccharides and starch have also considered as dietary fibers (Lattimer and Haub 2010).

CDs are cyclic oligosaccharides that are composed of glucopyranose molecules linked by α-1,4-glycosidic bonds (Li et al. 2014). There are three native CDs that are classified according to their number of glucopyranose units; α-CD, which has six units; β-CD, which has seven units; and γ-CD, which has eight units. CDs are used as food additives and in pharmaceutical formulations because they improve the stability and water solubility of guest molecules by forming complexes with them (Li et al. 2014). CDs are generally recognized as safe food additives, and although the acceptable daily intakes (ADIs) of α-CD and γ-CD have not been specified by

K. Chikamoto (✉)
CycloChem Bio Co., Ltd., Kobe, Hyogo, Japan
e-mail: keita.chikamoto@cyclochem.com

K. Terao
CycloChem Bio Co., Ltd., Kobe, Hyogo, Japan

Department of Social/Community Medicine and Health Science, Food and Drug Evaluation Science, Graduate School of Medicine, Kobe University, Kobe, Hyogo, Japan
e-mail: keiji.terao@cyclochem.com

T. M. Ho et al. (eds.), *Functionality of Cyclodextrins in Encapsulation for Food Applications*, https://doi.org/10.1007/978-3-030-80056-7_13

the Joint Food and Agriculture Organization/World Health Organization Expert Committee on Food Additives, the ADI of β-CD has been specified to be 5 mg/kg/day. α-CD and β-CD, unlike γ-CD, are not digested by human enzymes, and therefore reach the large intestine when ingested (Antenucci and Palmer 1984), where α-CD in particular is degraded by bacteria and has a beneficial effect on the local gut microbiota (Ommen et al. 2004).

In recent years, the use of α-CD as a type of dietary fiber has been studied because it can both form complexes with guest molecules and improve the composition of the gut microbiota. In this chapter, we describe new findings regarding the health-promoting effects of α-CD, as a type of dietary fiber, and consider its effects in the small and large intestine separately.

13.2 Effects of α-CD in the Small Intestine

In the small intestine, α-CD is not readily degraded by digestive enzymes and few bacteria can use α-CD. Therefore, the effects of α-CD in the small intestine are mediated by complex formation with guest molecules. Several studies have shown that α-CD can ameliorate or prevent metabolic disease by reducing the over-absorption of nutrients *via* the encapsulation of digestive enzymes, phospholipids, and food ingredients. In this section, we describe the effects of α-CD supplementation on the post-prandial increases in blood glucose, fat, and cholesterol concentrations, which are strongly associated with metabolic diseases.

13.2.1 The Prevention of Postprandial Hyperglycemia

Ingested carbohydrates are decomposed to form glucose by digestive enzymes, and the absorption of the liberated glucose increases the blood glucose concentration. Therefore, excessive carbohydrate intake causes postprandial hyperglycemia, which increases the risk of metabolic disease, including diabetes, obesity, and arteriosclerosis (Cavalot et al. 2006). Several studies have shown that dietary supplementation with α-CD suppresses postprandial hyperglycemia. For example, Buckley et al. (2006) reported a suppressive effect of α-CD on the increase in blood glucose following the consumption of boiled white rice by healthy individuals. Furthermore, we previously evaluated the effect of the pre-administration of α-CD on the increase in blood glucose that occurs following sucrose administration to healthy mice (Fig. 13.1), and found that it reduced this in a dose-dependent manner.

One mechanism by which α-CD may reduce the postprandial increase in blood glucose concentration might be an inhibition of the enzymes that digest carbohydrate in the small intestine. Ingested starch and sucrose are decomposed by enzymes including α-amylase and sucrase to liberate monosaccharides, which are absorbed.

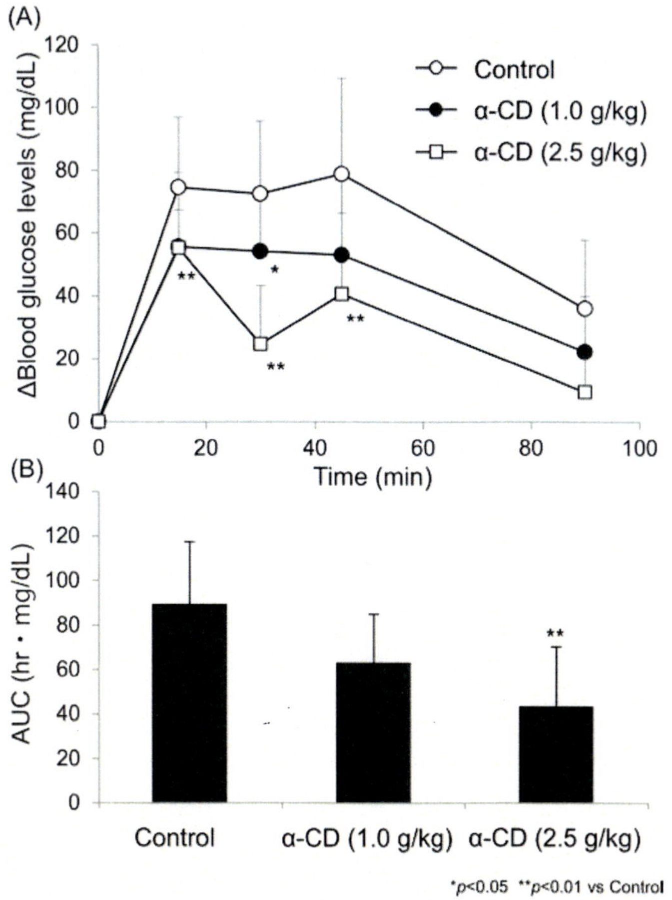

Fig. 13.1 Effects of the oral administration of sucrose and α-CD on the blood glucose concentration (**a**) and area under the blood glucose curve (**b**) in the healthy mice

Therefore, the inhibition of these enzymes would limit the postprandial increase in blood glucose concentration, and potentially therefore improve glycemic control (Agarwal and Gupta 2016). α-CD has been shown to inhibit the starch-degrading enzyme α-amylase. Additionally, Jo et al. (2011) reported that α-CD inhibits the sucrose-degrading enzyme sucrase. Larson et al. (2010) revealed that α-CD interacts with an amino acid residue in the active site of pig α-amylase by X-ray diffraction analysis of an α-amylase-α/CD mixture. Furthermore, Gentilcore et al. (2011) reported that high doses of α-CD inhibit carbohydrate absorption by delaying the rate of gastric emptying.

These findings imply that α-CD supplementation might have beneficial effects on glycemic control.

13.2.2 Prevention of Postprandial Hypertriglyceridemia

Obesity is associated with various other diseases, and in particular cardiovascular disease and diabetes, and dietary lipid restriction plays a very important role in body weight management (Astrup 2005). The inhibition of lipid absorption has the potential to reduce body weight, and several studies have shown that α-CD consumption inhibits post-prandial hypertriglyceridemia. Jarosz et al. (2013) reported that the consumption of 2 g of α-CD reduces the hypertriglyceridemia induced by the consumption of high-fat meal. Furthermore, Artiss et al. (2006) reported that 4 weeks of intake of α-CD prevented body weight gain, reduced serum triglyceride concentration, and increased fecal fat loss in high-fat diet (HFD)-fed rats.

Ingested triglycerides are degraded to liberate fatty acids by lipase in the small intestine. These fatty acids are incorporated into bile acid micelles, facilitating their absorption. The mechanism of the serum triglyceride-lowering effect of α-CD may involve three means of inhibiting lipid absorption. The first is a reduction in the formation of bile acid micelles because of the inclusion and precipitation of lecithin, the micelle-forming component of bile. Additionally, α-CD forms insoluble complexes by encapsulating fatty acids and the fatty acid components of triglyceride, thereby inhibiting their absorption. Furune et al. (2014) showed that α-CD reduced the solubility of cholesterol in artificial intestinal fluid that mimicked the post-prandial environment of the gut by forming a complex with lecithin. Interestingly, α-CD was the most effective form of water-soluble dietary fiber that was tested.

The second potential means of inhibiting lipid absorption was reported by Artiss et al. (2006). They showed that α-CDs inhibit the degradation of triglycerides by lipase because the micelles formed by α-CD and triglycerides are highly stable. Additionally, 1 g of α-CD was shown to bind 9 g of triglyceride, which may promote fecal triglyceride loss. Furthermore, we have previously shown that the particle size of micelles that are composed of α-CD and triglycerides increases according to the amount of triglyceride incorporated (Fig. 13.2).

The third potential means of inhibiting lipid absorption is the direct precipitation of fatty acids in bile acid micelles. In this way, α-CD may have a selective effect to

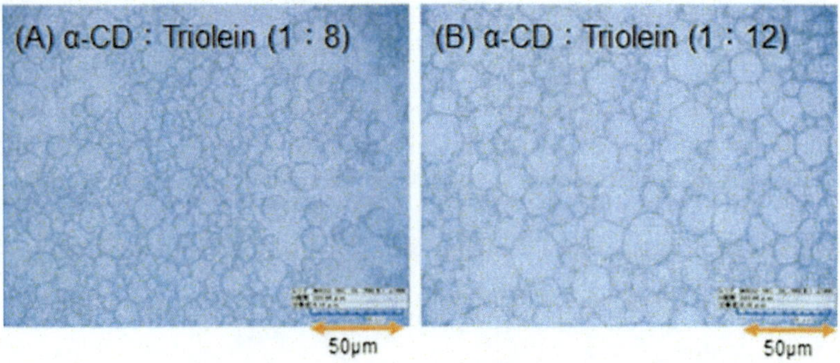

Fig. 13.2 Effect of α-CD on micelle formation by triolein 1:8 (**a**) and 1:12 (**b**)

promote the fecal loss of long-chain saturated fatty acids. Gallaher et al. (2007) showed that α-CD may selectively promote the loss of saturated triglycerides, rather than unsaturated triglycerides. To determine the mechanism of this effect, we evaluated the interaction of α-CD with various fatty acids in artificial fed state simulated intestinal fluid. We added α-CD to fluid containing 1.5 mM fatty acids at 37 °C and stirred this for 2 h. Then, after filtration of the mixture, the fatty acid composition was assessed using high-performance liquid chromatography. Study of the interactions of α-CD with the fatty acids showed that it reduced the solubility of stearic acid (C18:0) more than that of palmitic acid (C16:0) (Fig. 13.3a), and did not affect

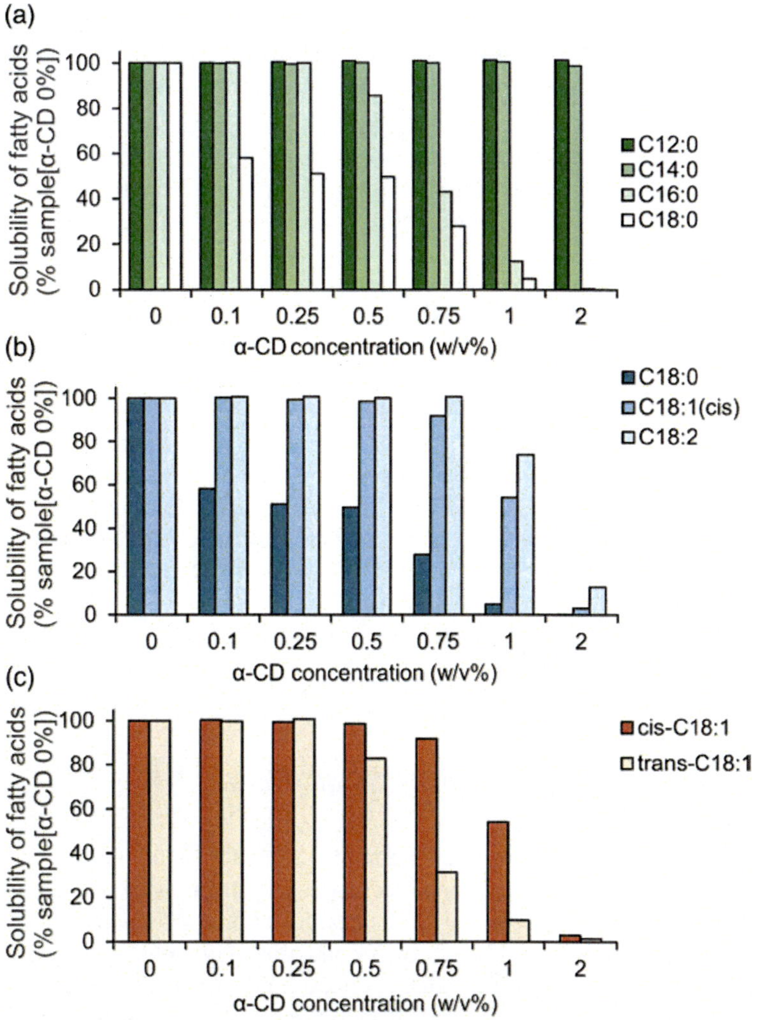

Fig. 13.3 Effect of α-CD on the solubility of fatty acids of differing length (**a**) saturation (**b**), and structure (**c**)

the solubilities of myristic acid (C14:0) or lauric acid (C12:0). Next, we compared the interactions of α-CD with saturated fatty acids and unsaturated fatty acids, and found that α-CD reduced the solubility of a saturated fatty acid (C18:0) more than those of unsaturated fatty acids (C18:1 and C18:2) (Fig. 13.3b). Finally, we compared the interactions of α-CD with trans-fatty acids and cis-fatty acids, and found that α-CD inhibited the solubility of elaidic acid (trans-C18:1) more than that of oleic acid (cis-C18:1) (Fig. 13.3c).

These results suggest that α-CD selectively interacts with fatty acids, depending on their structure.

13.2.3 Inhibition of Cholesterol Absorption

Serum low-density lipoprotein (LDL)-cholesterol concentration is clinically used as a predictor of hyperlipidemia and atherosclerosis. Several studies have shown that the intake of α-CD reduces serum cholesterol concentration. Comerford et al. (2012) reported that dietary supplementation with 6 g/day of α-CD reduced the serum total and LDL-cholesterol concentrations, as well as reducing body weight in healthy overweight individuals. Furthermore, Grunberger et al. (2007) reported that supplementation with α-CD at the same dose for 3 months reduced LDL-cholesterol concentration in individuals with obesity and diabetes.

Ezetimibe is a drug that suppresses the absorption of cholesterol in the small intestine, thereby reducing LDL-cholesterol in individuals with high LDL-cholesterol concentrations (Knopp et al. 2003). Therefore, the cholesterol-lowering effect of α-CD may be mediated through the inhibition of intestinal cholesterol absorption. α-CD reduces the solubility of lipids in small intestinal fluid (Sect. 13.2.2), but it is difficult for α-CD to directly encapsulate cholesterol because its pores are smaller than cholesterol molecules. Instead, α-CD reduces the solubility of cholesterol indirectly, by the inclusion of lecithin (Furune et al. 2014). These findings imply that α-CD reduces circulating LDL-cholesterol by inhibiting cholesterol absorption after a meal.

13.2.4 Reduction in Circulating Small, Dense LDL

Small, dense LDL particles are smaller than those of LDL, and higher circulating concentrations of this type have been identified in patients with atherosclerosis (Ivanova et al. 2017). Small, dense LDL is ideal for atheroma formation because it is easily oxidized, has a long plasma half-life, and can readily infiltrate vessel walls. A recent study showed that the circulating concentration of small, dense LDL is a more sensitive marker of atherosclerosis than LDL-cholesterol (Ivanova et al. 2017). Amar et al. (2016) reported a beneficial effect of α-CD on cholesterol particle size in healthy individuals. In their study, 75 participants had their diets supplemented with 6 g/day of α-CD and their circulating lipid concentrations were measured at baseline and after 3 months. Supplementation with α-CD reduced the number of small LDL particles more effectively than placebo, although there was no change in LDL-cholesterol concentration. With respect to the mechanism, Hirano et al. (2005) found that the small, dense LDL-cholesterol concentration is higher in patients with type 2 diabetes than in individuals with normal circulating lipid concentrations. Furthermore, Lemieux et al. (2000) found that the small, dense LDL concentration positively correlates with the postprandial triglyceride concentration of healthy individuals. These results suggest that α-CD may reduce the circulating small, dense LDL concentration in association with improving glycemic control and reducing the postprandial circulating triglyceride concentration.

13.3 Effects of α-CD in the Large Intestine

Some beneficial types of dietary fiber are not degraded by human digestive enzymes but are instead used by bacteria in the large intestine. These types of dietary fiber are referred to as "prebiotics", which are defined as "selectively fermented ingredients that allow specific changes, both in the composition and/or activity of the gastrointestinal microflora that confer benefit(s) upon host wellbeing and health" (Gibson et al. 2010). Well-known examples include inulin and fructo-oligosaccharide. α-CD is digested not by human digestive enzymes, but by gut bacteria (Antenucci and Palmer 1984). Therefore, in recent years, the prebiotic effects of α-CD have been extensively studied. Interestingly, the prebiotic effects of α-CD have been shown to differ from those of other prebiotics. In this section, we describe the various health-promoting effects of α-CD that are mediated by its prebiotic status and have been identified in recent studies.

13.3.1 Effects on the Gut Microbiota

Cyclodextrinase (EC 3.2.1.54), also known as cyclomaltodextrinase, is the type of α-amylase that can digest α-CD, and which is expressed by certain intestinal bacteria. α-CD is converted to maltooligosaccharides by this enzyme, and then pyruvate, *via* a modified Embden-Meyerhof pathway (Labes and Schönheit 2007). Antenucci and Palmer (1984) reported that 24 *Bacteroides* strains were able to degrade α-CD by analyzing their growth in the presence of CD. Chun et al. (2003) showed that *Bacillus* species can also utilize α-CD using Biolog® microplate assays. Furthermore, we have previously shown that some *Lactobacillus* species can utilize α-CD (Jo et al. 2007). However, *Bifidobacterium* cannot readily utilize α-CD because it does not express cyclodextrinase (Turroni et al. 2012). Several other studies have shown that the consumption of α-CD affects these constituents of the gut microbiota.

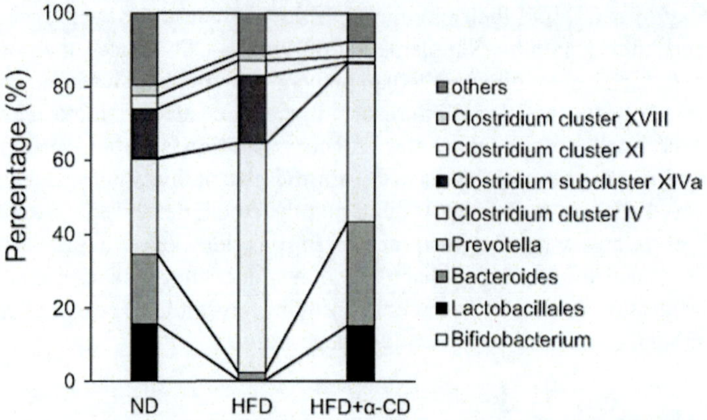

Fig. 13.4 Bacterial community dynamics in the feces of mice fed α-CD, a normal diet (ND), or a high-fat diet (HFD). (Adapted with permission from Nihei et al. 2018)

Sasaki et al. (2018) reported that α-CD may increase the metabolic activity of *Bacteroides* using an *in vitro* human gut microbiota model that was created by batch fermentation, starting from a fecal inoculum. Furthermore, Nihei et al. (2018) reported that α-CD increases the proportions of *Bacteroides* and *Lactobacillus* in the feces of HFD-fed mice (Fig. 13.4).

Short-chain fatty acids (SCFAs) are major bacterial metabolites of prebiotics, including α-CD (Besten et al. 2013). Therefore, we evaluated the production of SCFAs during a 24-h culture of *Bacteroides thetaiotaomicron* with α-CD, dextran (DXR), or dextrin (DEX), and found that α-CD is associated with the generation of more SCFAs than DXR or DEX (Fig. 13.5). We also administered healthy rats α-CD or another prebiotic, lactosucrose (LS), for 8 weeks, and found that α-CD increased the numbers of *Lactobacillus* and *Bacteroides*, whereas LS increased the number of *Bifidobacterium* in the cecal contents (Fig. 13.6). Furthermore, α-CD administration was associated with the production of larger amounts of lactate and SCFAs than LS (Table 13.1).

The mechanism whereby α-CD administration increases SCFAs production is related to the bacterial species that are targeted by prebiotics. Macfarlane and Macfarlane (2003) reported that *Bacteroides* spp. can produce SCFAs, especially acetate and propionate, while *Bifidobacterium* spp. mainly produce acetate from glucose. *Lactobacillus* spp. are well-known probiotics that are associated with better host immunity and less allergic disease. Lactate is produced by *Lactobacillus* in rats (Siebold et al. 1995) and is converted to butyrate by other bacteria (Bourriaud et al. 2005). This suggests that α-CD administration increases SCFA production by targeting SCFA-producing bacteria, such as *Bacteroides* and *Lactobacillus*.

In summary, recent findings indicate that the increase in SCFAs induced by α-CD administration has various health-promoting effects, as described in more detail below.

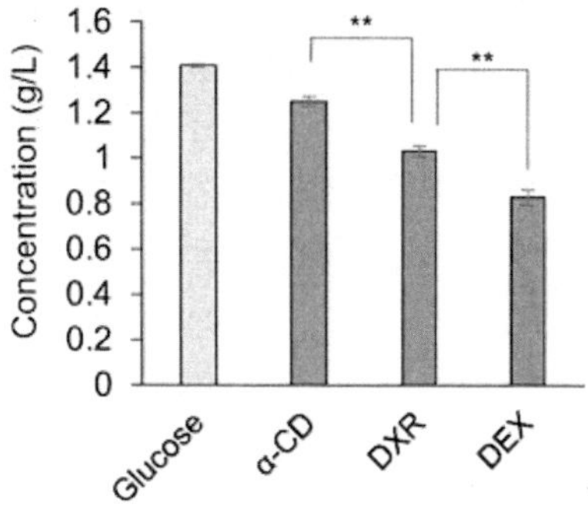

Fig. 13.5 Total SCFA concentration in cultures of *Bacteroides thetaiotaomicron* after a 24-h incubation with glucose, α-CD, DXR or DEX

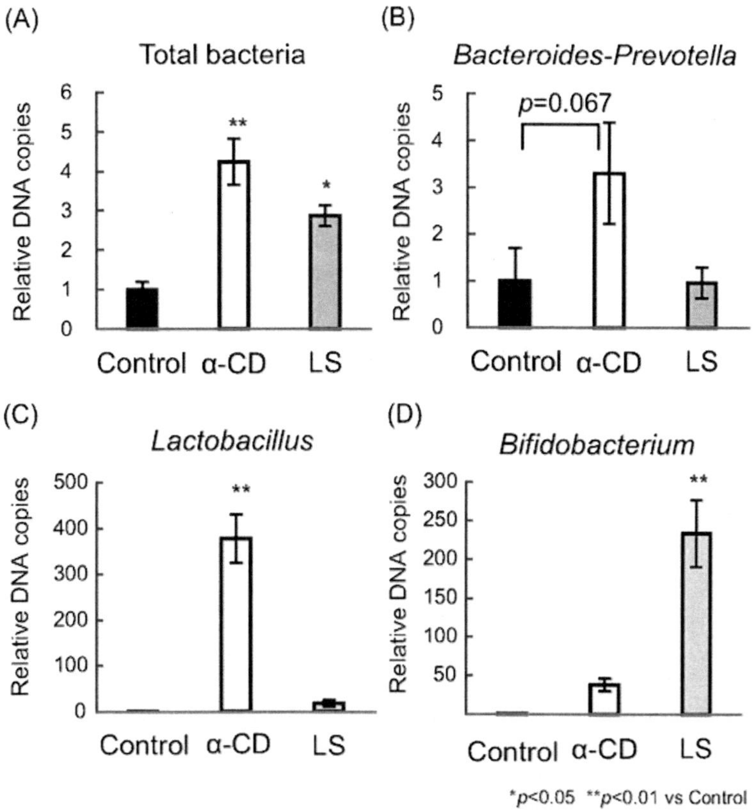

Fig. 13.6 Effect of α-CD on the total number of bacteria (**a**); and the numbers of *Bacteroides-Prevotella* (**b**), *Lactobacillus* (**c**), and *Bifidobacterium* (**d**) in the cecal contents of healthy rats

Table 13.1 Effect of α-CD on short-chain fatty acid production in the cecal content of healthy rats

	Control	α-CD	LS
Cecal content wet weight (g)	2.41 ± 0.10	5.29 ± 0.55**	3.45 ± 0.19
pH of cecal content	8.08 ± 0.24	6.47 ± 0.16**	8.38 ± 0.12
Organic acids (mg in the cecal content)			
Lactate	0.17 ± 0.03	2.02 ± 0.54**	0.24 ± 0.03
Acetate	2.44 ± 0.53	6.39 ± 1.01**	3.23 ± 0.45
Propionate	0.79 ± 0.18	2.91 ± 0.55**	1.15 ± 0.20
Butyrate	0.30 ± 0.19	1.10 ± 0.23**	0.18 ± 0.11
Total organic acids	3.75 ± 0.92	11.58 ± 1.74**	4.76 ± 0.66

**$p < 0.01$ vs. Control

13.3.2 Amelioration of Constipation

Constipation can be caused by lack of exercise, a lack of water intake, and/or a lack of fiber intake. The amelioration of constipation through dietary fiber intake is the best-known effect of its consumption. We previously studied the effect of α-CD intake on constipation by feeding a diet containing 5.5% α-CD for 8 weeks, which showed that this increased fecal bulk and the water content of the feces *versus* a control diet (Fig. 13.7). We hypothesized that this was the result of increases in the numbers of key bacterial species, or a stimulation of intestinal motility by the butyrate produced from the degradation of α-CD. Ge et al. (2017) reported that butyrate stimulates intestinal motility and ameliorates constipation by increasing the secretion of the hormones peptide-YY and glucagon-like peptide-1. We previously showed that α-CD supplementation is associated with high SCFA concentrations in normal rats (Sect. 13.3.1). Therefore, we considered that α-CD administration shortens fecal transit time and ameliorates constipation because of large production of SCFAs.

Protein is the most important nutrient for the maintenance of whole-body homeostasis. Most ingested proteins are digested by gastrointestinal proteases, but some proteins reach the large intestine, where they are metabolized by bacteria, generating putrefactive products such as phenol and indole (Fujikawa et al. 1991). These putrefactive products not only cause an unpleasant fecal odor but are also converted to uremic compounds such as phenyl sulfate that can exacerbate renal failure (Kikuchi et al. 2019; Evenepoel et al. 2017). We recently showed that the intake of α-CD may also reduce the production of putrefactive products via effects on the gut microbiota. We fed rats a diet containing 5.5% α-CD for 8 weeks, which reduced the cecal concentrations of putrefactive products *versus* control and LS-fed rats (Fig. 13.8). *Clostridium* spp. convert essential amino acids to putrefactive products in the colon (Fujikawa et al. 1991), and a previous study showed that α-CD supplementation reduces the proportion of *Clostridium* spp. in the feces of mice fed an HFD (Nihei et al. 2018). Thus, α-CD consumption may also reduce the production of putrefactive products by modulating the gut microbial composition.

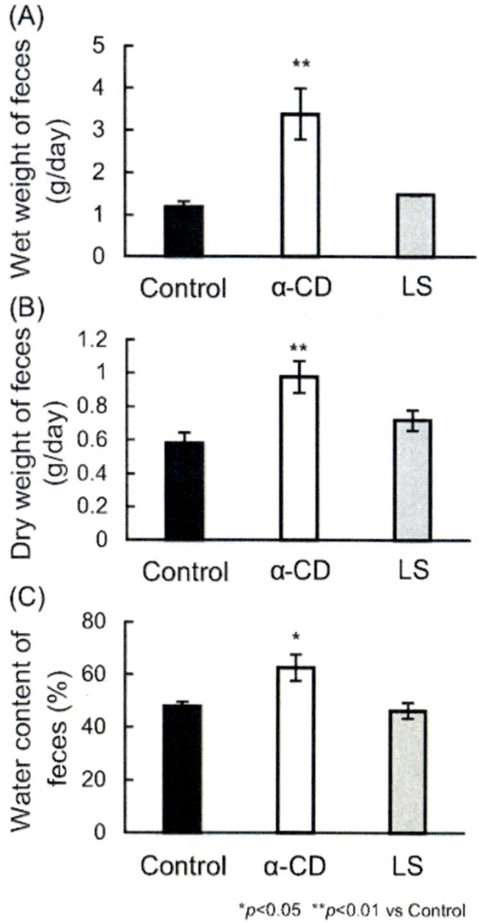

Fig. 13.7 Effect of α-CD on the wet mass (**a**), dry mass (**b**), and water content (**c**) of the feces in healthy rats

13.3.3 Prevention of Metabolic Disease

α-CD consumption is an effective means of controlling body weight because it inhibits the absorption of carbohydrate and fat (Sect. 13.2). Comerford et al. (2012) reported that dietary supplementation with 6 g/day α-CD causes weight loss in healthy people, and Grunberger et al. (2007) showed that the same level of α-CD supplementation for 3 months inhibits weight gain and increases the circulating adiponectin concentration of patients with obesity and type 2 diabetes. Although obesity is thought to be largely the result of overeating and adverse genetics, recent studies have shown that the gut microbial composition is also closely related to obesity (Cani et al. 2008). Nihei et al. (2018) reported that α-CD supplementation prevents obesity by modifying the gut microbiota. In this study, mice were fed an HFD containing 5.5% α-CD for 6 weeks, which prevented the expected body weight

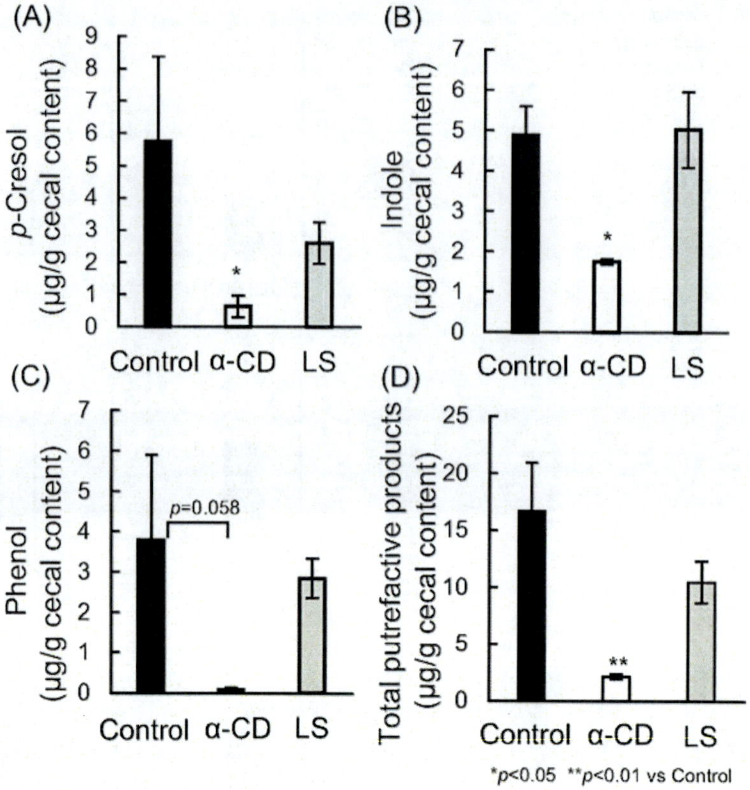

Fig. 13.8 Effect of α-CD on the concentrations of the phenolic metabolites *p*-cresol (**a**), indole (**b**), and phenol (**c**); and the total concentration of putrefactive products (**d**) in the cecal contents of healthy rats

gain and fat accumulation, as demonstrated by smaller adipocytes (Fig. 13.9a–e). Additionally, α-CD prevented the reduction in the proportion of *Bacteroides* that occurs with the development of obesity (Ley et al. 2005) and increased the SCFA content of the cecum. Furthermore, the expression of genes encoding molecules involved in lipid synthesis (fatty acid synthase and sterol regulatory element-binding protein-1c) was reduced and that of peroxisome proliferator-activated receptor (PPAR)α and PPARγ, which are important in adipocyte differentiation and are regulated by SCFAs (Yan and Ajuwon 2015), was increased by α-CD supplementation of the HFD (Fig. 13.9f). These results suggest that α-CD supplementation has an anti-obesity effect by modifying the gut microbiota and increasing SCFA production.

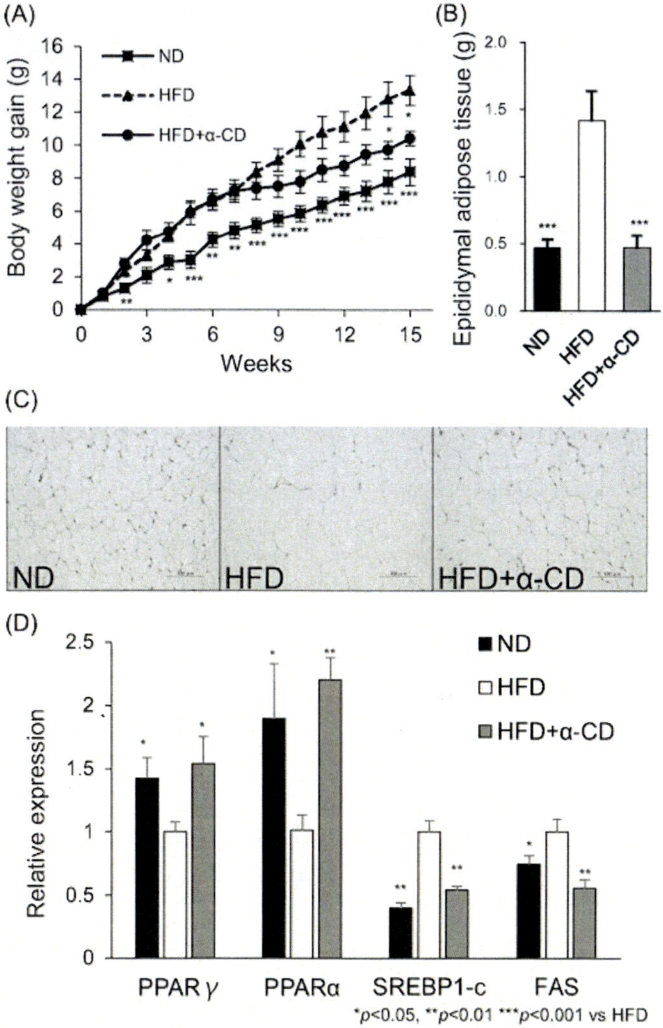

Fig. 13.9 Effect of α-CD on the body mass (**a**), fat mass (**b**), adipose tissue histology (**c**) and adipose gene expression (**d**) of mice fed a normal diet (ND), a high-fat diet (HFD), or an HFD supplemented with α-CD (HFD + α-CD). (Adapted with permission from Nihei et al. 2018)

13.3.4 Amelioration of Atherosclerosis

Atherosclerosis is a vascular disease that is characterized by the formation of atheromas, which are masses of cholesterol and triglycerides that accumulate in the vascular endothelium and represent a potent risk factor for serious cardiovascular disease, a major cause of morbidity and mortality in many countries (Frostegård 2013). Several recent studies have shown that the gut microbiota influences the development of atherosclerosis (Jonsson and Bäckhed 2017). Sakurai et al. (2017) showed that 1.5% α-CD supplementation reduces atherosclerosis in apolipoprotein E-deficient mice that were fed a Western diet for 11 weeks. However, interestingly, this effect was not observed in mice administered β-CD or inulin. Furthermore, there were correlations between the inhibition of atherogenesis and the changes in the gut microbiota and the mass of the cecal contents. These results suggest that α-CD consumption may prevent atherosclerosis by modifying the gut microbiota.

α-CD also reduces the circulating numbers of small LDL particles, which strongly correlate with atherosclerosis (Sect. 13.2.4). Therefore, α-CD may prevent atherosclerosis by two distinct mechanisms.

13.3.5 Improvement of Calcium Absorption and Bone Status

Calcium is the most abundant mineral in the human body and plays important roles as a second messenger in cells and in bone homeostasis (Brown and Macleod 2001). Calcium deficiency causes osteoporosis and a range of other diseases. Therefore, food ingredients that promote calcium absorption may be very useful. Several studies have shown that moderate dietary fiber intake increases calcium absorption (McCabe et al. 2015), and we have also shown that α-CD consumption may also increase calcium absorption and improve bone status. We fed a diet containing 5.5% α-CD to normal rats for 8 weeks and found that they showed greater calcium absorption than controls, based on the difference between the amounts of calcium consumed and lost in the feces (Fig. 13.10a). Interestingly, the effect of α-CD was larger than that of LS, which is recognized as a "food for a specified health use" and is used to promote calcium absorption in Japan (Fujita et al. 1999). Additionally, α-CD supplementation increased femoral mass and strength (Fig. 13.10b, c). The promotion of calcium absorbance by α-CD is the result of its effect to increase SCFA production by the microbiota. Younes et al. (1996) reported that dietary fiber supplementation increases calcium absorption by reducing the pH of the cecum, thereby increasing the soluble calcium concentration. Furthermore, Mineo et al. (2001) showed that SCFAs directly increase calcium transport *via* the paracellular route in the cecum and colon. As already discussed, α-CD consumption is associated with higher SCFA production than LS consumption (Sect. 13.3.1). These results suggest that α-CD is superior to more conventional sources of dietary fiber in its effects to promote calcium absorption and maintain bone homeostasis.

13.3.6 Improvement of Gut Immunity

Over 100 trillion bacteria live in a typical human intestinal tract. Therefore, the intestinal epithelium is constantly exposed to toxic substances and pathogens. The human intestinal tract has various barriers to the absorption of these toxic substances

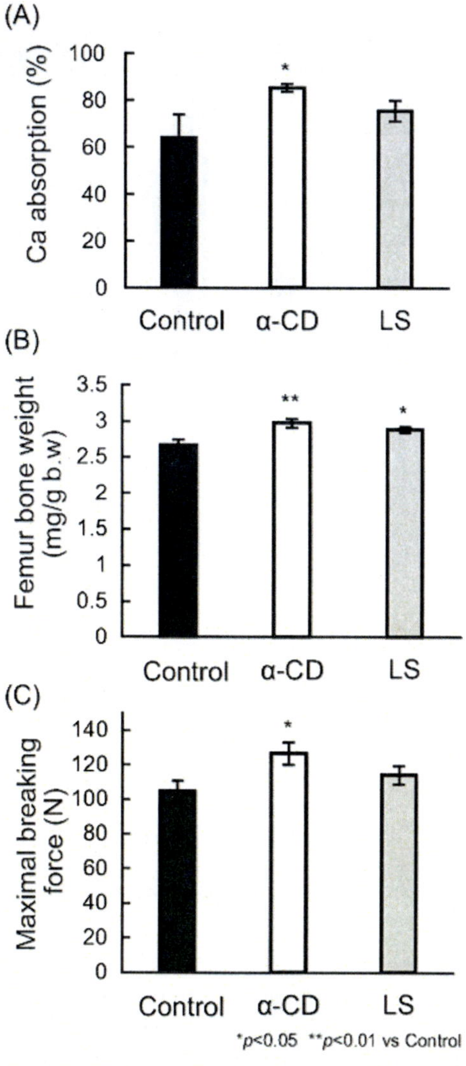

Fig. 13.10 Effect of α-CD on the calcium absorption (**a**), bone mass (**b**), and bone strength (**c**) of healthy rats

and pathogens. Mucin is a viscous glycoprotein that forms a mucosal barrier on epithelial cells and physically prevents toxic substances from entering the body (Odenwald and Turner 2017), and IgA is an immunoglobulin that is secreted by plasma cells, and labels and inactivates pathogens at the mucosal barrier (Macpherson et al. 2008). We have recently shown that α-CD consumption may also improve intestinal barrier function. We fed a diet containing 5.5% α-CD to normal rats for 8 weeks and found that they had a larger amount of mucin in their cecal contents than either control or LS-fed rats (Fig. 13.11a). Additionally, there were higher levels of secretory IgA in the cecal contents of the α-CD-fed rats (Fig. 13.11b). These potential improvements to the intestinal barrier function are likely to be related to

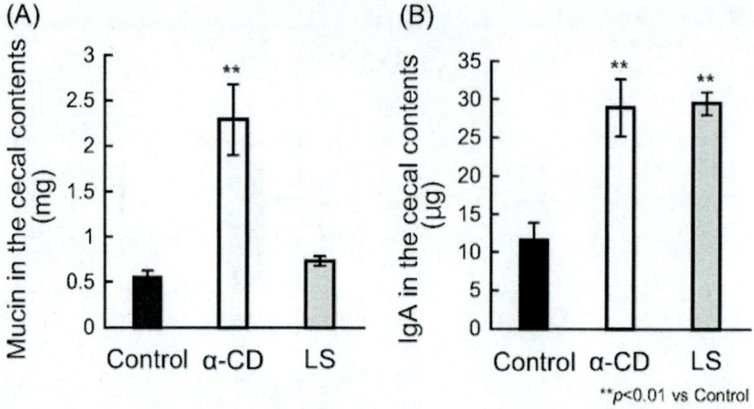

Fig. 13.11 Effect of α-CD on the mucin (**a**) and IgA (**b**) concentrations in the cecal contents of healthy rats

higher SCFA production (Sect. 13.3.1). Willemsen et al. (2003) reported that SCFAs, and especially acetate and propionate, dose-dependently induce the expression of *Muc2*, which encodes a mucin protein, by affecting prostaglandin release in a co-culture of intestinal epithelial cells and intestinal myofibroblasts. Furthermore, Wu et al. (2017) reported that acetate promotes secretory IgA production by B cells, which is mediated by G protein-coupled receptor 43. These results suggest that the intake of α-CD improves the intestinal barrier by increasing SCFA production.

13.3.7 Amelioration of Allergic Disease

Allergic diseases are caused by hypersensitivity of the immune system to environmental allergens, and include food allergies, atopic dermatitis, and allergic asthma. Common allergens include pollen, house dust, and food items. Allergies generally involve the release of pro-inflammatory substances, such as histamine and immunoglobulin E antibodies (IgE), in response to allergens (Gould and Sutton 2008; Thurmond et al. 2008). Recent studies have shown that dietary fiber may protect against allergic disease (Trompette et al. 2014; Tan et al. 2016). Therefore, we studied the effect of α-CD intake on allergic disease and found that it ameliorates features of allergic disease. First, we determined the effect of α-CD on the atopic dermatitis caused by picryl chloride in NC/Nga mice (Nakanishi et al. 2007). The oral administration of α-CD (5 mg/day) for 8 weeks reduced the dermatitis score and plasma IgE concentration of these mice. We also determined the effect of the consumption of 5 g of α-CD for 12 weeks in 14 patients with atopic dermatitis that had been diagnosed by a doctor. We found that α-CD significantly ameliorated their

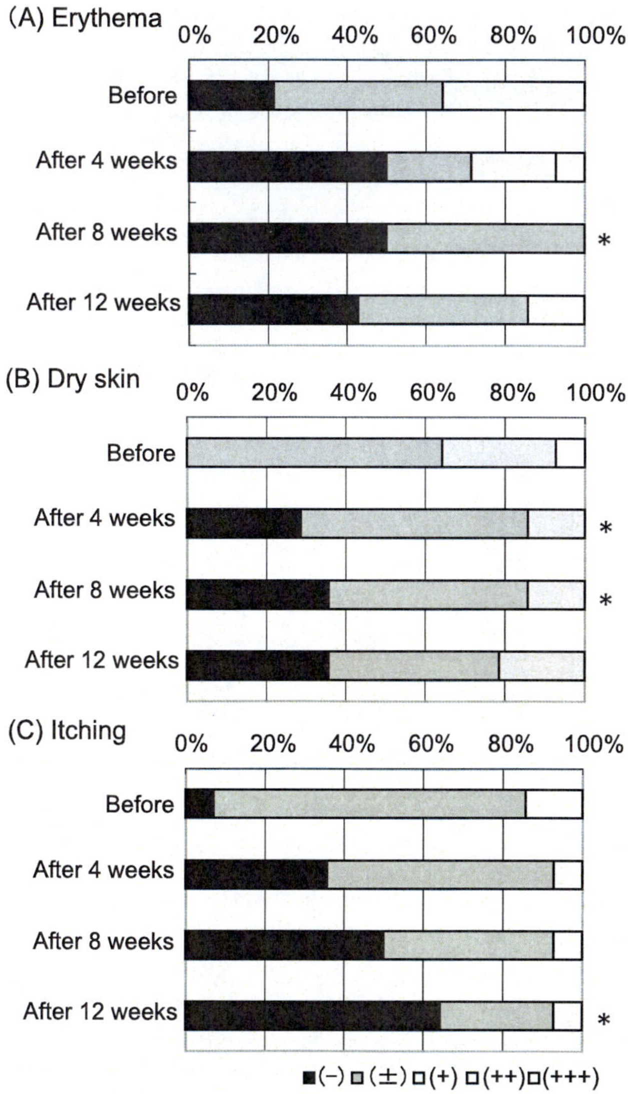

Fig. 13.12 Effect of α-CD on the severity of atopic symptoms in patients with atopic dermatitis

erythema, dry skin, and itching using five-point scales ((−):absence, (±), (+), (++), and (+++):severe) (Fig. 13.12). Moreover, the concentrations of specific circulating antigen-specific IgEs (to Dog dander, Japanese cedar, House dust 1, and Pityrosporum) were reduced by α-CD intake (Table 13.2).

We also studied the effect of α-CD on bronchial asthma. Seventeen patients with bronchial asthma took daily supplements of 5 g of α-CD for 12 weeks and then

Table 13.2 Antigen-specific IgE concentrations in patients with atopic symptoms before and after α-CD administration for 4, 8 or 12 weeks

Antigen-specific IgE (UA/ml)	Baseline	4 weeks	8 weeks	12 weeks
Dog dander	2.57 ± 0.9	2.46 ± 0.88	2.27 ± 0.77	2.06 ± 0.63*
Japanese cedar	27.9 ± 7.6	26.0 ± 7.4**	23.0 ± 7.1**	21.8 ± 7.0**
Housedust1	61.6 ± 11.3	59.2 ± 11.6	57.1 ± 11.5*	57.4 ± 11.5**
Pityrosporum	10.8 ± 5.4	12.2 ± 6.0	7.3 ± 3.6*	7.8 ± 3.9*

*$p < 0.05$, **$p < 0.01$ vs. Baseline

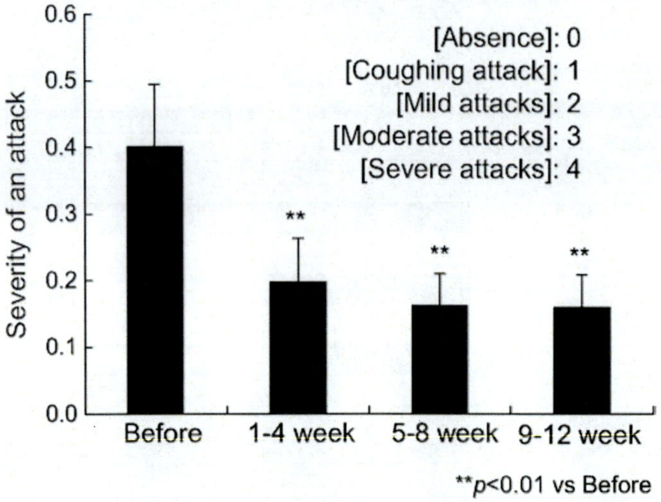

Fig. 13.13 Effect of α-CD on the severity of an attack in patients with bronchial asthma by the questionnaire regarding the subjective severity of their bronchial asthma attacks using the classification Absence: 0, Coughing attacks: 1, Mild attacks: 2, Moderate attacks: 3, and Severe attacks: 4

completed a questionnaire regarding the subjective severity of their bronchial asthma attacks. This showed that 4 weeks of α-CD intake was associated with less severe bronchial asthma attacks, using the classification Absence: 0, Coughing attacks: 1, Mild attacks: 2, Moderate attacks: 3, and Severe attacks: 4 (Fig. 13.13). Furthermore, the circulating Cypress pollen IgE concentration was lower after 12 weeks of α-CD intake.

The beneficial effects of α-CD on allergic diseases may also be related to the higher intestinal SCFA concentration. Furusawa et al. (2013) studied the effects of dietary fiber intake on the immune system and found that it increased the activity of the intestinal bacteria and the amount of SCFA produced, and especially the amount of butyrate. This promoted the differentiation of regulatory T cells, which regulate the immune response. Furthermore, Chiu et al. (2019) reported that gut microbe-derived butyrate inhibits IgE responses to allergens in childhood asthma. Additionally, α-CD may inhibit the entry of allergens by improving the integrity of the gastrointestinal barrier (Sect. 13.3.6). These results suggest that α-CD ameliorates allergic symptoms by increasing microbial SCFA production.

13.3.8 Improvement of Exercise Performance

Exercise promotes muscle strength and health. Exercise performance can be improved by daily training and diet, but recent studies have also demonstrated a relationship between the gut microbiota and exercise performance. Morita et al. (2020) showed that α-CD promotes exercise performance in mice and humans. In this study, they evaluated the effect of α-CD on exercise performance and the intestinal population of *Bacteroides uniformis*, which have been shown to correlate. They found that 5% α-CD administration increased the time taken to reach exhaustion while swimming and the number of *Bacteroides uniformis*. Furthermore, a daily supplement of 200 mg of α-CD for 8 weeks reduced the time required for volunteers to pedal 10 km on an exercise bike and their subsequent fatigue, and also increased the number of intestinal *Bacteroides uniformis* that were present. The authors have speculated that α-CD might improve exercise performance by increasing SCFA production by *Bacteroides uniformis*.

13.4 Conclusion and Further Perspectives

In this chapter, we have described the beneficial effects of α-CD as a source of dietary fiber. α-CD has effects in both the small and large intestine because it has the ability to form an inclusion complex with guest molecules and also to be used by the gut microbiota as a prebiotic. In the small intestine, α-CD inhibits the absorption of carbohydrate, fat, and cholesterol by reducing the activity of digestive enzymes and lipid solubility. In the large intestine, α-CD modulates the gut microbial composition and increases microbial SCFA production. In these ways, α-CD supplementation has various beneficial effects, including anti-obesity and anti-atherosclerotic effects, the amelioration of constipation, the strengthening of bone, the improvement of gut immunity, the amelioration of allergic disease, and the enhancement of exercise performance. Interestingly, the prebiotic effect of α-CD differs from that of conventional sources of dietary fiber, which target *Bifidobacterium*, because α-CD preferentially targets *Lactobacillus* and *Bacteroides*, which are more capable of producing SCFAs.

References

Agarwal P, Gupta R (2016) Alpha-amylase inhibition can treat diabetes mellitus. Res Rev J Med Health Sci 5(4):1–8

Aller R, de Luis DA, Izaola O et al (2004) Effect of soluble fiber intake in lipid and glucose levels in healthy subjects: a randomized clinical trial. Diabetes Res Clin Pract 65(1):7–11

Amar MJA, Kaler M, Courville AB et al (2016) Randomized double blind clinical trial on the effect of oral α-cyclodextrin on serum lipids. Lipids Health Dis 15(1):115

Anderson JW, Baird P, Davis RH et al (2009) Health benefits of dietary fiber. Nutr Rev 67(4):188–205

Antenucci RN, Palmer JK (1984) Enzymatic degradation of alpha- and beta-cyclodextrins by Bacteroides of the human colon. J Agric Food Chem 32(6):1316–1321

Artiss JD, Brogan K, Brucal M et al (2006) The effects of a new soluble dietary fiber on weight gain and selected blood parameters in rats. Metab Clin Exp 55(2):195–202

Astrup A (2005) The role of dietary fat in obesity. Semin Vasc Med 5(1):40–47

Besten GD, Eunen KV, Groen AK et al (2013) The role of short-chain fatty acids in the interplay between diet, gut microbiota, and host energy metabolism. J Lipid Res 54:2325–2340

Bourriaud C, Robins RJ, Martin L et al (2005) Lactate is mainly fermented to butyrate by human intestinal microfloras but inter-individual variation is evident. J Appl Microbiol 99(1):201–212

Brown EM, Macleod RJ (2001) Extracellular calcium sensing and extracellular calcium signaling. Physiol Rev 81:239–297

Buckley JD, Thorp AA, Murphy KL et al (2006) Dose-dependent inhibition of the post-prandial glycaemic response to a standard carbohydrate meal following incorporation of alpha-cyclodextrin. Ann Nutr Metab 50:108–114

Cani PD, Bibiloni R, Knauf C et al (2008) Changes in gut microbiota control metabolic endotoxemia-induced inflammation in high-fat diet-induced obesity and diabetes in mice. Diabetes 57(6):1470–1481

Cavalot F, Petrelli A, Traversa M et al (2006) Postprandial blood glucose is a stronger predictor of cardiovascular events than fasting blood glucose in type 2 diabetes mellitus, particularly in women: lessons from the San Luigi Gonzaga Diabetes Study. J Clin Endocrinol Metab 91(3):813–819

Chiu CY, Cheng ML, Chiang MH et al (2019) Gut microbial-derived butyrate is inversely associated with IgE responses to allergens in childhood asthma. Pediatr Allergy Immunol 30(7):689–697

Chun SC, Schneider RW, Chung IM (2003) Determination of carbon source utilization of Bacillus and Pythium species by biolog® microplate assay. J Microbiol 41(3):252–258

Comerford KB, Artiss JD, Jen KLC et al (2012) The beneficial effects α-cyclodextrin on blood lipids and weight loss in healthy humans. Obesity 19(6):1200–1204

Evenepoel P, Poesen R, Meijers B (2017) The gut–kidney axis. Pediatr Nephrol 32:2005–2014

Frostegård J (2013) Immunity, atherosclerosis and cardiovascular disease. BMC Med 11:117

Fujikawa S, Okazaki M, Matsumoto N et al (1991) Effect of xylooligosaccharide on growth of intestinal bacteria and putrefaction products. J Jpn Soc Nutr Food Sci 44:37–40

Fujita K, Hara K, Nakayama S et al (1999) Effect of 4G-beta-galactosylsucrose (lactosucrose) on bone in weanling rats. J Jpn Soc Nutr Food Sci 52:343–348

Furune T, Ikuta N, Ishida Y et al (2014) A study on the inhibitory mechanism for cholesterol absorption by α-cyclodextrin administration. J Org Chem 10:2827–2835

Furusawa Y, Obata Y, Fukuda S et al (2013) Commensal microbe-derived butyrate induces the differentiation of colonic regulatory T cells. Nature 504:446–450

Gallaher D, Gallaher C, Plank D (2007) Alpha-cyclodextrin selectively increases fecal excretion of saturated fats. FASEB J 21:A730

Ge X, Zhao W, Ding C et al (2017) Potential role of fecal microbiota from patients with slow transit constipation in the regulation of gastrointestinal motility. Sci Rep 7:441

Gentilcore D, Vanis L, Teng J et al (2011) The oligosaccharide a-cyclodextrin has modest effects to slow gastric emptying and modify the glycaemic response to sucrose in healthy older adults. Br J Nutr 106(4):583–587

Gibson GR, Scott KP, Rastall RA et al (2010) Dietary prebiotics: current status and new definition. Food Sci Technol Bull Funct Foods 7:1–19

Gould H, Sutton B (2008) IgE in allergy and asthma today. Nat Rev Immunol 8:205–217

Grunberger G, Jen KLC, Artiss JD (2007) The benefits of early intervention in obese diabetic patients with FBCx™—new dietary fibre. Diabetes Metab Res Rev 23:56–62

Hirano T, Ito Y, Koba S et al (2005) Clinical significance of small dense low-density lipoprotein cholesterol levels determined by the simple precipitation method. Arterioscler Thromb Vasc Biol 24(3):558–563

Holscher HD (2017) Dietary fiber and prebiotics and the gastrointestinal microbiota. Gut Microbes 8(2):172–184

Ivanova EA, Myasoedova VA, Melnichenko AA et al (2017) Small dense low-density lipoprotein as biomarker for atherosclerotic diseases. Oxidative Med Cell Longev 2017:1273042

Jarosz PA, Fletcher E, Elserafy E et al (2013) The effect of α-cyclodextrin on postprandial lipid and glycemic responses to a fat-containing meal. Metabolism 62(10):1443–1447

Jo A, Nakata D, Terao K et al (2007) New synbiotics by the combination of α-cyclodextrin and lactic acid bacteria. In: Proceedings of 25th Cyclodextrin symposium in Japan, Tochigi, pp 144–145

Jo A, Nakata D, Terao K et al (2011) Inhibition effect by alpha Cyclodextrin on an elevation of blood glucose level after administration of sucrose. In: Proceedings of 33rd Annual meeting of the Japanese Society of Clinical Nutrition, Tokyo, p 124

Jonsson A, Bäckhed F (2017) Role of gut microbiota in atherosclerosis. Nat Rev Cardiol 14:79–87

Kikuchi K, Saigusa D, Kanemitsu Y et al (2019) Gut microbiome-derived phenyl sulfate contributes to albuminuria in diabetic kidney disease. Nat Commun 10:1835

Knopp RH, Gitter H, Truitt T et al (2003) Effects of ezetimibe, a new cholesterol absorption inhibitor, on plasma lipids in patients with primary hypercholesterolemia. Eur Heart J 24(8):729–741

Labes A, Schönheit P (2007) Unusual starch degradation pathway via cyclodextrins in the hyperthermophilic sulfate-reducing archaeon archaeoglobus fulgidus strain 7324. J Bacteriol 189(24):8901–8913

Larson SB, Day LS, McPherson A (2010) X-ray crystallographic analyses of pig pancreatic α-amylase with limit dextrin, oligosaccharide, and α-cyclodextrin. Biochemistry 49(14):3101–3115

Lattimer JM, Haub MD (2010) Effects of dietary fiber and its components on metabolic health. Nutrients 2:1266–1289

Lemieux I, Couillard C, Pascot A et al (2000) The small, dense LDL phenotype as a correlate of postprandial lipemia in men. Atherosclerosis 153:423–432

Ley RE, Bäckhed F, Turnbaugh P et al (2005) Obesity alters gut microbial ecology. Proc Natl Acad Sci U S A 102(31):11070–11075

Li Z, Chen S, Gu Z et al (2014) Alpha-cyclodextrin: enzymatic production and food applications. Trends Food Sci Technol 35(2):151–160

Macfarlane S, Macfarlane GT (2003) Regulation of short-chain fatty acid production. Proc Nutr Soc 62(1):67–72

Macpherson A, McCoy K, Johansen FE et al (2008) The immune geography of IgA induction and function. Mucosal Immunol 1:11–22

McCabe L, Britton RA, Parameswaran N (2015) Prebiotic and probiotic regulation of bone health: role of the intestine and its microbiome. Curr Osteoporos Rep 13:363–371

Mineo H, Hara H, Tomita F (2001) Short-chain fatty acids enhance diffusional Ca transport in the epithelium of the rat cecum and colon. Life Sci 69:517–526

Morita H, Kano C, Ishii C et al (2020) Bacteroides uniformis enhances endurance exercise performance through gluconeogenesis. BioRxiv 2020:975730

Nakanishi K, Nakata D, Konishi M et al (2007) Effect of cyclodextrin on allergic action of the PiCl-induced in NC/Nga mice. J Incl Phenom Macrocycl Chem 57:61–64

Nihei N, Okamoto H, Furune T et al (2018) Dietary α-cyclodextrin modifies gut microbiota and reduces fat accumulation in high-fat-diet-fed obese mice. BioFactors 44(4):336–347

Odenwald MA, Turner JR (2017) The intestinal epithelial barrier: a therapeutic target? Nat Rev Gastroenterol Hepat 14:9–21

Ommen BV, de Bie ATHJ, Bär A (2004) Disposition of 14C-α-cyclodextrin in germ-free and conventional rats. Regul Toxicol Pharmacol 39:57–66

Sakurai T, Sakurai A, Chen Y et al (2017) Dietary α-cyclodextrin reduces atherosclerosis and modifies gut flora in apolipoprotein E-deficient mice. Mol Nutr Food Res 61(8):1600804

Sasaki D, Sasaki K, Ikuta N et al (2018) Low amounts of dietary fibre increase in vitro production of short-chain fatty acids without changing human colonic microbiota structure. Sci Rep 8:435

Siebold M, Frieling PV, Joppien R et al (1995) Comparison of the production of lactic acid by three different lactobacilli and its recovery by extraction and electrodialysis. Process Biochem 30(1):81–95

Tan J, McKenzie C, Vuillermin PJ et al (2016) Dietary fiber and bacterial SCFA enhance oral tolerance and protect against food allergy through diverse cellular pathways. Cell Rep 15(12):2809–2824

Thurmond R, Gelfand E, Dunford P (2008) The role of histamine H1 and H4 receptors in allergic inflammation: the search for new antihistamines. Nat Rev Drug Discov 7:41–53

Trompette A, Gollwitzer E, Yadava K et al (2014) Gut microbiota metabolism of dietary fiber influences allergic airway disease and hematopoiesis. Nat Med 20:159–166

Tungland BC, Meyer D (2002) Nondigestible oligo- and polysaccharides (dietary fiber): their physiology and role in human health and food. Compr Rev Food Sci Food Saf 1(3):90–109

Turroni F, Strati F, Foroni E et al (2012) Analysis of predicted carbohydrate transport systems encoded by Bifidobacterium bifidum PRL2010. Appl Environ Microbiol 78(14):5002–5012

Willemsen LEM, Koetsier MA, van Deventer SJH et al (2003) Short chain fatty acids stimulate epithelial mucin 2 expression through differential effects on prostaglandin E1 and E2 production by intestinal myofibroblasts. Gut 52:1442–1447

Wu W, Sun M, Chen F et al (2017) Microbiota metabolite short-chain fatty acid acetate promotes intestinal IgA response to microbiota which is mediated by GPR43. Mucosal Immunol 10:946–956

Yan H, Ajuwon KM (2015) Mechanism of butyrate stimulation of triglyceride storage and adipokine expression during adipogenic differentiation of porcine stromovascular cells. PLoS One 10(12):e0145940

Younes H, Demigné C, Rémésy C (1996) Acidic fermentation in the caecum increases absorption of calcium and magnesium in the large intestine of the rat. Br J Nutr 75(2):301–314

Chapter 14
Complexation of Ingredients in Foods by Alpha-Cyclodextrin to Improve Their Functions

Takahiro Furune and Keiji Terao

14.1 Introduction

Fresh and processed foods contain various functional ingredients, but some of them have low thermal stability and/or poor water solubility, or must be consumed in large quantities to exert their function. Cyclodextrins (CDs) are used as inclusion agents in a variety of food applications to stabilize certain compounds, improve water solubility, mask odors, and aid in the selective extraction of functional compounds. Among the various CDs, α-CD is also used as a dietary fiber. The acceptable daily intakes (ADIs) for α-CD have not been specified by the Joint FAO/WHO Expert Committee on Food Additives (JECFA). Moreover, α-CD have been reported to have multiple functions including anti-bacterial activity (Zhang et al. 2008) and as a dietary fiber (see Chap. 13).

Recent studies have shown that α-CD cannot only stabilize and improve the water solubility of functional ingredients by complexation, but also combine with other food ingredients to improve their functions. In this chapter, we introduce the concept of combining α-CD with functional components in foods to improve their functionality.

T. Furune (✉)
CycloChem Bio Co., Ltd., Kobe, Hyogo, Japan
e-mail: takahiro.furune@cyclochem.com

K. Terao
CycloChem Bio Co., Ltd., Kobe, Hyogo, Japan

Department of Social/Community Medicine and Health Science, Food and Drug Evaluation Science, Graduate School of Medicine, Kobe University, Kobe, Hyogo, Japan
e-mail: keiji.terao@cyclochem.com

T. M. Ho et al. (eds.), *Functionality of Cyclodextrins in Encapsulation for Food Applications*, https://doi.org/10.1007/978-3-030-80056-7_14

14.2 Manuka Honey α-Cyclodextrin Complex Improves Anti-bacterial Activity

14.2.1 Manuka Honey

Manuka honey is a special honey that can be harvested for only 4 weeks in a year in New Zealand and Australia. This honey has an elegant herb-like flavor and contains minerals, vitamins, amino acids, and enzymes. In addition, manuka honey contains a large amount of methylglyoxal (MGO, Fig. 14.1), which is a unique and special anti-bacterial compound. Consequently, the antibacterial properties of manuka honey are much stronger than those of other general honeys (Mavric et al. 2008; Almasaudi et al. 2017). Manuka honey has been demonstrated to have anti-bacterial activity against oral pathogenic bacteria; *Staphylococcus aureus* (*S. aureus*), which causes skin infections; and *Helicobacter pylori*, the bacterium that causes gastric cancer (Safii et al. 2017; Johnston et al. 2018; al Somal et al. 1994). Furthermore, manuka honey has been shown to have anti-viral activity (Watanabe et al. 2014), skin wound healing properties (Carter et al. 2016), and an ameliorating effect on inflammatory bowel disease (Medhi et al. 2008).

14.2.2 Enhancement of Anti-bacterial Activity of Manuka Honey by Complexation with α-CD

It is known that α-CD has anti-bacterial activity against *S. aureus* and *Bacillus cereus* (Zhang et al. 2008). Therefore, it is speculated that the antibacterial activity of a complex of manuka honey and α-CD might be stronger and wider than that of each component alone.

The complex of manuka honey and α-CD is prepared as follows: manuka honey and α-CD (45:55 w/w) were added to water and the mixture was stirred at 40 °C. After drying the mixture, a fine powder (manuka honey and α-CD powder, MAP) was obtained.

The anti-bacterial activity of MAP against *S. aureus in vitro* is tested. For this analysis, *S. aureus* was cultured in heart infusion broth medium with MAP, the physical mixture of manuka honey and α-CD, manuka honey, α-CD, or no additives (control) at 37 °C. Then, bacterial viability was evaluated using the conventional dilution plate method. Compared with each component alone, MAP drastically inhibited the growth of *S. aureus* (Fig. 14.2). Interestingly, the anti-bacterial activity

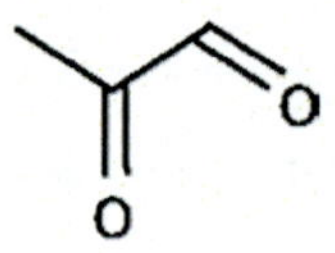

Fig. 14.1 Chemical structure of methylglyoxal

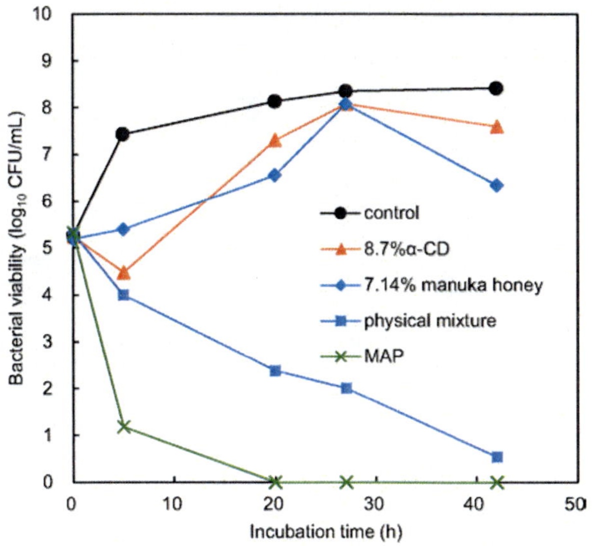

Fig. 14.2 Antibacterial activity of MAP against *S. aureus*. The test sample of MAP or physical mixture contained 8.7% α-CD and 7.14% manuka honey

of MAP was stronger than that of the physical mixture of manuka honey and α-CD. Swift et al. (2014) reported that the anti-bacterial activity of MAP was stronger and showed a broader range than that of each single component. These results suggest that MAP is an excellent broad-range anti-bacterial compound.

14.2.3 Application of Manuka Honey α-Cyclodextrin Complex to Oral Care

Some oral bacteria produce odor-causing substances including volatile sulfur compounds (VSCs) such as hydrogen sulfide and methyl mercaptan, these bacteria are related to poor oral health and periodontal disease (Ratcliff and Johnson 1999). *Porphyromonas gingivalis* is a key periodontal disease-causing bacterium that produces biofilm and VSCs (Mysak et al. 2014; Nakano et al. 2002). The results of a recent study suggest that *P. gingivalis* is also related to Alzheimer's disease (Dominy et al. 2019).

The anti-bacterial activity of MAP against *P. gingivalis in vitro* was tested (Ueno et al. 2012). *P. gingivalis* was cultured in brain heart infusion yeast (BHIY) broth medium with MAP, manuka honey, α-CD, or no additives. After cultivation at 37 °C for 54 h under anaerobic conditions, bacterial viability was evaluated using the conventional dilution plate method. Compared with each compound alone, MAP inhibited the growth of *P. gingivalis*. The inhibitory effect of MAP on biofilm formation

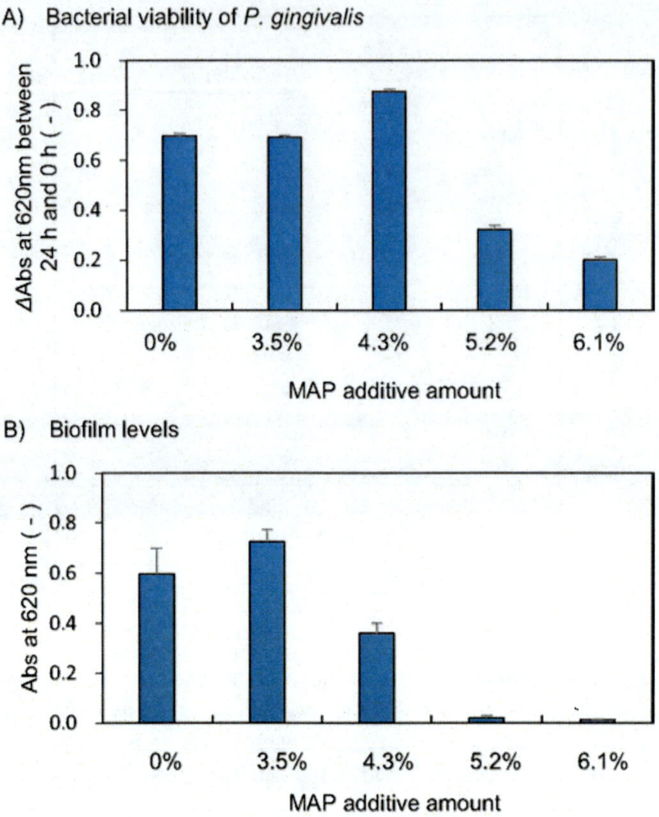

Fig. 14.3 Effect of MAP on the biofilm formation by *P. gingivalis*

(Fig. 14.3) and VSCs production (Fig. 14.4) by *P. gingivalis* was also tested. To investigate the effect of MAP on biofilm formation, *P. gingivalis* was cultured in BHIY broth medium with MAP or no additives at 37 °C for 24 h under anaerobic conditions. The amount of biofilm was determined by the conventional crystal violet method as described by Wakabayashi et al. (2009). To investigate the effect of MAP on VSCs production, *P. gingivalis* (2.2×10^6 CFU/mL) was cultured in BHIY broth medium with MAP or no additives at 37 °C for 46 h under anaerobic conditions. The amount of VSCs produced was determined as described by Suzuki et al. (2016), with slight modifications. The culture samples containing methionine were sealed in a sampling bag and incubated at 37 °C for 90 min. After incubation, the amount of methyl mercaptan that had been converted from methionine was measured by a detector tube. The results of these analyses showed that MAP inhibited biofilm formation and VSCs production *via* its anti-bacterial activity. These results suggest that MAP is an excellent oral care material.

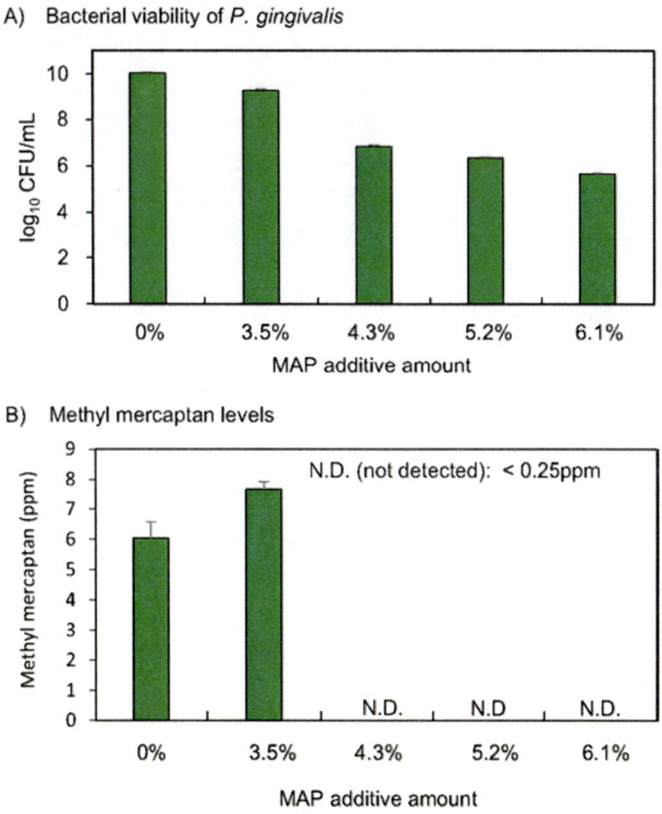

Fig. 14.4 Effect of MAP on the methyl mercaptan production by *P. gingivalis*

14.2.4 Other Benefits of the Manuka Honey α-Cyclodextrin Complex

Several studies have reported that MAP has an anti-inflammatory effect (Chepulis and Francis 2012), reduces glucose absorption (Okamoto et al. 2017), improves symptoms of osteoporosis (Katsumata et al. 2015), stimulates the gut microbiota (Katsumata et al. 2015; Parkar et al. 2017), and has antioxidant activity (Kimura et al. 2017).

Parkar et al. (2017) reported that MAP has a beneficial effect on the gut microbiota. To explore its effect on digestion, MAP, manuka honey, or α-CD was treated with artificial gastric and intestinal juice containing digestive enzymes, and then the digested samples were dialyzed in a 100–500 Da molecular weight cut-off membrane to simulate the absorption of small sugars in the small intestine. The remaining gastrointestinal digesta were applied to a human feces fermentation system and cultured at 37 °C for 16 h under anaerobic conditions. The pH and organic acid content of each fermented sample are shown in Fig. 14.5a, b. The pH was

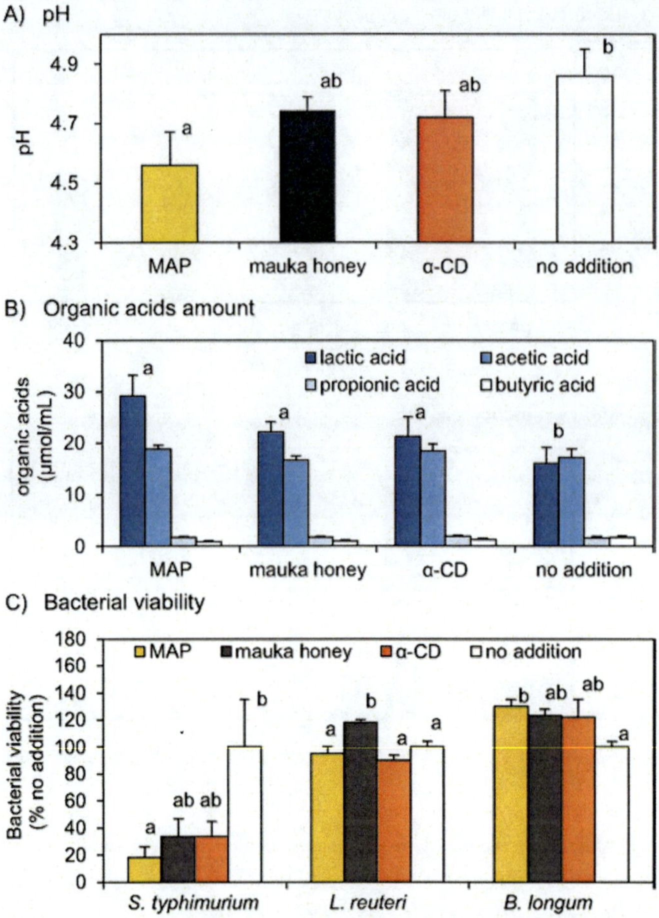

Fig. 14.5 Effect of MAP on the pH, organic acids and bacterial viability by the fermentation system. Letters after means indicate results of post hoc tests; means that do not have a letter in common are significantly different (Tukey's test at p = 0.05). (Adapted with permission from Parkar et al. 2017)

significantly lower and the lactic acid content was significantly higher in the MAP sample than in the sample with no additives. Furthermore, MAP significantly reduced the abundance of *Salmonella typhimurium* (*S. typhimurium*) and significantly increased the abundance of *Lactobacillus reuteri* (*L. reuteri*) (Fig. 14.5c). These results are reasonable because *Lactobacillus* species produce organic acids to lower gut pH and reduce the number of harmful bacteria. Together, these results suggest that MAP has potential applications as a prebiotic material.

Manuka honey has strong scavenging activity for superoxide anion radicals because of the presence of methyl syringate (Inoue et al. 2005). Kimura et al. (2017) tested the antioxidant activity of MAP using a conventional DPPH-free radical

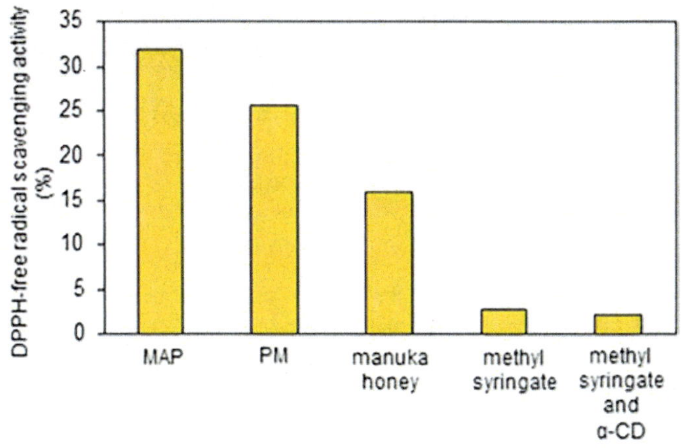

Fig. 14.6 DPPH-free radical scavenging activity of MAP. *PM* physical mixture of manuka honey and α-CD

scavenging assay, and found that the antioxidant activity of MAP was stronger than that of each component alone. When samples with equal methyl syringate concentrations were tested in DPPH analyses, the radical scavenging rate of MAP was higher than that of the physical mixture of manuka honey and α-CD and that of manuka honey (Fig. 14.6). This result suggested that complexation of manuka honey and α-CD enhances the antioxidant effect, as well as improving anti-bacterial activity. In contrast, the complex of methyl syringate with α-CD did not show stronger radical scavenging activity than that of methyl syringate alone.

14.3 "Daikon" Radish α-Cyclodextrin Complex Improves 4-Methylthio-3-butenyl isothiocyanate Stability and Anti-obesity Effect

14.3.1 Radish 4-Methylthio-3-butenyl isothiocyanate

Radish (*Raphanus sativus* L.) is in the *Brassicaceae* family, and its root is mainly used as a vegetable in Asia and Europe. Some species of radish produce pungent compounds, especially 4-methylthio-3-butenyl isothiocyanate (MTBI, Fig. 14.7) (Kjajr et al. 1978). In the absence of stress, MTBI exists as a stable glycoside (MTBG), but when the radish root is cut or ground before cooking or during processing, MTBG is converted to MTBI by the endogenous enzyme myrosinase. Recent studies have reported that MTBI has anti-bacterial (Uda and Shitara 2013) and anti-cancer activities (Hirata et al. 2019) and aids in the prevention of nonalcoholic fatty liver disease (You et al. 2015). However, MTBI is unstable as it is easily hydrolyzed (Li et al. 2015).

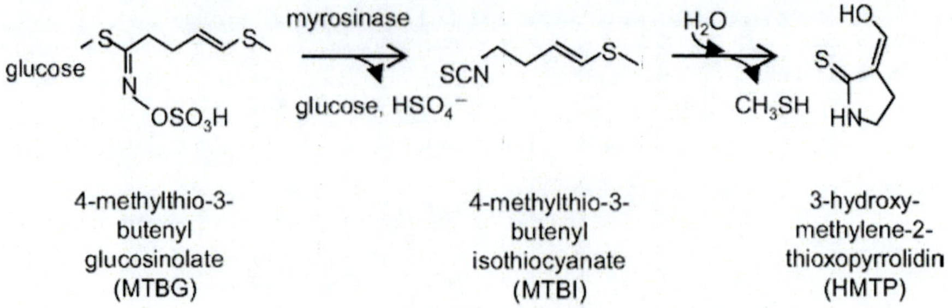

Fig. 14.7 MTBI biosynthesis and degradation. (Adapted with permission from Okamoto et al. 2019)

14.3.2 *Stabilization of MTBI in Radish by Complexation with α-Cyclodextrin*

It is difficult to maintain the MTBI content in radish during and after food processing. Because CDs can control the release of allyl isothiocyanate (AITC), which has a similar structure to that of MTBI (Li et al. 2007). The effect of CDs on MTBI stability was investigated (Ueno et al. 2013). Each CD was added to freshly grated radish at 10% (w/w). After filtering the mixture through cotton gauze, the filtrate was kept at room temperature and its MTBI content was measured using a gas chromatography—flame ionization detection method as described by Okamoto et al. (2019). As shown in Fig. 14.8a, the stability of MTBI was higher in the sample containing α-CD than in the other samples. The interaction between α-CD and MTBI was also investigated in an ultraviolet spectrum analysis, and found that the binding constant of MTBI and α-CD was 129 M^{-1} that is thought to be enough to explain the stability improvement of MTBI by complexation with α-CD.

Next, a radish and α-CD complex powder was developed to improve the stability of MTBI. The mixture of grated radish and α-CD was stirred, freeze-dried, and then ground to a fine powder (MTBI-αCD). During the manufacturing process, α-CD suppressed the degradation of MTBI, and the storage stability of MTBI was very high in the obtained powder. As shown in Fig. 14.8b, the MTBI content in the sample containing α-CD remained at almost 100%, but that in the sample without additives was reduced to less than 20% after processing.

14.3.3 *Anti-obesity Effect of MTBI Stabilized by α-Cyclodextrin*

The anti-obesity effect of MTBI-αCD in obese mice fed with a high-fat diet was investigated (Okamoto et al. 2019). Male C57BL/6J mice were randomly divided into four groups: a normal diet (ND) group, a high-fat diet (HFD) group, a high-fat

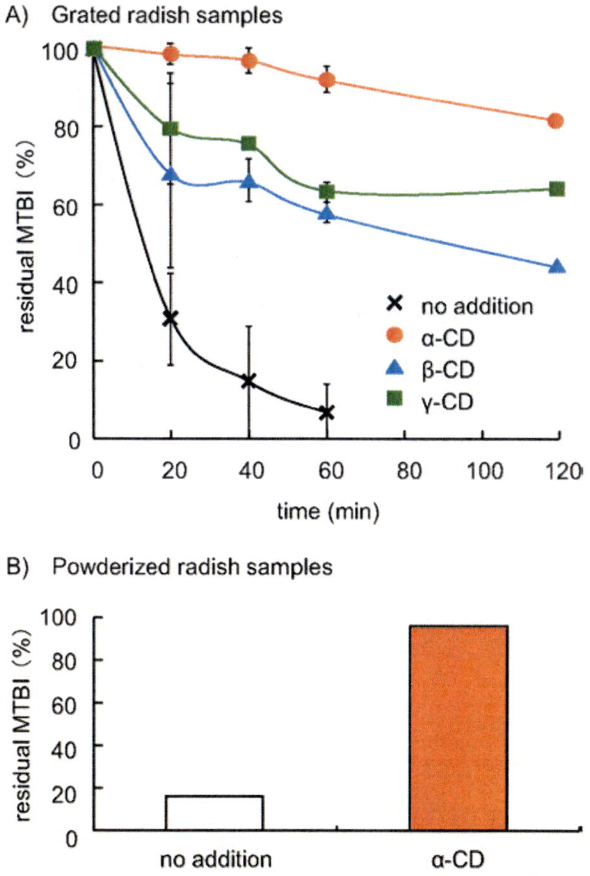

Fig. 14.8 Effect of CDs on the MTBI stability

diet with α-CD group, and a high-fat diet with MTBI-αCD (MTBI-αCD) group. The groups of mice were fed with these diets for 16 weeks. The body weight gain in the MTBI-αCD group was significantly suppressed compared with that in the HFD group, and was similar to that in the ND group (Fig. 14.9a). After 16 weeks of feeding, adipocytes were significantly smaller in the MTBI-αCD group and the α-CD group than in the HFD group, and were a similar size in the MTBI-αCD group and the ND group (Fig. 14.9b). Thus, compared with α-CD alone, MTBI-αCD had a stronger anti-obesity effect. The serum triglyceride and cholesterol levels and insulin sensitivity were also significantly lower in the MTBI-αCD group than in the HFD group. Furthermore, the analyses for mRNA transcription revealed that MTBI-αCD suppressed lipid accumulation by altering the transcription of genes encoding antioxidant enzymes and anti-inflammatory factors.

Together, all of these results show that the anti-obesity effect of MTBI-αCD is due to its ability to inhibit fat accumulation in animals fed with a high-fat diet, and is stronger than that of α-CD alone.

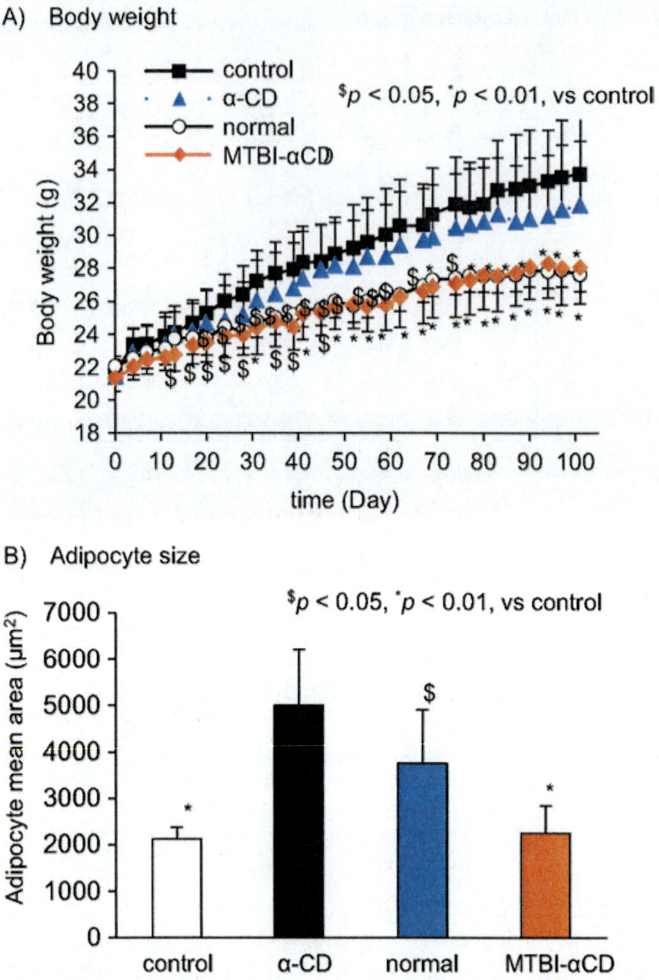

Fig. 14.9 Effect of MTBI-αCD on the body weight gain and adipocyte size in mice. (Adapted with permission from Okamoto et al. 2019)

14.4 Kiwi Fruit α-Cyclodextrin Complex Improves Protease Stability and Bioactivity

14.4.1 Actinidin, the Kiwi Fruit Protease

Kiwi fruit (*Actinidia deliciosa*) belongs to the Actinidiaceae family and is a popular fruit in many countries. It contains a specific protease, actinidin, which enhances digestion (Kaur et al. 2010) and reduces the formation of tongue plaque (Mugita et al. 2017). When kiwi fruit is marinated with meat, the meat becomes tender *via* the activity of actinidin (Lewis and Luh 1988). However, actinidin has low thermal stability, so it is easily destroyed during food processing.

14.4.2 Stabilization of Kiwi Fruit Protease by Complexation with α-Cyclodextrin

To improve the thermal stability of actinidin, a powder of a kiwi fruit juice and α-CD complex was prepared and its effect on the thermal stability of protease was investigated (Asada et al. 2009). To obtain the powder, 2 g α-CD was added to 18 g fresh kiwi fruit juice, and then the mixture was stirred and then freeze-dried. The resulting fine powder of kiwi fruit juice and α-CD was designated as KAP. Figure 14.10 shows the thermal stability of KAP. For this analysis, each sample was added to 25 mmol/L citric acid buffer solution (pH 6.0) containing 12.5 mmol/L EDTA disodium salt, and then heated at 70 °C for 6 min. The protease activity was measured using the artificial substrate (Pyr-Phe-Lue-pNa) method described by Dubois et al. (1988) with slight modifications. The protease activity of kiwi fruit juice was reduced by heat treatment, but the protease activity of KAP was maintained (Fig. 14.10). Furthermore, we found that the protease activity of KAP was maintained after storage at 4 °C for 1 year.

Linde et al. (2009) reported that α-CD forms an inclusion complex with hydrophobic amino acids. Therefore, the mechanism of stabilization may involve the formation of an inclusion complex *via* the interactions between α-CD and the hydrophobic amino acid residues of actinidin.

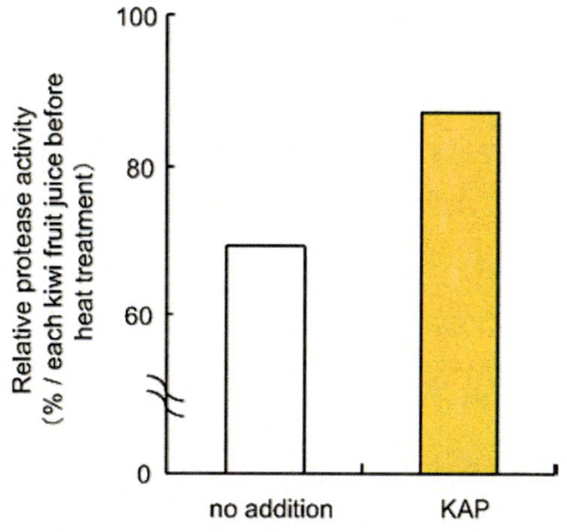

Fig. 14.10 Protease activity of kiwi fruit juice samples after heat treatment (70 °C, 6 min)

14.4.3 Enhancement of Kiwi Fruit Bioactivity by Complexation with α-Cyclodextrin

Actinidin has a tongue plaque removing effect. Because α-CD is used for oral care products, KAP has potential applications in oral care products such as chewables, candy, and gum.

In addition, because kiwi fruit contain high amount of pectin, which is a water-soluble dietary fiber, KAP is expected to increase its prebiotic effect.

14.5 Improvement of Phospholipase C Stability and Bioactivity by Cucumber α-Cyclodextrin Complex

14.5.1 Phospholipase C in Cucumber

Cucumber (*Cucumis sativus* L.) belongs to the Cucurbitaceae family and is a popular vegetable in many countries. It has many health benefits including antioxidant activity, an anti-diabetic effect, and anti-bacterial activity (Mukherjee et al. 2013). Cucumber has also been reported to lower blood lipid levels in people with hyperlipidemia (Molly et al. 2017). Cucumber contains various active ingredients, among which phospholipase C is very abundant (Sugimori et al. 2011). Phospholipase C catalyzes the cleavage of the phosphate group and glycerol moiety to reduce lipid emulsification properties. Because lecithin enhances lipid solubility and absorption to bile salts in the small intestinal fluid (Tamesue et al. 1973), the lipid-reducing effect of cucumber may result from a decrease in lipid micellar solubility *via* lecithin degradation in the gut. However, cucumber must be consumed in large quantities daily to achieve this effect, and the activity of phospholipase C is generally diminished or destroyed during food processing.

14.5.2 Improvement of Phospholipase C Activity by Complexation with α-Cyclodextrin

Previous studies have shown that α-CD not only stabilizes enzyme activity (Asada et al. 2009), but also decreases lipid micellar solubility *via* lecithin precipitation (Furune et al. 2014). Therefore, we investigated the effect of α-CD on the stability of phospholipase C and the effect of a cucumber α-CD complex on lipid micellar solubility in artificial small intestinal fluid.

For the phospholipase C activity test, a complex powder of the crude cucumber enzyme and α-CD was prepared. The preparation the crude cucumber enzyme and the phospholipase C activity assay were conducted as described by Sugimori et al. (2011) with minor modifications. Cucumber crude enzyme and α-CD (1:1 w/w) were mixed in water, then the mixture was freeze-dried to obtain a fine powder. The same procedure was conducted with micro-crystalline cellulose (MCC) instead of α-CD for comparison. To evaluate the phospholipase C activity, phosphatidylcholine was hydrolyzed to choline phosphate by each test sample in Tris-HCl buffer (pH 7.2) with 2 mM EDTA at 37 °C for 30 min. After heating at 95 °C for 10 min, the reaction mixture was treated at 37 °C for 20 min by 0.2 U of alkaline phosphatase in 50 mM glycine-NaOH buffer (pH 9.6) with 1 mM MgCl and 0.1 mM $ZnCl_2$. The amount of released phosphate in the mixture was evaluated using the commercial kit (Phospha-C test Wako, Japan). For the heat treatment, each sample was added to Tris-HCl buffer (pH 7.2), and the mixture was heated at 95 °C for 10 min. Figure 14.11 shows the relative phospholipase C activity of each sample (Furune et al. 2019). The phospholipase C activity was higher in the α-CD sample than in the samples with cucumber crude enzyme only or MCC. After heat treatment, the phospholipase C activity of cucumber was retained in the α-CD sample, but completely eliminated in the samples with enzyme alone or MCC.

Previous studies have shown that α-CD can form inclusion complexes with hydrophobic amino acid residues of proteins (Linde et al. 2009) and lecithin (Furune et al. 2014) in mixtures. Thus, it was thought that improved stability of phospholipase C was caused by complexation of phospholipase C with α-CD.

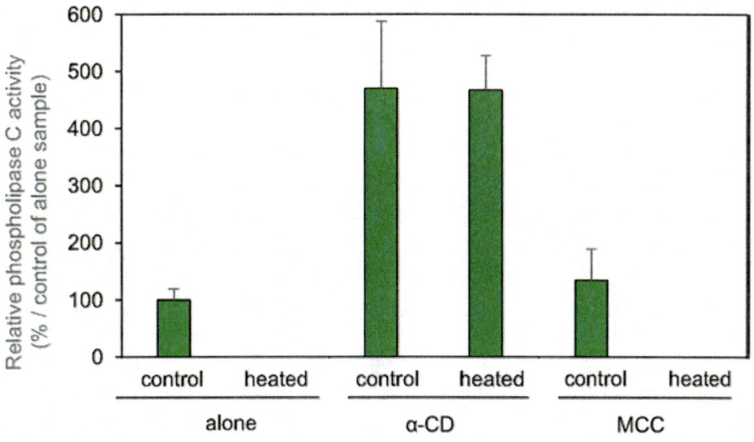

Fig. 14.11 Effect of α-CD on the phospholipase C activity of crude cucumber enzyme samples

14.5.3 Decrease of Cholesterol Solubility by Cucumber α-Cyclodextrin Complex

Next, the effects of the cucumber and α-CD complex on cholesterol micellar solubility was explored. The cucumber and α-CD (1:1 w/w) were mixed in water, and then dried to obtain a fine powder (Cuc50% α-CD). For comparison, the same process was performed with MCC instead of α-CD (Cuc50% MCC). Figure 14.12 shows the effect of the various samples on cholesterol micellar solubility in fed-state simulated intestinal fluid (FeSSIF) (Furune et al. 2019). The components of FeSSIF used in this study are shown in Table 14.1. $CaCl_2$ was added for activation of enzyme. Excess cholesterol and each cucumber powder were added to FeSSIF, and then the mixture was shaken for 2 h at 37 °C. After filtering through a 0.2-μm PTFE filter, the filtrate was analyzed by high performance liquid chromatography according to the method described by Furune et al. (2014). At the same amount of α-CD (6% Cuc50% α-CD and 3% α-CD), Cuc50% α-CD had a stronger effect to decrease cholesterol micellar solubility than did α-CD alone. In contrast, Cuc50% MCC did not affect cholesterol micellar solubility.

14.5.4 Other Benefits of the Cucumber and α-Cyclodextrin Complex

Cucumber has a specific odor, and one of the odor compounds is (E,Z)-2,6-nonadienal (Kemp et al. 1974). To investigate the deodorizing effect of (E,Z)-2,6-nonadienal by the complexation with α-CD, a sensory test to evaluate the odor of

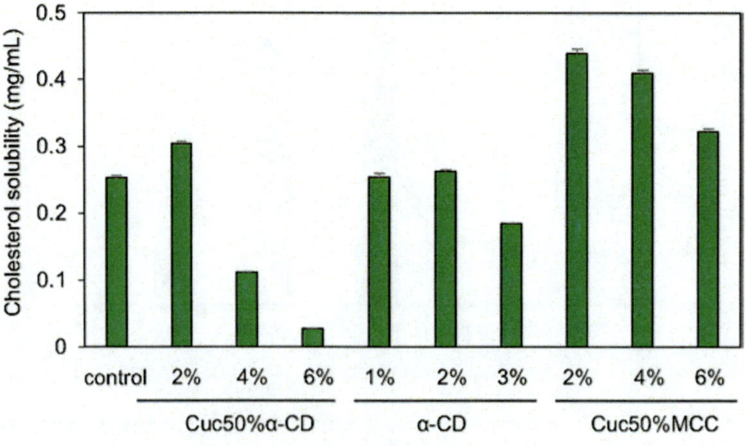

Fig. 14.12 Cholesterol micellar solubility of each cucumber samples

Table 14.1 Components of FeSSIF

Components	mM
Sodium taurocholate	15
Lecithin	3.75
Acetic acid	144
NaCl	173
NaOH	~101 (adjust to pH 5.0)
$CaCl_2$	10

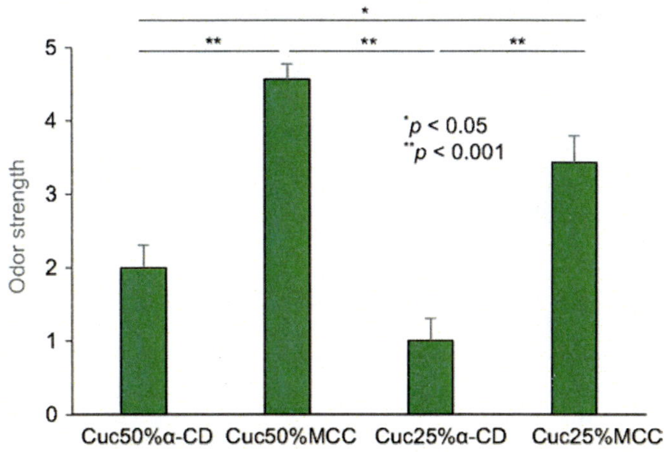

Fig. 14.13 Odor strength of cucumber α-CD powders

each cucumber powder using a single-blind technique was conducted. The sample volumes were equalized on a cucumber solids content basis. The scale was 0 (no odor) to 5 (strong cucumber odor) and the odor of each sample was evaluated relative to that of cucumber solids alone, with a score of 5. The results showed that cucumber with α-CD had a weaker odor than that of cucumber with MCC (Fig. 14.13).

The green color stability of each cucumber powder sample was evaluated after standing for 7 days at room temperature without shading. To determine the chlorophyll level, 40 mg cucumber powder sample was added to 20% aqueous acetone with 0.2% calcium carbonate. After filtration through a 0.2-μm PTFE filter, the absorbance of the filtrate was measured at a wavelength of 663 nm.

Compared with the cucumber powder with MCC, that with α-CD showed higher color stability and chlorophyll retention (Fig. 14.14). It has been reported that CDs stabilize chlorophyll *via* complex formation (Seto et al. 1990). These results suggest that α-CD not only stabilizes and enhances the functionality of phospholipase C, but also improves the color quality of cucumber powder.

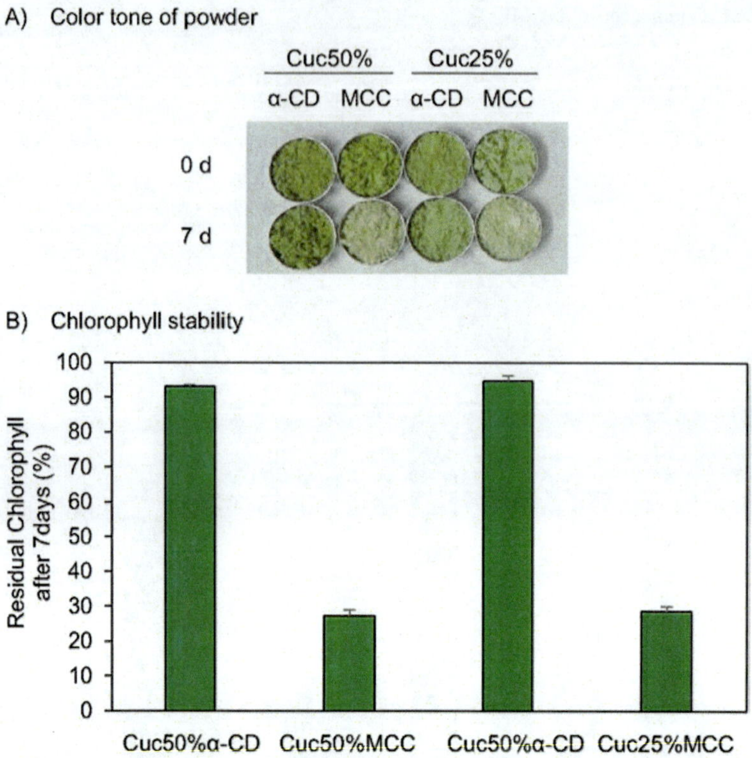

Fig. 14.14 Green color stability of cucumber α-CD powders

14.6 Potential for Other Applications

Since α-CD has multiple functions such as a prebiotic effect and an ability to inhibit the absorption of sugar and lipids, it has the potential to be used in combination with many different materials. In this section, we introduce the possibility of combining it with a probiotic, camel milk, and flaxseed oil.

14.6.1 Combination with Probiotics to Create a Synbiotic

A "synbiotic" is the combination of probiotics, which are beneficial gut bacteria, and prebiotics, which are utilized by gut bacteria and improve the health of the host (Pandey et al. 2015). Synbiotics are thought to be more beneficial for host health than each component alone. However, if the probiotic bacteria are able to utilize the prebiotics in the product, then the levels of prebiotics decrease before the product reaches the consumer. It is preferable for prebiotics to be retained in

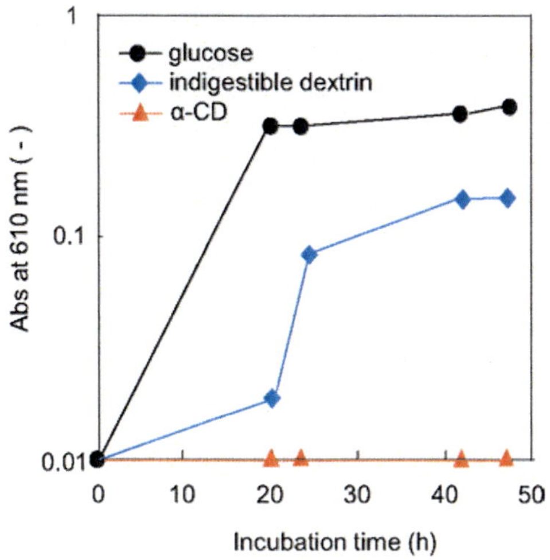

Fig. 14.15 Utilization of glucose, indigestible dextrin or α-CD by *L. casei*

the product before intake so that they are available for use by good bacteria in the gut after being consumed.

The potential use of α-CD as a prebiotic for preparing synbiotic was studied (Jo et al. 2007). The growth of *Lactobacillus casei* (*L. casei*) strain shirota YIT 9029, a probiotic bacterium, was tested using carbohydrate-free medium (peptone–yeast extract broth medium) with α-CD, glucose, or commercial indigestible dextrin. The optical density at 610 nm was measured as an indicator of bacterial viability. As the results, the bacterium utilized glucose and indigestible dextrin, but not α-CD which indicate that α-CD can be stably present in processed foods such as yogurt containing *L. casei* (Fig. 14.15). Pranckute et al. (2014) also reported that α-CD can be used in combination with *Lactococcus* bacteria. These results suggested that α-CD has potential as a component of synbiotic products. Moreover, butyric acid bacteria (*Clostridium butyricum*), which specifically produces butyric acid and stimulates mucosal immune activity (Murayama et al. 1995), and α-CD can be used as a synbiotic because α-CD also enhances the production of mucin and immunoglobulin A (IgA), which contributes to immune protection (see Chap. 13).

14.6.2 Improvement of Anti-diabetic Effect of Camel Milk by Its Combination with α-CD

Camel milk has anti-diabetic, anti-hypertensive, and anti-carcinogenic properties (Yadav et al. 2015). Mirmiran et al. (2017) systematically reviewed the anti-diabetic effects of camel milk. The proposed mechanism of the anti-diabetic action of camel milk is its high insulin content. In addition, a protein derived from camel milk was shown to inhibit DPP-4, exhibit anti-oxidant activity, and have protective effects on beta cells (Ayoub et al. 2018). Another advantage of camel milk is that it does not contain β-lactoglobulin, one of the major allergens in cow milk (Yadav et al. 2015).

It is known that α-CD can suppress the absorption of dietary digestible carbohydrates (Buckley et al. 2006). The combination α-CD with camel milk is expected to have synergically affect its anti-diabetic properties.

14.6.3 Reduction of Small Dense LDL-Cholesterol Levels in Serum by Combination of α-CD with Flaxseed Oil

Flaxseed oil is a functional vegetable oil rich in polyunsaturated fatty acids, especially α-linolenic acid. Flaxseed oil has various health benefits, such as lowering blood pressure and improving hyperlipidemia (Goyal et al. 2014). Recently, it has been reported to reduce the levels of small dense LDL in serum (Kawakami et al. 2015). Among the smaller LDL particles, small dense LDL is a better biomarker of atherosclerotic diseases than is LDL-cholesterol (Ivanova et al. 2017).

Because α-CD has been reported to reduce the levels of small LDL in serum (Amar et al. 2016), it is expected to act synergistically when combined with flaxseed oil. Although unsaturated fatty acids are easily oxidized, α-CD can improve their oxidation stability *via* the formation of an inclusion complex (Regiert 2007).

14.7 Conclusion and Future Perspectives

In this chapter, we discuss the combination of α-CD with various functional foods not only in terms of its inclusion effect, but also in terms of its effects to promote anti-bacterial activity and dietary fiber function. Studies have shown that the combination of α-CD with manuka honey improves the honey's anti-bacterial effect; with radish, it can improve the stability of MTBI and the anti-obesity effect; and with kiwi fruit, it can stabilize the protease activity of actinidin. In combination with cucumber, α-CD can stabilize phospholipase C activity, thereby reducing lipid micellar solubility. In addition, α-CD can be used as a prebiotic for preparing synbiotic when combined with probiotics, and it has the potential to improve the anti-diabetic effect and reduce small dense LDL-cholesterol when combined with camel milk or flaxseed oil.

The functional components of various foods and their actions have been studied in detail over the years, while more recent studies have explored the health benefits of α-CD itself. Further studies should explore the high-functionality combination of α-CD with various functional foods beyond its inclusion effect.

References

al Somal N, Coley KE, Molan PC et al (1994) Susceptibility of *Helicobacter pylori* to the antibacterial activity of manuka honey. J R Soc Med 87(1):9–12

Almasaudi SB, Al-Nahari AAM, El Sayed M et al (2017) Antimicrobial effect of different types of honey on *Staphylococcus aureus*. Saudi J Biol Sci 24(6):1255–1261

Amar MJA, Kaler M, Courville AB et al (2016) Randomized double blind clinical trial on the effect of oral α-cyclodextrin on serum lipids. Lipids Health Dis 15(1):115

Asada Y, Jo A, Shiga H et al (2009) Protective effect of α-cyclodextrin on fruits protease activity. In: Proceedings of 26th Cyclodextrin symposium in Japan, Tochigi, pp 148–149

Ayoub MA, Palakkott AR, Ashraf A et al (2018) The molecular basis of the anti-diabetic properties of camel milk. Diabetes Res Clin Pract 146:305–312

Buckley JD, Thorp AA, Murphy KJ et al (2006) Dose-dependent inhibition of the post-prandial glycaemic response to a standard carbohydrate meal following incorporation of alpha-cyclodextrin. Ann Nutr Metab 50(2):108–114

Carter DA, Blair SE, Cokcetin NN et al (2016) Therapeutic manuka honey: no longer so alternative. Front Microbiol 7:569

Chepulis LM, Francis E (2012) An initial investigation into the anti-inflammatory activity and antioxidant capacity of alpha-cyclodextrin-complexed manuka honey. J Complement Integr Med 9(1):25

Dominy SS, Lynch C, Ermini F et al (2019) *Porphyromonas gingivalis* in Alzheimer's disease brains: evidence for disease causation and treatment with small-molecule inhibitors. Sci Adv 5(1):eaau3333

Dubois T, Jacquet A, Schnek AG et al (1988) The thiol proteinases from the latex of *Carica papaya L.* I. Fractionation, purification and preliminary characterization. Biol Chem 369:733–740

Furune T, Ikuta N, Ishida Y et al (2014) A study on the inhibitory mechanism for cholesterol absorption by α-cyclodextrin administration. Beilstein J Org Chem 10:2827–2835

Furune T, Ishida Y, Terao K et al (2019) A study on enhancement of lipid lowering effect of cucumber by α-cyclodextrin. In: Proceedings of 26th Cyclodextrin symposium in Japan, Tochigi, pp 165–166

Goyal A, Sharma V, Upadhyay N et al (2014) Flax and flaxseed oil: an ancient medicine & modern functional food. J Food Sci Technol 51(9):1633–1653

Hirata T, Cho YM, Suzuki I et al (2019) 4-Methylthio-3-butenyl isothiocyanate (MTBITC) induced apoptotic cell death and G2/M cell cycle arrest via ROS production in human esophageal epithelial cancer cells. J Toxicol Sci 44(2):73–81

Inoue K, Murayama S, Seshimo F et al (2005) Identification of phenolic compound in manuka honey as specific superoxide anion radical scavenger using electron spin resonance (ESR) and liquid chromatography with coulometric array detection. J Sci Food Agric 85:872–878

Ivanova EA, Myasoedova VA, Melnichenko AA et al (2017) Small dense low-density lipoprotein as biomarker for atherosclerotic diseases. Oxidative Med Cell Longev 2017:1273042

Jo A, Nakata D, Terao K et al (2007) New synbiotics by the combination of α-cyclodextrin and lactic acid bacteria. In: Proceedings of 25th Cyclodextrin symposium in Japan, Tochigi, pp 144–145

Johnston M, McBride M, Dahiya D et al (2018) Antibacterial activity of manuka honey and its components: an overview. AIMS Microbiol 4(4):655–664

Katsumata S, Wolber FM, Tadaishi M et al (2015) Effects of manuka honey combined with α-cyclodextrin on bone metabolism and caecal bacterial contents in ovariectomized mice. J Food Nutr Sci 2(3):1–6

Kaur L, Rutherfurd SM, Moughan PJ et al (2010) Actinidin enhances protein digestion in the small intestine as assessed using an *in vitro* digestion model. J Agric Food Chem 58(8):5074–5080

Kawakami Y, Yamanaka-Okumura H, Naniwa-Kuroki Y et al (2015) Flaxseed oil intake reduces serum small dense low-density lipoprotein concentrations in Japanese men: a randomized, double blind, crossover study. Nutr J 14:39

Kemp TR, Knavel DE, Stoltz LP (1974) Identification of some volatile compounds from cucumber. J Agric Food Chem 22(4):717–718

Kimura M, Ueno C, Uekaji U et al (2017) Improvement of antioxidant effect of manuka honey by α-cyclodextrin. In: Proceedings of 34th Cyclodextrin symposium in Japan, Aichi, pp 290–291

Kjajr A, Madsen JO, Maeda Y et al (1978) Volatiles in distillates of fresh radish of Japanese and Kenyan origin. Agric Biol Chem 42(9):1715–1721

Lewis D, Luh BS (1988) Application of actinidin from kiwi fruit to meat tenderization and characterization of beef muscle protein hydrolysis. J Food Biochem 12(3):147–158

Li X, Jin Z, Wang J (2007) Complexation of allyl isothiocyanate by α- and β-cyclodextrin and its controlled release characteristics. Food Chem 103(2):461–466

Li J, Xie B, Yan S et al (2015) Extraction and determination of 4-methylthio-3-butenyl isothiocyanate in Chinese radish (*Raphanus sativus L.*) roots. LWT 60:1080–1087

Linde GA, Junior AL, de Faria EV et al (2009) Taste modification of amino acids and protein hydrolysate by α-cyclodextrin. Food Res Int 42(7):814–818

Mavric E, Wittmann S, Barth G et al (2008) Identification and quantification of methylglyoxal as the dominant antibacterial constituent of manuka (*Leptospermum scoparium*) honeys from New Zealand. Mol Nutr Food Res 52(4):483–489

Medhi B, Prakash A, Avti PK et al (2008) Effect of manuka honey and sulfasalazine in combination to promote antioxidant defense system in experimentally induced ulcerative colitis model in rats. Indian J Exp Biol 46(8):583–590

Mirmiran P, Ejtahed HS, Angoorani P (2017) Camel milk has beneficial effects on diabetes mellitus: a systematic review. Int J Endocrinol Metab 15(2):e42150

Molly J, Edison S, Vijajaraghavan R et al (2017) Effect of curry leaves and cucumber fruit on lipid profile in menopausal women with hyperlipidaemia: a randomized controlled pilot study. Int J Clin Trials 4(1):7–13

Mugita N, Nambu T, Takahashi K et al (2017) Proteases, actinidin, papain and trypsin reduce oral biofilm on the tongue in elderly subjects and in vitro. Arch Oral Biol 82:233–240

Mukherjee PK, Nema NK, Maity N et al (2013) Phytochemical and therapeutic potential of cucumber. Fitoterapia 84:227–236

Murayama T, Mita N, Tanaka M et al (1995) Effects of orally administered Clostridium butyricum MIYAIRI 588 on mucosal immunity in mice. Vet Immunol Immunopathol 48(3–4):333–342

Mysak J, Podzimek S, Sommerova P et al (2014) *Porphyromonas gingivalis*: major periodontopathic pathogen overview. J Immunol Res 2014:476068

Nakano Y, Yoshimura M, Koga T (2002) Correlation between oral malodor and periodontal bacteria. Microbes Infect 4(6):679–683

Okamoto H, Nihei N, Furune T et al (2017) Development of a sweetener, manuka honey α-dextrin powder, for improving in dietary habit. In: Proceedings of 34th Cyclodextrin symposium in Japan, Aichi, pp 288–289

Okamoto H, Ino S, Nihei N et al (2019) Anti-obesity effects of α-cyclodextrin-stabilized 4-methylthio-3-butenyl isothiocyanate from daikon (*Raphanus sativus var. longipinnatus*) in mice. J Clin Biochem Nutr 65(2):99–108

Pandey KR, Naik SR, Vakil BV (2015) Probiotics, prebiotics and synbiotics - a review. J Food Sci Technol 52(12):7577–7587

Parkar SG, Jobsis CMH, Herath TD et al (2017) Metabolic and microbial responses to the complexation of manuka honey with α-cyclodextrin after simulated gastrointestinal digestion and fermentation. J Funct Foods 31:266–273

Pranckute R, Kaunietis A, Kuisiene N et al (2014) Development of synbiotics with inulin, palatinose, alpha-cyclodextrin and probiotic bacteria. Pol J Microbiol 63(1):33–41

Ratcliff PA, Johnson PW (1999) The relationship between oral malodor, gingivitis, and periodontitis. A review. J Periodontol 70:485–489

Regiert R (2007) Oxidation-stable linoleic acid by inclusion in α-cyclodextrin. J Incl Phenom Macrocycl Chem 57:471–474

Safii SH, Tompkins GR, Duncan WJ (2017) Periodontal application of manuka honey: antimicrobial and demineralising effects *in vitro*. Int J Dent 2017:9874535

Seto Y, Nakanishi Y, Saeki T et al (1990) The stabilizing effect of cyclodextrin on the photodegradation of chlorophyll. Sci Cookery 23(4):392–396. (Japanese)

Sugimori D, Nishina A, Chiba Y (2011) Japanese unexamined patent application publication no. P2011-10608A

Suzuki N, Higuchi T, Nakajima M et al (2016) Inhibitory effect of *Enterococcus faecium* WB2000 on volatile sulfur compound production by *Porphyromonas gingivalis*. Int J Dent 2016:8241681

Swift S, Chepulis LM, Uy B et al (2014) Enhanced antibacterial activity of MGO™ manuka honey complexed with a-cyclodextrin (manuka honey with CycloPower™). FFHD 4(5):172–181

Tamesue N, Inoue T, Juniper K Jr (1973) Solubility of cholesterol in bile salt-lecithin model systems. Am J Dig Dis 18(8):670–678

Uda Y, Shitara T (2013) Effects of benzyl isothiocyanate and (E,Z)-4-methylthio-3-butenyl isothiocyanate on *in vitro* growth of human intestinal bacteria and on in vivo formation of short-chain fatty acids in the rat intestine. In: Méndez-Vilas A (ed) Worldwide research efforts in the fighting against microbial pathogens: from basic research to technological developments. Brown Walker Press, Florida, pp 150–154

Ueno C, Jo A, Nakata D et al (2012) Enhancement of antibacterial activity of manuka honey on periodontal pathogenic bacteria with α-cyclodextrin. In: Proceedings of 29th Cyclodextrin symposium in Japan, Tokyo, pp 198–199

Ueno C, Nakata D, Ishida Y et al (2013) Stabilization of pungent principle of daikon (white radish) by cyclodextrin. In: Proceedings of 30th Cyclodextrin symposium in Japan, Kumamoto, pp 278–279

Wakabayashi H, Yamauchi K, Kobayashi T et al (2009) Inhibitory effects of lactoferrin on growth and biofilm formation of *Porphyromonas gingivalis* and *Prevotella intermedia*. Antimicrob Agents Chemother 53(8):3308–3316

Watanabe K, Rahmasari R, Matsunaga A et al (2014) Anti-influenza viral effects of honey in vitro: potent high activity of manuka honey. Arch Med Res 45(5):359–365

Yadav AK, Kumar R, Priyadarshini L (2015) Composition and medicinal properties of camel milk: a review. Asian J Dairy Food Res 34(2):83–91

You H, Hao R, Li R et al (2015) The effect of radish sourced 4-(Methylthio)-3-butenyl isothiocyanate on ameliorating the severity of high fat diet inducted nonalcoholic fatty liver disease in rats. Int J Clin Exp Med 8(9):15910–15919

Zhang HM, Li Z, Uematsu K et al (2008) Antibacterial activity of cyclodextrins against *Bacillus* strains. Arch Microbiol 190:605–609

Chapter 15
Fruit Packaging with 1-Methylcyclopropene Included in Alpha-Cyclodextrin

Hermawan D. Ariyanto and Hidefumi Yoshii

15.1 Introduction

As a vulnerable commodity, fruit requires special attention and treatment to prevent a decline in quality during postharvest processing, distribution and handling. Fruits quality is greatly affected by ripening, decaying, and senescence due to ethylene action. In 2011, the Food Agriculture Organization (FAO) estimated that 32% of all fruit produced in the world was lost or wasted during the shipping and distribution process, equivalent to approximately 1.3 billion tons of fruit (Lipinski et al. 2013). This loss occurs mostly in climacteric fruit, whose behaviour is associated with ethylene production and respiration rate. Ethylene is a plant growth hormone that promotes and triggers ripening processes in fruit, including physiological changes in color, texture, flavor (taste and aroma), and nutritional quality. Figure 15.1 shows physiological changes in apple during storage. These changes indicate the maturation and ripening of fruit organs at maximum stage of growth and development. Many studies have shown that ethylene biosynthesis and responses have an impact on fruit maturation and play a key role in the ripening process in most climacteric fruits, including banana, apple, pear, melon, and tomato (Barry and Giovannoni 2007; Liu et al. 2015). The ripening process in fruit frequently occurs during long shipment and unloading periods. Environmental conditions such as temperature and relative humidity (RH), which change drastically during unloading, also promote a decrease in fruit quality.

H. D. Ariyanto (✉)
Department of Industrial Chemical Engineering, Vocational School of Diponegoro University, Semarang, Indonesia
e-mail: hd.ariyanto@live.undip.ac.id

H. Yoshii
Department of Food Science and Human Nutrition, Setsunan University, Osaka, Japan
e-mail: hidefumi.yoshii@setsunan.ac.jp

T. M. Ho et al. (eds.), *Functionality of Cyclodextrins in Encapsulation for Food Applications*, https://doi.org/10.1007/978-3-030-80056-7_15

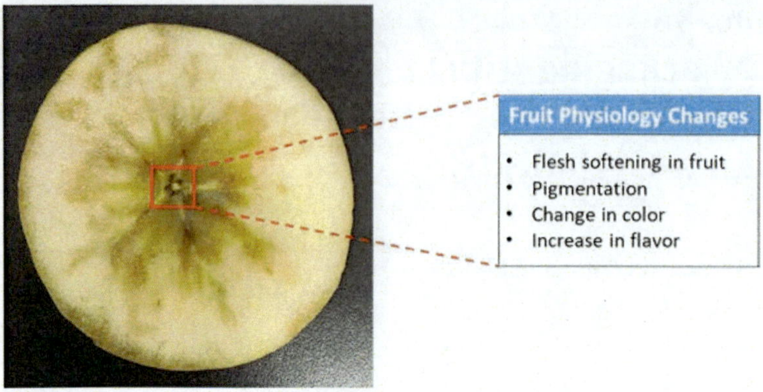

Fig. 15.1 Physiological changes of fruit during ripening

1-Methylcyclopropene (1-MCP), originally discovered by Sisler and Serek (1997), is the last generation of ethylene antagonist agents. 1-MCP demonstrates superiority and became popular among ethylene physiologists, horticulturists, food scientists, and postharvest technologists. Blankenship and Dole (2003), and Watkins (2006) reviewed the responses of 1-MCP and its use on fruits and vegetables. They reported that 1-MCP can be used as an ethylene antagonist agent to delay the ripening process of fruits and vegetables by blocking the ethylene receptor. 1-MCP was made commercially available in the form of an inclusion complex with α-cyclodextrin (α-CD) as the carrier material, registered by Rohm and Haas Company (AgroFresh, Inc.) and sold under the trade name SmartFresh in 2002. Encapsulation of 1-MCP to produce inclusion complex powders has been developed in several agents including molecular sieves, modified starch, sawdust, and α-CD which α-CD as a wall material is most preferred (Neoh et al. 2007; Zhang et al. 2011).

Over the last decade, the use of 1-MCP inclusion complexes in α-CD has become popular to potentially prolong the shelf life of fruit during postharvest processing. For this reason, the development of fruit packaging containing 1-MCP inclusion complex shows a great potential for the preservation of fresh products. The additional functions of fruit packaging for prolonged shelf life and enhanced consumer convenience are useful during postharvest processing, distribution and handling. This chapter aims to summarise the development of fruit packaging containing a 1-MCP inclusion complex. The topics in this chapter include effectiveness of 1-MCP in delaying of fruit ripening, formation of packaging material with 1-MCP inclusion complex, and release behaviour.

15.2 1-MCP Effectiveness for Delaying Fruit Ripening

1-MCP is an advanced compound that can maintain and prolong the shelf life of fresh fruit by inhibiting ethylene action (Sisler and Serek 1997; Blankenship and Dole 2003; Nanthachai et al. 2007). 1-MCP has been registered for use with a wide variety of crops and food products in many countries (Watkins 2006). In 2009, 1-MCP was registered by the Ministry of Health, Labour, and Welfare in Japan for use as an agrochemical for the preservation of apples, pears, and persimmons. According to the safety, toxicity, and environmental profile of 1-MCP, application of 1-MCP is strongly favourable. At low concentrations, 1-MCP is non-toxic and similar to naturally occurring substances. A small amount of 1-MCP, in the range of 1 ppm, is necessary to prevent ethylene response.

The conventional application or delivery system of 1-MCP to fresh fruits and vegetables involves dissolving the inclusion complex into water or other solvents. 1-MCP gas is released into the atmosphere and applied to fruits and vegetables. Fruits and vegetables absorb 1-MCP gas through the epidermis of the host tissue, binding with the ethylene receptor sites of the host, delaying the ripening process and ethylene action. 1-MCP responds positively on climacteric fruit, but does not have the same reaction on non-climacteric fruit (Watkins 2006; Fan et al. 1999; Huber 2008). Moretti et al. (2002) investigated 1-MCP to delay tomato ripening. They found that applying 1-MCP in the early stage of tomato ripening immediately after harvest made the treatment ineffective; the tomatoes spoiled before becoming viable for sale. Ariyanto et al. (2019a, b) also reported the effect of 1-MCP release from inclusion complex powder on the color change of tomatoes in storage. They indicated that 1-MCP treatment significantly affected the color change during storage at room temperature, as shown in Fig. 15.2.

Fig. 15.2 Effect of different amounts of 1-MCP on the color change of tomato in storage

The effectiveness of 1-MCP treatment depends on the concentration, treatment time, and treatment temperature. The concentration of 1-MCP gas should be sufficient to saturate the receptors and compete with any ethylene that is present. The treatment time also affects the penetration of 1-MCP into the plant tissue. The temperature at which treatment takes place influences the length of time necessary for treatment. The ability of 1-MCP to delay ripening, senescence, and decay of climacteric fruit has been demonstrated in many studies (Liu et al. 2015; Barry and Giovannoni 2007; Lelièvre et al. 1997; Saltveit 1999). Bouzayen et al. (2010) reviewed the mechanism of fruit ripening and reported that ethylene effects include fruit ripening, induction of flowering, loss of chlorophyll, abortion of plant parts, stem shortening, abscission (dropping) of plant parts, epinasty (stem bend), and dormancy. Ethylene can be produced when plants are injured, either by mechanical actions or by disease. During ethylene biosynthesis, plant cells contain ethylene binding receptors, which are proteins on the cell membrane. The ethylene receptor has ethylene binding sites with similar binding affinities for ethylene, and triggers ripening when it chemically reacts with ethylene (O'Malley et al. 2005). Ethylene binding and action can be suppressed by the potential inhibitor 1-MCP. In fact, 1-MCP has been shown to bind competitively to the ethylene receptors with a greater affinity for the receptors than ethylene (Sisler et al. 2003; Hall et al. 2000). Thewes et al. (2018) investigated 1-MCP suppression in the anaerobic metabolism of apples in a dynamic controlled atmosphere. They reported that 1-MCP application resulted in lower 1-aminocyclopropane-1-carboxylic acid (ACC) oxidase enzyme activity and ethylene production.

15.3 Development of Packaging Material Containing 1-MCP Inclusion Complexes

As a water-soluble powder, inclusion complexes of 1-MCP into α-CD are unstable under certain conditions, such as high temperature and RH. At 80% RH, the fine particles of 1-MCP inclusion complex powder aggregate and slow the 1-MCP release. Moreover, the increase in RH triggered the collapse of the crystalline structure (Neoh et al. 2008, 2017). Ariyanto and Yoshii (2019) also reported that the collapse behaviour of 1-MCP inclusion complex powder in α-CD occurred at greater than 60% RH and decreased the 1-MCP release from the powder. Thus, the new approach for gas composition with 1-MCP controlled release improves 1-MCP efficiency in the packaging. Controlled release packaging (CRP), is the latest technology focusing on the release of an active compound under certain conditions to maintain the quality and safety of food during storage (Lacoste et al. 2005). This technology has become popular in the last few years for the preservation of fresh products. Among active packaging technologies, CRP is one of the most innovative and challenging because of its diversity of materials, methods of application, and the variety of food products that can be protected (Malhotra et al. 2015;

Table 15.1 Packaging material containing 1-MCP inclusion complex in cyclodextrin

Carrier materials	Carrier form	References
Tyvek®, paper, low-density polyethylene (LDPE), and polyvinyl acetate (PVAc)	Release sachet	Lee et al. (2006)
Polymer resins (LDPE, PVC, PP, PS, and EVOH)	Polymer film	Hotchkiss et al. (2007)
Soy protein	Releasing pad	Ortiz et al. (2013)
Polystyrene	Electrospun fibre	Neoh et al. (2017)
Nano-Ag polyethylene	Nano packaging bag	Xu et al. (2018)
Shellac	Coated paper	Ariyanto et al. (2019a, b)

Khaneghah et al. 2018). Active packaging is a promising technology that enables the package to interact with the food to improve food quality, safety, and convenience. Appendini and Hotchkiss (2002) discussed active packaging materials and their research development in food packaging applications and other food contact surfaces. They highlighted the different types of polymers developed for food contact, commercial applications, and testing methods. Table 15.1 shows some of the recent developments in packaging materials containing 1-MCP inclusion complex in CD.

In active packaging, CD inclusion complexes have been used to incorporate active agents such as oxygen scavengers, carbon dioxide emitters, moisture absorbers, ethylene absorbers/inhibitors, ethanol emitters, flavor-releasing systems, and antimicrobial agents. By mixing CD inclusion complex with active agents, packaging materials such as films, containers, laminates, and trays can be produced. In these products, the complexed substances are homogeneously dispersed and released slowly from the polymer matrix (Szente and Fenyvesi 2018). Active packaging materials can be produced for specific purposes, incorporating and slowly releasing active substances (antimicrobials, insecticides, insect repellents, fragrant components, plant hormones, etc.). The application of active substances in CD-complexed form has several advantages:

- Reduced loss of biological compounds or active agents by decomposition or volatilization at high temperatures during the production of material packaging.
- Incorporation of CD-complexed active substances into the polymer matrix without reducing their chemical stability.
- Effective blending of cyclodextrin inclusion complexes containing active agents into the polymer phase.
- Improved release of the complex and polymer incorporation of active substances.

The 1-MCP delivery system was proposed by Lee et al. (2006) and Hotchkiss et al. (2007). They investigated the delivery of 1-MCP with controlled release in sachet release systems and materials such as polyvinyl acetate (PVAc), low density polyethylene (LDPE), and Tyvek®. They found that PVAc sachets containing silica gel can potentially be used for the slow release of 1-MCP. The release of 1-MCP from a sachet can be influenced by a number of factors including film or material characteristics, the adsorption abilities of the adsorbing agents, the concentration of sorbate, the size of the sachet, and environmental factors. Temperature and RH conditions during storage are critical in developing a controlled-release system for 1MCP. Using constant and linear RH, Neoh et al. (2017) investigated the controlled release of 1-MCP from electrospun fibres. Electrospun fibres 1.5–4.4 μm in diameter were produced with 12.5–20.0% (w/w) polystyrene (PS) in electrospun solution and an equal amount of the inclusion complex. Neoh et al. (2017) indicated that in an electrospun fibre system containing 1-MCP, RH also affected the release rate constant of 1-MCP through a moisture–adsorption mechanism.

In 2013, Ortiz et al. developed and tested soy protein biodegradable films releasing an ethylene inhibitor. Soy protein pads were cast from formulations containing different glycerol concentrations (20%, 40%, or 60% on a protein basis) and pH (2.0, 7.0, or 10.0). They found the best performing films had a pH of 7.0 and 20% glycerol and used them to characterise the influence of the 1-MCP-release pads on tomato quality. With application to the tomatoes, the pads containing 1-MCP delayed ripening without causing negative quality changes. Although the release of 1-MCP can be controlled using α-CD in the inclusion complex, soy protein can also improve the release of 1-MCP from pads, and can be useful for postharvest treatment during transit and distribution. The results showed that soy protein 1-MCP release pads delayed tomato softening and pectin solubilisation, reduced decay and lycopene accumulation, and were useful in postharvest treatments.

Recent advances in the development of a 1-MCP controlled-release system using coated paper in functional packaging for fruits and vegetables have been proposed by Ariyanto et al. (2019a, b), as shown in Fig. 15.3. They developed an additional function of 1-MCP inclusion complex in α-CD with a controlled-release system using coated paper that delays ripening and prolongs the shelf life of fruits and vegetables. In this study, a 1-MCP controlled release system on coated paper was produced using shellac as the coating material. The results showed that the dynamic release of 1-MCP from the coated paper was affected mainly by RH; the release rate constant can be correlated with the moisture concentration in humid air. Figure 15.4 shows the structure and morphology of 1-MCP coated paper at 80% RH. At low RH

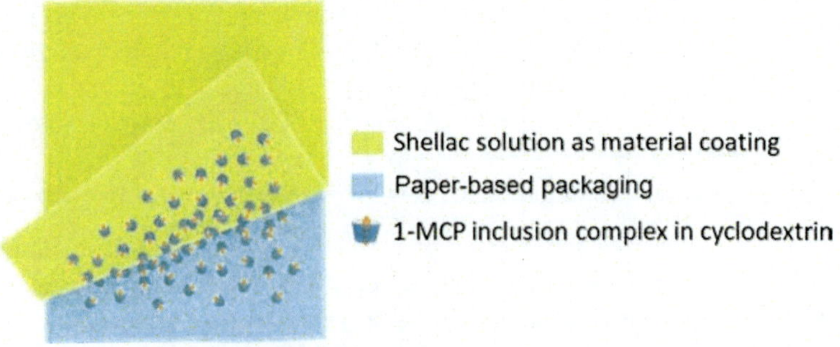

Fig. 15.3 Coated paper containing 1-MCP inclusion complex in CD as functional packaging

Fig. 15.4 SEM images of (**a**) initial 1-MCP coated paper before treatment and 1-MCP coated paper after treatment at 80% RH and different temperatures: (**b**) at 40 °C; (**c**) at 50 °C; (**d**) at 60 °C. (Adapted with permission from Ariyanto et al. 2019a, b)

(40–60%) in all temperature conditions, fine particles of the 1-MCP inclusion complex powder with shellac coating were clearly observed. However, the particles became aggregated at 80% RH, especially at 60 °C. The aggregation induced the collapse phenomenon; moisture penetrated the 1-MCP inclusion complex powder through the shell wall of the shellac and changed the structure of the powder. The release of 1-MCP can be improved by coating the paper with shellac, with potential for further application as functional controlled-release packaging.

In recent decades, research on edible films and coatings on foods has been undertaken by food engineers, driven by high consumer's demand for longer shelf life, better fresh food quality, and environmentally friendly packaging. Coatings can extend the shelf life of many foods, controlling moisture and solute migration, gas exchange, and oxidative reaction rates. In addition, edible coatings can be used as carriers for bioactive compounds such as antioxidants, antimicrobials, flavors, and probiotics to improve the quality of food products. These approaches can be useful for extending shelf life and improving product functionality. The use of polymer blends as coating materials for controlled systems offers major advantages, including: (1) adjustment of active compound release patterns, mechanical properties, and release mechanisms, (2) improvement of film formation and storage stability, and

(3) possibility of developing novel strategies for specific purposes. Quirós-Sauceda et al. (2014) reviewed edible coatings as encapsulating matrices for bioactive compounds. In their review, the most important factors affecting the release of active compounds are storage temperature and RH, which change the stability and characteristics of the polymer coating material.

15.4 Release of 1-MCP from the Packaging System

Active compounds introduced in packaging can act through direct or indirect contact with the product. Furthermore, active compounds be used in packaging system with different ways: as a coating layer directly applied to the product surface, as a self-supported film or coated paper, or as an object inserted in the package. The release of active compound from packaging materials depends on the mechanisms influencing the stability of the active compounds. In all cases, the same series of unit operations with mass and heat transfers, diffusion, and reaction must be considered. There are several release mechanisms for active compounds from polymeric systems, including diffusion, swelling, and polymer chain relaxation. However, most active compounds are released from polymeric systems by diffusion. The release of active compounds from a polymeric packaging film involves three steps: (1) molecular diffusion within the film toward the film/food interface, (2) mass transfer across the interface, and (3) dispersion into food or desorption into the package headspace (Irkin and Esmer 2015).

The diffusion step, which controls the release rate and release amount, may change based on the duration of release and other external factors such as temperature, moisture, solubility, and diffusivity of the active compound. The diffusivity and the partition coefficient are two model parameters commonly used to describe the release behaviour of an active compound from a polymer film to a food or food simulant. Diffusivity indicates how fast the active compound moves within the film, and the partition coefficient indicates how much of the active compound is released from the film to the food at equilibrium (Chen et al. 2019). The release kinetics can be defined by the rate-controlling step of diffusion.

Moisture absorption is an important factor in the release of active compounds from inclusion complexes with cyclodextrin as wall material (Reineccius et al. 2002). The build-up of a dynamic vapor sorption (DVS) system was used to conduct a real-time evaluation of the release mechanism of inclusion complex powder under linearly ramped or stepwise RH changes at constant temperature (Yamamoto et al. 2012). Ariyanto et al. (2019a, b) adopted a DVS system to determine the release mechanism of 1-MCP gas from coated paper under stepwise RH changes at different storage temperatures. Figure 15.5 shows the release characteristics of 1-MCP subjected to stepwise RH changes at constant temperature. RH strongly increased the release of 1-MCP from coated paper. Different storage temperatures also affected the release flux of 1-MCP from the coated paper. As shown in Fig. 15.5,

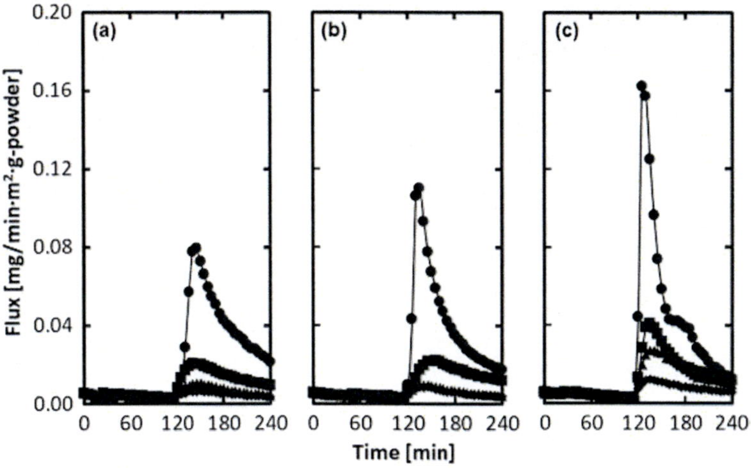

Fig. 15.5 Release profile of 1-MCP gas from coated paper with stepwise RH changes of 40% RH (◆), 50% RH (▲), 60% RH (■), and 80% RH (●) at different storage temperatures: (**a**) 40 °C; (**b**) 50 °C; (**c**) 60 °C. (Adapted with permission from Ariyanto et al. 2019a, b)

Fig. 15.6 Correlation of moisture concentration (C_w) and release rate constant (k) of 1-MCP from coated paper at different storage temperatures. (Adapted with permission from Ariyanto et al. 2019a, b)

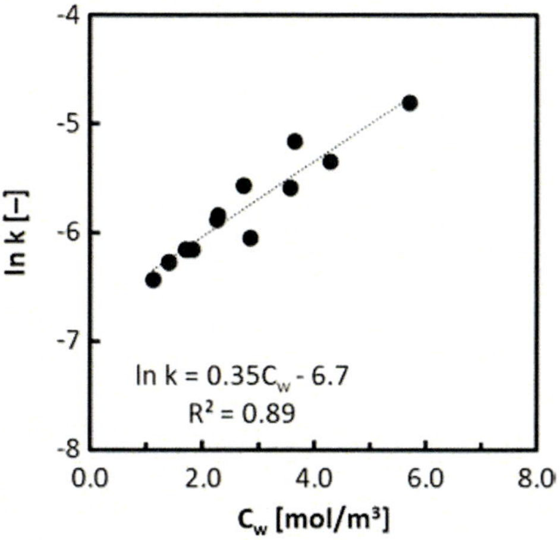

storage temperatures from 40 to 60 °C significantly increased the 1-MCP release flux from the coated paper under all RH conditions (40–80% RH). Figure 15.6 confirms the relationship between the release rate constant of 1-MCP and the moisture concentration. A higher moisture concentration increased the release rate constant of 1-MCP from the coated paper. This relationship is assumed to result from the shellac coating of the 1-MCP inclusion complex powder, which improved the stability and release rate of 1-MCP during storage.

15.5 Kinetic Model of Controlled Release of Active Compound

Mathematical models are useful for summarizing kinetic release data, predicting release behaviour, and providing mechanistic insights. Different mathematical models were used to investigate the release kinetics of the active compound based on its release mechanism and release behaviour. Some commonly applied models for the release of active compounds are summarized.

15.5.1 Crank's Diffusion Model

The most popular mathematical models to describe release behaviour are derived from differential equations based on one-dimensional Fickian diffusion with appropriate initial and boundary conditions (Crank 1957). These models typically require the following assumptions: the partition coefficient and diffusivity are constant at a given temperature.

Partition coefficient (K) estimation:

$$K_{F,P} = \frac{C_F}{C_P} \text{or} K_{P,F} = \frac{C_P}{C_F} \tag{15.1}$$

where C_F and C_P are the concentrations of the active compound in the food and the package at equilibrium, respectively.

Diffusion coefficient (diffusivity) estimation

Crank's diffusion model:

$$\frac{M_{f,t}}{M_\infty} = 1 - \sum_{n=1}^{\infty} \exp\left(\frac{2\alpha(1+\alpha)}{1+\alpha+\alpha^2 q_n{}^2}\right)\left(-\frac{Dq_n{}^2 t}{L_p{}^2}\right) \tag{15.2}$$

$$\alpha = K_{FP}\frac{V_F}{V_P}$$

where V_F is the volume of food, V_P is the volume of the package, K_{FP} is the partition coefficient, q_ns are non-zero positive roots of tan $q_n = -\alpha q_n$, $M_{f,t}$ is the mass of active compound migrated to the food at time t, M_∞ is the mass of the active compound in the food at equilibrium, L_P is the film thickness, D is the diffusivity of the active compound in the film, and t is time. The estimation of D can be obtained using computer programs such as Microsoft Excel and MATLAB.

When the food simulant is infinite compared to the film, Eq. (15.1) can be simplified to Eq. (15.2)

$$\frac{M_{f,t}}{M_{\alpha}} = 1 - \sum_{n-1}^{\infty} \frac{8}{(2n-1)^2 \pi^2} \exp\left(\frac{-D(2n+1)^2 \pi^2 t}{L_P^2}\right) \tag{15.3}$$

When $M_{f,t}/M_{P,0}$ is <0.6, Eq. (15.3) can be simplified to Eq. (15.4)

$$\frac{M_{f,t}}{M_{P,0}} = \frac{4}{L_P}\left(\frac{D_t}{\pi}\right)^{0.5} \tag{15.4}$$

where $M_{P,0}$ is the initial loading of the active compounds in the package. D can be estimated from the slope of the plot of $M_{f,t}/M_{P,0}$ versus $t^{0.5}$.

15.5.2 Zero-Order Diffusion Model

A zero-order model is typically used for volatile drug release (Vazquez-Landaverde et al. 2007; Zhang et al. 2018). In this model, diffusion is not dependent on the concentration of the active compound. This model is applicable when deliberation is slow compared to the actives loaded in the packaging matrix. A true zero-order is desirable in the food industry for its short release time, solubility of active compounds in the matrix, surface activity, and burst release from the capsule (Vasisht 2014).

$$Q_t = Q_0 + k_0 t \tag{15.5}$$

where Q_t is the amount of active compound released in time t; Q_0 is the amount of initial release of the active compound; k_0 is the zero-order release rate constant.

15.5.3 First-Order Diffusion Model

A first-order model is typically used for volatile release by erosion or diffusion mechanisms. The volatile release rate decreases exponentially with time (Meiners 2012). The diffusion is proportionally related to the concentration of active compounds. The model can be expressed as

$$Q_t = Q_0 e^{-k_1 t} \tag{15.6}$$

where Q_t is the amount of active compound released in time t; Q_0 is the amount of initial release of the active compound; k_1 is the zero-order release rate constant.

Zero-order and first-order kinetic models are used to describe controlled release (Lakkis 2007). The integrated first-order rate is expressed as

$$\ln Q_t = \ln Q_0 - k_1 t \tag{15.7}$$

15.5.4 *Weibull Distribution Function or Avrami's Model*

Many controlled release systems do not follow a pure diffusional process; several other approaches have been developed to consider other mechanisms. One alternative to describe release behaviour is the empirical Weibull function, which is valid for the entire release curve.

$$\frac{M_t}{M_\infty} = 1 - \exp\left[-(kt)^n\right] \tag{15.8}$$

where M_t is the total amount of active compound released; M_∞ is the total amount of active compound at equilibrium, t is the storage time; k is the release rate constant; n is the release mechanism number/parameter. A value of n in the range of 0.54–0.75 indicates that Fick diffusion is the ruling mechanism (Papadopoulou et al. 2006), and an n value close to 0.89 indicates Case II transport (polymer chain relaxation rules the release). If the value of n is >1, a complex release mechanism is indicated (Estevez-Areco et al. 2018).

15.6 Conclusion and Future Perspective

1-MCP has been successfully applied to extend the shelf life of fruits and vegetables. Thus, fruit and vegetable packaging based on 1-MCP encapsulated in cyclodextrin has great potential for commercial application. 1-MCP is useful in packaging materials during postharvest processing and distribution to maintain freshness for a wide range of fruits and vegetables. A novel packaging material containing 1-MCP encapsulated in cyclodextrin may have great commercial potential, and is still being researched. The release of 1-MCP depends on the RH and storage temperature conditions. Future packaging can inhibit collapse behaviour and the release of 1-MCP inclusion complex in high RH conditions, impacting food preservation (maintaining the quality and/or decision support systems), leading to further improvement in postharvest processing of fruits and vegetables. In the future, packaging material containing 1-MCP included in cyclodextrin can be coated as a film or applied to labels, paper, or inside the packaging. These structures are highly desirable for promoting the release of 1-MCP during storage.

References

Appendini P, Hotchkiss JH (2002) Review of antimicrobial food packaging. Innov Food Sci Emerg Technol 6:113–126

Ariyanto HD, Yoshii H (2019) Effect of stepwise humidity change on the release rate constant of 1-methylcyclopropene (1-MCP) in a cyclodextrin inclusion complex powder. Food Packag Shelf Life 21:100322. https://doi.org/10.1016/j.fpsl.2019.100322

Ariyanto HD, Bercis MN, Yoshii H (2019a) Effect of 1-methylcyclopropene release from inclusion complexes powder on the quality of tomato by short-term treatment. IOP Conf Ser Earth Environ Sci 309:012065. https://doi.org/10.1088/1755-1315/309/1/012065

Ariyanto HD, Chiba M, Oguma K et al (2019b) Release behavior of 1-methylcylopropene coated paper-based shellac solution in response to stepwise humidity changes to develop novel functional packaging for fruit. Packag Technol Sci 32(10):523–533. https://doi.org/10.1002/pts.2468

Barry CS, Giovannoni JJ (2007) Ethylene and fruit ripening ethylene and fruit ripening. J Plant Growth Regul 26:143–159. https://doi.org/10.1007/s00344-007-9002-y

Blankenship SM, Dole JM (2003) 1-Methylcyclopropene: a review. Postharvest Biol Technol 28(5):1–25. https://doi.org/10.1016/S0925-5214(02)00246-6

Bouzayen M, Latché A, Nath P et al (2010) Mechanism of fruit ripening. In: Pua EC, Davey MR (eds) Plant developmental biology-biotechnological perspectives, 1st edn. Springer, Berlin, pp 119–340. https://doi.org/10.1007/978-3-642-02301-9

Chen X, Chen M, Xu C et al (2019) Critical review of controlled release packaging to improve food safety and quality. Crit Rev Food Sci Nutr 59(15):2386–2399. https://doi.org/10.1080/10408398.2018.1453778

Crank J (1957) The mathematics of diffusion. Oxford University Press, Oxford

Estevez-Areco S, Guz L, Candal R et al (2018) Release kinetics of rosemary (Rosmarinus officinalis) polyphenols from polyvinyl alcohol (PVA) electrospun nanofibers in several food simulants. Food Packag Shelf Life 18(4):42–50. https://doi.org/10.1016/j.fpsl.2018.08.006

Fan X, Blankenship SM, Mattheis JP (1999) 1-Methylcyclopropene inhibits apple ripening. J Am Soc Hortic Sci 124(6):690–695

Hall AE, Findell JL, Schaller GE et al (2000) Ethylene perception by the ERS1 protein in Arabidopsis. Plant Physiol 123(4):1449–1457. https://doi.org/10.1104/pp.123.4.1449

Hotchkiss JH, Watkins CB, Sanchez DG (2007) Release of 1-methylcyclopropene from heat-pressed polymer films. J Food Sci 72(5):1–5. https://doi.org/10.1111/j.1750-3841.2007.00391.x

Huber DJ (2008) Suppression of ethylene responses through application of 1-methylcyclopropene: a powerful tool for elucidating ripening and senescence mechanisms in climacteric and nonclimacteric fruits and vegetables. HortScience 43(1):106–111

Irkin R, Esmer OK (2015) Novel food packaging systems with natural antimicrobial agents. J Food Sci Technol 52(10):6095–6111. https://doi.org/10.1007/s13197-015-1780-9

Khaneghah AM, Hashemi SMB, Limbo S (2018) Antimicrobial agents and packaging systems in antimicrobial active food packaging: an overview of approaches and interactions. Food Bioprod Process 111:1–19. https://doi.org/10.1016/J.FBP.2018.05.001

Lacoste A, Schaich KM, Zumbrunnen D et al (2005) Advancing controlled release packaging through smart blending. Packag Technol Sci 18(2):77–87. https://doi.org/10.1002/pts.675

Lakkis JM (2007) Encapsulation and controlled release technology in food systems. Blackwell, Iowa

Lee YS, Beaudry R, Kim JN et al (2006) Development of a 1-methylcyclopropene (1-MCP) sachet release system. J Food Sci 71(1):C1–C6. https://doi.org/10.1111/j.1365-2621.2006.tb12380.x

Lelièvre JM, Latchè A, Jones B et al (1997) Ethylene and fruit ripening. Physiol Plant 101:727–739

Lipinski B, Hanson C, Lomax J et al. (2013) Reducing food loss and waste. https://www.wri.org/publication/reducing-food-loss-and-waste. Accessed 14 Dec 2020. https://doi.org/10.2499/9780896295827_03

Liu M, Pirrello J, Chervin C et al (2015) Ethylene control of fruit ripening: revisiting the complex network of transcriptional regulation. Plant Physiol 169(4):2380–2390. https://doi.org/10.1104/pp.15.01361

Malhotra B, Keshwani A, Kharkwal H (2015) Antimicrobial food packaging: potential and pitfalls. Front Microbiol 6(6):1–9. https://doi.org/10.3389/fmicb.2015.00611

Meiners JA (2012) Fluid bed microencapsulation and other coating methods for food ingredient and nutraceutical bioactive compounds. In: Garti N, McClements DJ (eds) Encapsulation technologies and delivery systems for food ingredients and nutraceuticals, 1st edn. Woodhead Publishing, Cambrige, pp 151–176. https://doi.org/10.1533/9780857095909.2.151

Moretti CL, Araujo AL, Marouelli WA (2002) 1-Methylcyclopropene delays tomato fruit ripening. Hortic Bras 20(4):659–663

Nanthachai N, Ratanachinakorn B, Kosittrakun M et al (2007) Absorption of 1-MCP by fresh produce. Postharvest Biol Tec 43(3):291–297. https://doi.org/10.1016/j.postharvbio.2006.10.003

Neoh TL, Yamauchi K, Yoshii H et al (2007) Kinetics of molecular encapsulation of 1-methylcyclopropene into alpha cyclodextrin. J Agric Food Chem 55(26):11020–11026. https://doi.org/10.1021/jf072357t

Neoh TL, Yamauchi K, Yoshii H et al (2008) Kinetic study of thermally stimulated dissociation of inclusion complex of 1-methylcyclopropene with α-cyclodextrin by thermal analysis. J Phys Chem B 112:15914–15920. https://doi.org/10.1021/jp806233c

Neoh TL, Ariyanto HD, Menéndez Galvan P et al (2017) Controlled release of 1-methylcyclopropene from its functionalized electrospun fibres under constant and linearly ramped humidity. Food Addit Contam Part A Chem Anal Control Expo Risk Assess 34(10):1690–1702. https://doi.org/10.1080/19440049.2017.1325520

O'Malley RC, Rodriguez FI, Esch JJ et al (2005) Ethylene-binding activity, gene expression levels, and receptor system output for ethylene receptor family members from Arabidopsis and tomato. Plant J 41(5):651–659. https://doi.org/10.1111/j.1365-313X.2004.02331.x

Ortiz CM, Mauri AN, Vicente AR (2013) Use of soy protein based 1-methylcyclopropene-releasing pads to extend the shelf life of tomato (Solanum lycopersicum L.) fruit. Innov Food Sci Emerg Technol 20:218–287. https://doi.org/10.1016/j.ifset.2013.07.004

Papadopoulou V, Kosmidis K, Vlachou M et al (2006) On the use of the Weibull function for the discernment of drug release mechanisms. Int J Pharm 309(1–2):44–50. https://doi.org/10.1016/j.ijpharm.2005.10.044

Quirós-Sauceda AE, Ayala-Zavala JF, Olivas GI et al (2014) Edible coatings as encapsulating matrices for bioactive compounds: a review. J Food Sci Technol 51(9):1674–1685. https://doi.org/10.1007/s13197-013-1246-x

Reineccius TA, Reineccius GA, Peppard TL (2002) Encapsulation of flavors using cyclodextrins: comparison of flavor retention in alpha, beta, and gamma types. J Food Sci 67(9):3271–3279. https://doi.org/10.1111/j.1365-2621.2002.tb09577.x

Saltveit ME (1999) Effect of ethylene on quality of fresh fruits and vegetables. Postharvest Biol Technol 15:279–292

Sisler EC, Serek M (1997) Inhibitors of ethylene responses in plants at the receptor level: recent developments. Physiol Plant 100(3):577–582. https://doi.org/10.1034/j.1399-3054.1997.1000320.x

Sisler EC, Dupille E, Serek M et al (2003) Compounds interacting with the ethylene receptor. Plant Biol 5:473–480. https://doi.org/10.17660/ActaHortic.2001.553.32

Szente L, Fenyvesi È (2018) Cyclodextrin-enabled polymer composites for packaging. Molecules 23:1556–1574. https://doi.org/10.3390/molecules23071556

Thewes FR, Brackmann A, de Oliveira Anese R et al (2018) 1-Methylcyclopropene suppresses anaerobic metabolism in apples stored under dynamic controlled atmosphere monitored by respiratory quotient. Sci Hortic 227:288–295. https://doi.org/10.1016/j.scienta.2017.09.028

Vasisht N (2014) Factors and mechanisms in microencapsulation. In: Gaonkar AR, Vasisht N, Khare AR et al (eds) Microencapsulation in the food industry, 1st edn. Academic, Cambridge, pp 15–24. https://doi.org/10.1016/b978-0-12-404,568-2.00002-9

Vazquez-Landaverde PA, Qian MC, Torres JA (2007) Kinetic analysis of volatile formation in milk subjected to pressure-assisted thermal treatments. J Food Sci 72(7):E389–E398. https://doi.org/10.1111/j.1750-3841.2007.00469.x

Watkins CB (2006) The use of 1-methylcyclopropene (1-MCP) on fruits and vegetables. Biotechnol Adv 24(4):389–409. https://doi.org/10.1016/j.biotechadv.2006.01.005

Xu F, Liu Y, Shan X et al (2018) Evaluation of 1-methylcyclopropene (1-MCP) treatment combined with nano-packaging on quality of pleurotus eryngii. J Food Sci Technol 55(11):4424–4431. https://doi.org/10.1007/s13197-018-3354-0

Yamamoto C, Neoh TL, Honbou H et al (2012) Kinetic analysis and evaluation of controlled release of d-limonene encapsulated in spray-dried cyclodextrin powder under linearly ramped humidity. Dry Technol 30(11–12):1283–1291. https://doi.org/10.1080/07373937.2012.681089

Zhang Q, Zhen Z, Jiang H et al (2011) Encapsulation of the ethylene inhibitor 1-methylcyclopropene by cucurbit[6]uril. J Agric Food Chem 59(19):10539–10545. https://doi.org/10.1021/jf2019566

Zhang L, Alfano J, Race D et al (2018) Zero-order release of poorly water-soluble drug from polymeric films made via aqueous slurry casting. Eur J Pharm Sci 117(2):245–254. https://doi.org/10.1016/j.ejps.2018.02.029

Chapter 16
Encapsulation of Fruit Ripening Controlling Compounds

Chalida Cholmaitri, Apiradee Uthairatanakij, Natta Laohakunjit, and Bhesh R. Bhandari

16.1 Introduction

Fresh fruits are highly perishable agricultural commodities still living and respiring after harvested. Ethylene is a plant hormone responsible for triggering the ripening process causing the fruit to change color, lose firmness, increase sugar content, decrease acidity and develop aroma compounds (Kader 2002). There are several compounds used that can accelerate ripening or delay ripening. The use of ethylene gas can achieve faster and more uniform ripening of fruit. One of the exogenous ethylene producing products in a liquid formulation is known as 2-chloroethylphosphonic acid with the brand name 'Ethers®'. Ethylene released from ethrel was effective in stimulating mango ripening (Mohamed and Abu-Goukh 2003), enhancing color development and de-greening citrus fruits (Saltveit 1999). Conversely, 1-methylcyclopropene (1-MCP) is an inhibitor of ethylene activity by inhibiting the ethylene receptor (Tassoni et al. 2006). 1-MCP has been reported to delay the ripening in bananas (Harris et al. 2000), avocados (Pesis et al. 2002), apples (Apollo Arquiza et al. 2005) and plums (Candan et al. 2006). In addition, both methyl salicylate (MeSA) and methyl jasmonate (MeJA) have also been reported to effectively inhibit ethylene formation from 1-aminocyclopropane-1-carboxylate (ACC) synthase in tomatoes (Ding et al. 2001) delay the postharvest ripening process and maintain quality in kiwi fruit (Aghdam et al. 2011), sweet cherries (Giménez et al. 2015), and peaches (Jin et al. 2009). However, uses of fruit

C. Cholmaitri · A. Uthairatanakij (✉) · N. Laohakunjit
Division of Postharvest Technology, School of Bioresources and Technology, King Mongkut's University of Technology Thonburi, Bangkok, Thailand
e-mail: apiradee.uth@kmutt.ac.th; nutta.lao@kmutt.ac.th

B. R. Bhandari
School of Agriculture and Food Sciences, University of Queensland, Brisbane, QLD, Australia
e-mail: b.bhandari@uq.edu.au

T. M. Ho et al. (eds.), *Functionality of Cyclodextrins in Encapsulation for Food Applications*, https://doi.org/10.1007/978-3-030-80056-7_16

ripening controlling compounds in postharvest are somewhat limited due to rapid evaporation during fumigation and contamination of intact fruit after spraying or immersion.

The encapsulation technique is a process of coating or entrapping active compounds within other materials to enhance the stability of active compounds with controlled release. The core material is an active compound which will be released in a controlled manner in a given condition. The coating material is also called a shell, carrier material or wall material, which do not chemically react with the core material like adsorb and protect the core material from environmental factors, such as heat, light and moisture (Reineccius 1995). Moreover, the encapsulation can convert liquids or gases into powder forms possessing longer storage times and allow easier use and handling. One of the materials used as encapsulating agents is cyclodextrins (CDs), which is used in the food industry because they are very effective encapsulant materials although they are relatively expensive. CDs are approved as Generally Recognized As Safe (GRAS) substances by the US Food and Drug Administration (FDA). CDs are natural cyclic oligosaccharides derived from bacterial degradation of starch. The common natural CDs, consisting of six, seven, and eight D-glucopyranose units, are called alpha (α-), beta (β-), and gamma (γ-) cyclodextrins (CD), respectively whose properties are presented in Table 16.1 and also elsewhere in this book. The hydrogen bond network affects the CDs water solubility. The α-CD has a higher water solubility than β-CD because it has a weaker hydrogen bond network compared with β-CD. One property of CDs is the hydrophobic inner cavity which can encapsulate hydrophobic molecules to form inclusion complexes with a wide variety of guest molecules, while their outer surface is hydrophilic (Alvarez-Parrilla et al. 2005). The α-CDs encapsulate well for low molecular weight molecules or aliphatic polymer chains. The β-CDs are suitable for accommodating aromatic and heterocyclic compounds, and γ-CDs are able to incorporate large molecules such as macrocycles and steroids (Del Valle 2004). The binding of active compounds within the CD cavity constitutes a dynamic equilibrium in that some factors can affect the CD inclusion complex formation, such as type of CDs, cavity size, kind of active compounds, moisture content and temperature (Marques 2010). Interestingly, CDs can encapsulate gases and liquids converting them to powder form, such as 1-MCP, ethylene, MeSA, MeJA and

Table 16.1 Molecular formula and properties of cyclodextrins (adapted with permission from Maksimowski and Rumianowski 2016)

Properties	α-Cyclodextrin	β-Cyclodextrin	γ-Cyclodextrin
Number of glucopyranose units	6	7	8
Molecular formula	$C_{36}H_{60}O_{30}$	$C_{42}H_{70}O_{35}$	$C_{48}H_{80}O_{40}$
Molecular weight (g/mol)	972.85	1134.99	1297.14
Solubility in water at 25 °C (g/100 mL)	14.5	1.85	23.2
Cavity diameter (nm)	0.47–0.53	0.60–0.65	0.75–0.83
Height of torus (nm)	0.79	0.79	0.79
Cavity volume (nm^3)	0.174	0.262	0.427

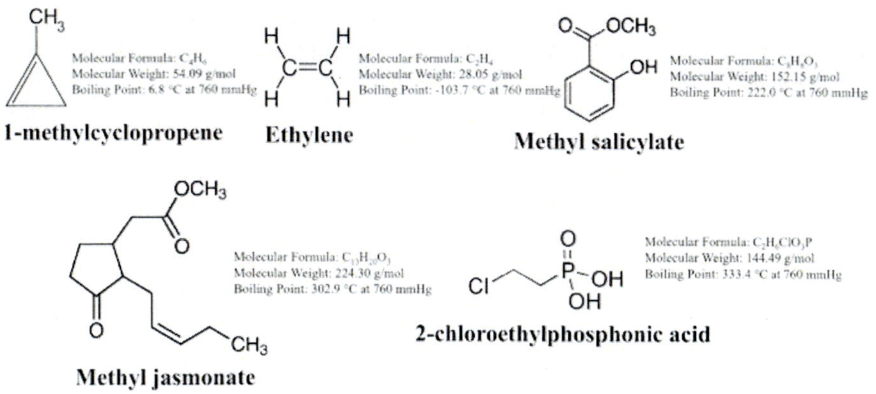

Fig. 16.1 Molecular structure of ripening controlling compounds

2-chloroethylphosphonic acid. The molecular structure of these compounds is presented in Fig. 16.1. This chapter reviews potential encapsulations of various fruit ripening controlling compounds by CDs.

16.2 1-MCP Encapsulation Using Alpha and Modified Beta-Cyclodextrins

The gaseous molecule 1-MCP is a known cyclopropene derivative with the molecular weight of 54.09 g/mol at standard temperature and pressure. Its chemical formula is C_4H_6 (Fig. 16.1). The 1-MCP molecule has the ability to inhibit ethylene synthesis by competing for the binding site of ethylene receptors (Tassoni et al. 2006). The 1-MCP molecule has been successfully used to extend shelf life by delaying the ripening process and also to improve postharvest quality in climacteric and nonclimacteric fruit, including apples (Fan and Mattheis 2001), peaches (Kluge and Jacomino 2002), avocados (Jeong et al. 2003), bananas (Jiang and Joyce 2003), pears (Trinchero et al. 2004), mangoes (Sakhale et al. 2018), oranges (Porat et al. 1999), strawberries (Jiang et al. 2001) and jujubes (Zhang et al. 2012). However, 1-MCP is present as a gaseous compound which is chemically unstable and presents a risk as an explosive hazard when compressed. To resolve these potential troubles, various methods have been developed to incorporate 1-MCP in molecular encapsulation agents, such as modified starches (Tester and Debon 2000), molecular sieves (Daly and Kourelis 2001) and CDs (Loftsson et al. 2005), of which CDs are the most preferred one.

The properties of CDs are widely recognized and used in industry to improve stability, solubility and delivery of many products. The CDs are employed to form an inclusion complex and are selected to fit the molecular size of the active compounds (Neoh et al. 2007). The 1-MCP molecule has a size that is suitable to form an inclusion complex with α-CD (Zhang et al. 2011). The 1-MCP molecule is hydrophobic that can easily be entrapped in the cavity of α-CD by hydrophobic interactions as well as physical entrapment such as dipole-dipole bonds, nonpolar and polar interactions, hydrophobic associations or Van der Waals interactions (Pinho et al. 2014).

The method of encapsulation of 1-MCP with α-CD usually occurs in aqueous solutions. The reaction can occur in two-steps comprising of dissolution of gaseous 1-MCP into aqueous phase, and the encapsulation of the dissolved molecules by α-CD. The α-CD molecular weight is 972.85 g/mol, the 1-MCP molecular weight is 54.09 g/mol, and the maximum complexation ratio of α-CD and 1-MCP is 1:1 (Neoh et al. 2007). There are several reports on the efficiency of the 1-MCP/α-CD inclusion complex on delaying the ripening and extending the shelf life of fresh products. For example, Jinwei et al. (2019) reported that the inclusion compound of 1-MCP/α-CD delayed the ripening of sweet cherries.

The release property of 1-MCP-α-CD complex in different temperature and humidity environment conditions is presented in Fig. 16.2. Ariyanto et al. (2019) reported that the slow release of 1-MCP from α-CD delayed the ripening of tomatoes during storage time at 25 °C for 18 days. The 1-MCP molecule is released from the α-CD complex when exposed to water or water vapor (Neoh et al. 2007). Fresh products can generate moisture by itself via the transpiration process. The moisture from the fresh products can accelerate the release of 1-MCP from the α-CD complex to inhibit the action of ethylene (Watkins 2015). Presently, 1-MCP/α-CD complex powders are commercially available and a commercial product having the trade

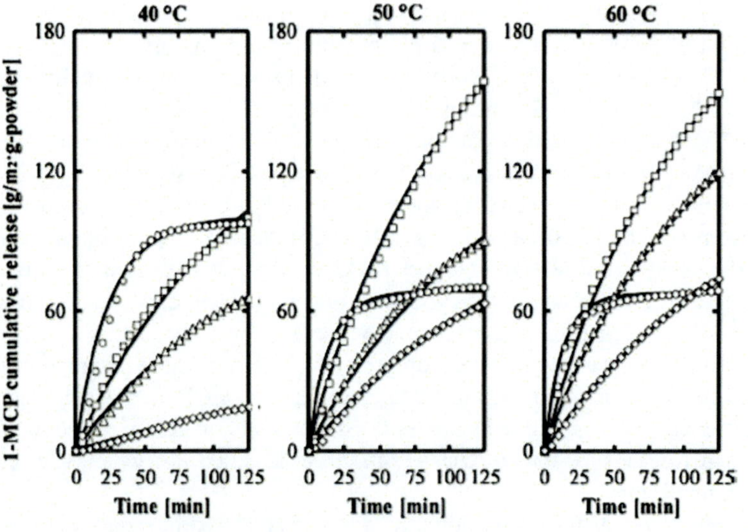

Fig. 16.2 Controlled release property of 1-MCP from 1-MCP/α-CD inclusion complex under relative humidity conditions of 40% (◇), 50% (△), 60% (□), and 80% (○) at different temperatures. (Adapted with permission from Ariyanto and Yoshii 2019)

name of SmartFresh® and EthylBloc® (Blankenship and Dole 2003). Even though α-CD is quite successful in the market, some remaining problems limit its applications. First of all, the cost of α-CD is relative expensive when compared to the β-CD. Secondly, α-CD is quite moisture sensitive (Ho et al. 2011). During storage of 1-MCP/α-CD complex powders, α-CD rapidly absorb water, 1-MCP will be unexpectedly released as they are kept in open environment. Moreover, α-CD has a high solubility possibly affecting the rapid release of 1-MCP. These limitations may lead to a problem of controlled release technique for applying in the field or the open environment.

However, β-CD has a lower cost and moisture sensitivity compared with α-CD. It also has a similar structure as α-CD but with a ring of seven units of glucose showing a larger size inner cavity compared with α-CD with six units of glucose (Table 16.1). Due to the increase of cavity size, 1-MCP cannot be tightly encapsulated in β-CD anymore as it used to be in α-CD. As a result, the loading of 1-MCP in β-CD is less than 0.1% (Zhang 2016). Similarly, γ-CD has the largest internal cavity compared with β-CD and α-CD. The encapsulating a small molecule such as 1-MCP in β-CD or γ-CD has not been successful (Neoh et al. 2007). Therefore, a modified β-CD (Mβ-CD) improved the loading capacity of 1-MCP and achieved the control release has been proposed.

The concept of the Mβ-CD is essentially to cross link individual β-CD molecules with proper cross-linkers to create a three-dimensional sponge-like polymer structure, which can accommodate more 1-MCP compounds due to the changes of porous networks during the modification process (Seglie et al. 2011). The synthesis of Mβ-CD commonly uses *N,N′*-carbonyldiimidazole as reagent for the activation reaction in an anhydrous solvent, such as tetrahydrofuran, chloroform, benzene, dimethylformamide and dichloromethane (Hu et al. 2014). The factors needed to be considered in the encapsulation process of 1-MCP/Mβ-CD are temperature, pH, concentration of Mβ-CD, concentration of 1-MCP and encapsulation time that may affect the final loading of 1-MCP in Mβ-CD (Zhang 2016). As a result, the pore size of Mβ-CD has been reduced significantly from 3.2 nm to 1.4 nm whereas 1-MCP loading was 0.529% after modification (Zhang 2016). Similarly, Seglie et al. (2011) also reported that Mβ-CD nanosponges can store high amounts of 1-MCP. Moreover, 1-MCP released from Mβ-CD can extend the shelf life of bananas for 5 days at 23 °C (Chen 2015).

16.3 Ethylene Encapsulation Using Alpha-Cyclodextrin

Ethylene is a gaseous natural plant hormone having the formula C_2H_4. Ethylene plays an important role in seed germination, vegetative growth, leaf abscission, flowering, senescence and fruit ripening (Saltveit 1999). Excessive amounts of ethylene will lead to changes in color, flavor, texture, aroma, nutrition and appearance of fruits (Kader 2002). In commercial practice, ethylene gas is often applied at high concentrations of 100–150 μL/L (ppm) in tightly closed rooms or chambers to promote uniform ripening of various fresh products such as bananas, melons, mangoes, tomatoes, and avocados (Beaudry and Kays 1988). In addition, ethylene enhances the red color of 'Gala' apples (Brighenti et al. 2017) and promotes de-greening of 'Oroblanco' citrus fruit (Porat et al. 2001). The mechanism for the biosynthesis of ethylene starts from the amino acid methionine through the intermediary S-adenosyl methionine (SAM) by SAM synthase. Then SAM is converted to 1-aminocyclopropane-1-carboxylic acid (ACC) by ACC synthetase. Finally, ACC oxidase converts ACC to ethylene (Yang and Hoffman 1984). However, ethylene is commonly used in compressed gas form which is unsafe and is not available in small quantity. Therefore, ethylene gas encapsulation in powder form would potentially be a more practical for use in fruit ripening and other applications. The α-CDs have encapsulation potential with ethylene (C_2H_4) (Ho et al. 2016). The α-CD has an interior cavity 0.47–0.53 nm in diameter and is suitable for inclusion complexes with molecules comprising five carbon atoms or less. Ethylene gas has a hydrophobic characteristic, low molecular weight and small size with a molecule consisting of two carbon atoms. Ethylene gas can be encapsulated in the cavity α-CD molecules more easily than β-CD and γ-CD (Cramer and Henglein 1956).

Molecular encapsulation techniques have been used to trap ethylene inside α-CD cavity by liquid method. The method involved the dissolution of α-CD powder in water followed by exposure to ethylene gas under high pressure conditions and long time periods. After that, the crystalline complex is collected and air dried at room temperature. The ethylene/α-CD inclusion complex showing an approximate 1:1 molar ratio and the yield increased with increasing pressures and time (Ho et al. 2011). Ho and Bhandari (2016) also developed a solid encapsulation method in which the ethylene is directly complexed with amorphous α-CD powder under pressure. Ethylene/α-CD inclusion complex powder has been successfully used for its applicability in horticulture to enhance the ripening process of mangoes. The ethylene/α-CD powder (100 g) was able to stimulate the ripening of mangoes (20 tons) in a container during transit from farm to market (Ho et al. 2016). Ethylene gas released from the α-CD powder triggered mango ripening similar to conventional ethylene gas (Ho et al. 2012).

Environmental factors such as temperature and humidity have affected the controlled release rate of ethylene (Neoh et al. 2007). An increased in temperature and humidity can accelerate the release rates of ethylene gas from the α-CD complex. The release property of ethylene-α-CD powder is presented in Fig. 16.3. Ho et al. (2011) reported that the release rates of ethylene were accelerated by increases in temperature and relative humidity. Moreover, placing the mixture of ethylene-α-CD complexes with deliquescent salts, e.g., $CaCl_2$ and $MgCl_2$, can dissolve the complex powder and control the release of ethylene gas from α-CD complex (Ho et al. 2012).

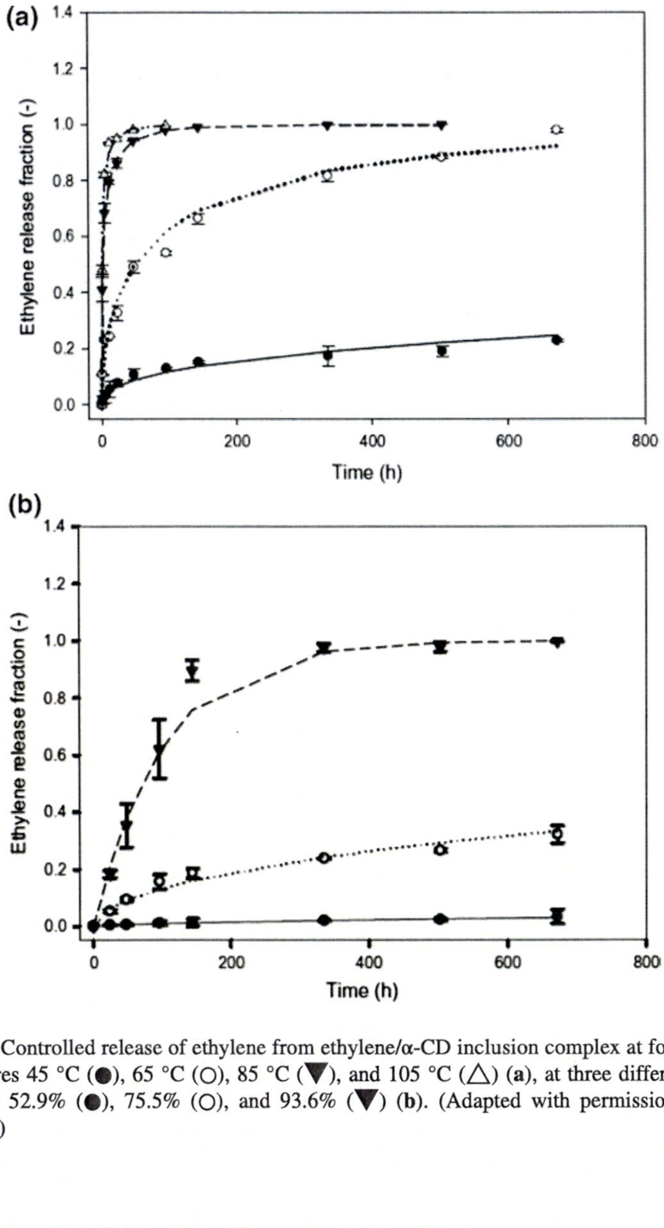

Fig. 16.3 Controlled release of ethylene from ethylene/α-CD inclusion complex at four different temperatures 45 °C (●), 65 °C (○), 85 °C (▼), and 105 °C (△) (**a**), at three different relative humidities 52.9% (●), 75.5% (○), and 93.6% (▼) (**b**). (Adapted with permission from Ho et al. 2011)

16.4 Methyl Salicylate Encapsulation Using Beta-Cyclodextrin

Methyl salicylate (MeSA) or salicylic acid methyl ester is a volatile organic compound with the formula $C_6H_4(OH)(CO_2CH_3)$ and molecular weight of 152.15 g/mol, a colorless liquid with a sweet, fruity odor, and miscible in organic solvents. MeSA is used as a food additive that is classified as a GRAS substance. In postharvest technology, MeSA has been used to delay the ripening and extend the shelf life of fresh produces (Ding et al. 2002). The mechanism of MeSA on delaying fruit ripening is not fully understood. However, it has been assumed that MeSA affects ripening and senescence processes by suppressing the expression of 1-aminocyclopropane-1-carboxylate (ACC) synthase and ACC oxidase activity in ethylene biosynthesis (Yin et al. 2013). The results are supported by a recent report that MeSA inhibited the activity of ACC synthase and ACC oxidase in tomatoes and plums (Ding et al. 2002). Similarly, Aghdam et al. (2012) reported that MeSA delayed the ripening of tomatoes by inhibiting ACC synthase. Ding and Wang (2003) found that MeSA vapor at high concentration (0.5 mmol/L) delayed ripening of tomatoes by reducing ethylene production and respiration rate. Min et al. (2018) reported that MeSA at 0.05 mmol/L delayed fruit ripening of tomatoes. MeSA also delayed weight loss and total soluble solids, and inhibited the activity of ascorbate peroxidase (APX) and catalase (CAT) in 'Hayward' kiwifruit (Aghdam et al. 2011). Chotikakham et al. (2019) reported that bananas dipped in MeSA solution at concentrations of 2 and 4 mmol/L delayed fruit ripening for 6 days at 25 ± 2 °C and 75 ± 5% relative humidity. Moreover, MeSA vapor at 1 mmol/L delayed ripening of sweet cherries for 20 days at 2 °C (Giménez et al. 2016).

MeSA is usually applied to fruit either by fumigation or by dipping them in a solution, but this technique is limited by its quick release and low solubility in water. Thus, one drawback has been solved by entrapping MeSA in the internal cavity of CDs. MeSA is an aromatic compound containing benzene ring which is substituted by a hydroxyl and ester groups. The molecular volume of a MeSA molecule is 0.209 nm^3, but the cavity volume of β-CD is 0.262 nm^3 whereas α-CD and γ-CD is 0.174 nm^3 and 0.427 nm^3, respectively (Del Valle 2004). Thus, the β-CD is more suitable to entrap MeSA molecules than α-CD and γ-CD. Entrapping MeSA inside the β-CD cavity depends on the dipole-dipole and van der Waals interactions (Herrera et al. 2019).

The inclusion complexes of MeSA and β-CD have been prepared by the co-precipitation method. This method involved the dissolution of β-CD in a warm ethanol and water (1:2) mixture followed by slow addition of MeSA to the solution. Afterwards, the resultant mixture is stirred for 4 h to allow complex formation. The final solution is kept overnight at 4 °C. The precipitated crystalline MeSA/β-CD complex is obtained by filtration and then dried in a convection oven at 50 °C for 24 h (Lee et al. 2020). MeSA/β-CD inclusion complex at 1:1 showed the MeSA entrapment efficiency of 59.0%and Fourier-transform infrared spectroscopy (FTIR) indicated that the functional groups of MeSA/β-CD were different from the MeSA spectrum ^{1}H nuclear magnetic resonance (^{1}H NMR) spectra showed the chemical shift of the protons of β-CD, MeSA, and MeSA/β-CD. In addition, thermal analysis (TGA) showed the thermal degradation patterns of the MeSA/β-CD differ from that of β-CD, and scanning electron microscopy (SEM) confirmed that the surface morphology of MeSA/β-CD inclusion complex smaller than that of β-CD (Lee et al. 2020).

The controlled release property of encapsulated MeSA is depicted in Fig. 16.4. The release rate of MeSA from β-CD complex increased with increasing temperature and relative humidity. Moreover, the controlled release rate followed a diffusive and first-order kinetic reaction (Lee et al. 2020). This complex has the potential to be applied during fumigation in cold storage or insertion in active packaging to prolong the shelf life of fresh products (Del Valle 2004).

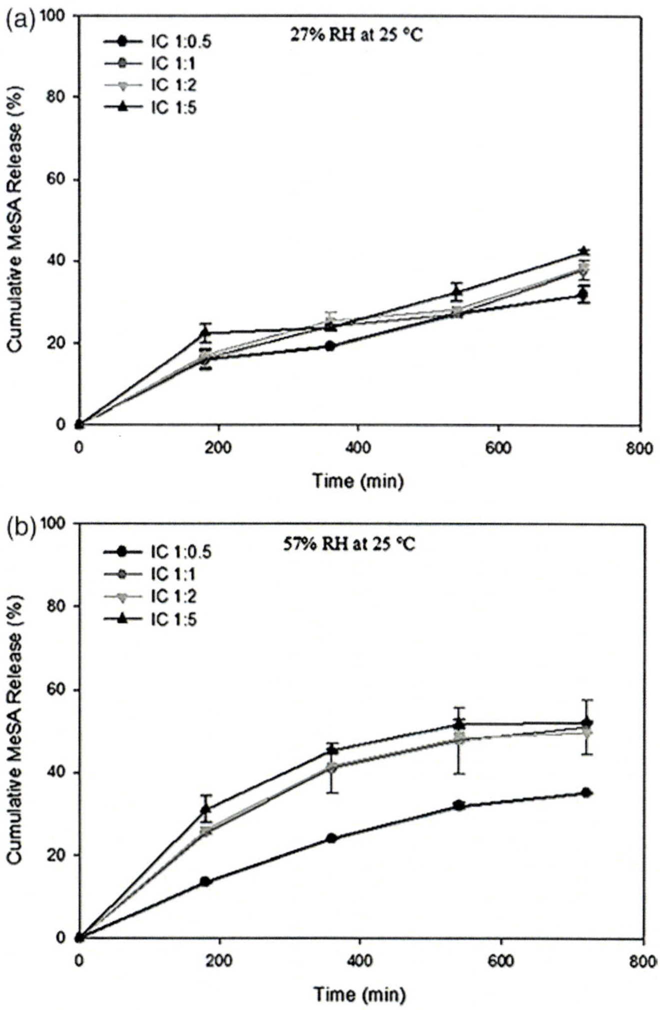

Fig. 16.4 Controlled release properties of MeSA/β-CD inclusion complex at different ratios of β-CD:MeSA (1:0.5, 1:1, 1:2, and 1:5) under temperatures of 7 °C and 25 °C and relative humidities of 27%, 57%, and 95% (Adapted with permission from Lee et al. 2020)

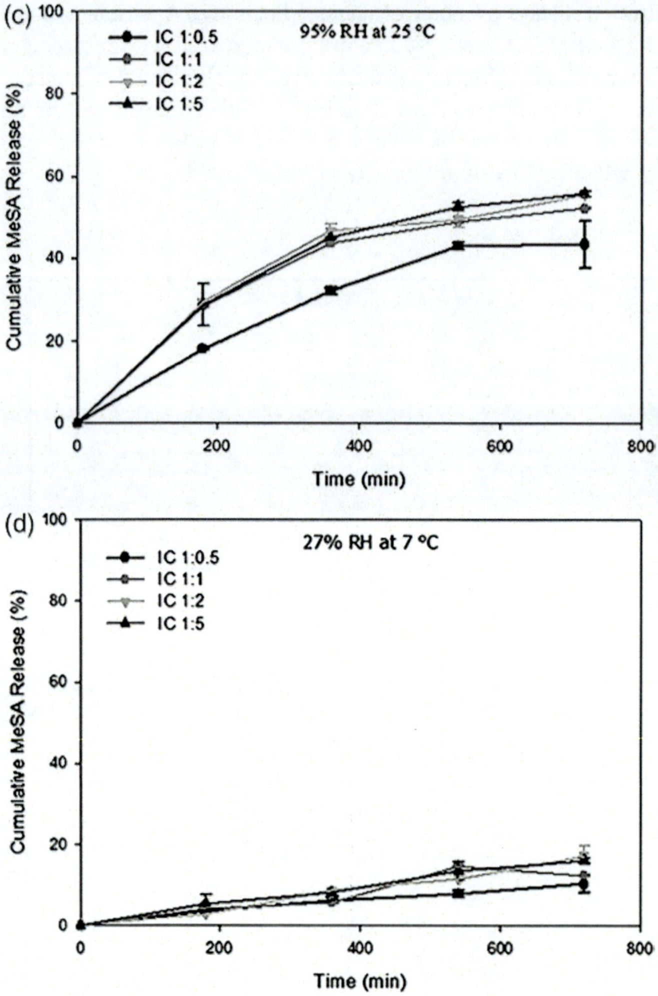

Fig. 16.4 (continued)

16.5 Methyl Jasmonate Encapsulation Using Beta-Cyclodextrin

Methyl jasmonate (MeJA) is a volatile organic compound that affects plant growth including seed germination, growth of roots and plant senescence (Wasternack et al. 2006). MeJA has also been applied in fruit ripening. Min et al. (2020) reported that MeJA significantly promoted the ripening of tomatoes by increasing ethylene biosynthesis. MeJA vapors accelerated ethylene production in kiwifruit (Wu et al. 2020). MeJA treatment enhanced ripening of apples by activating the synthesis of anthocyanin pigments (Mattheis et al. 2004). The use of 0.05 mmol/L of MeJA

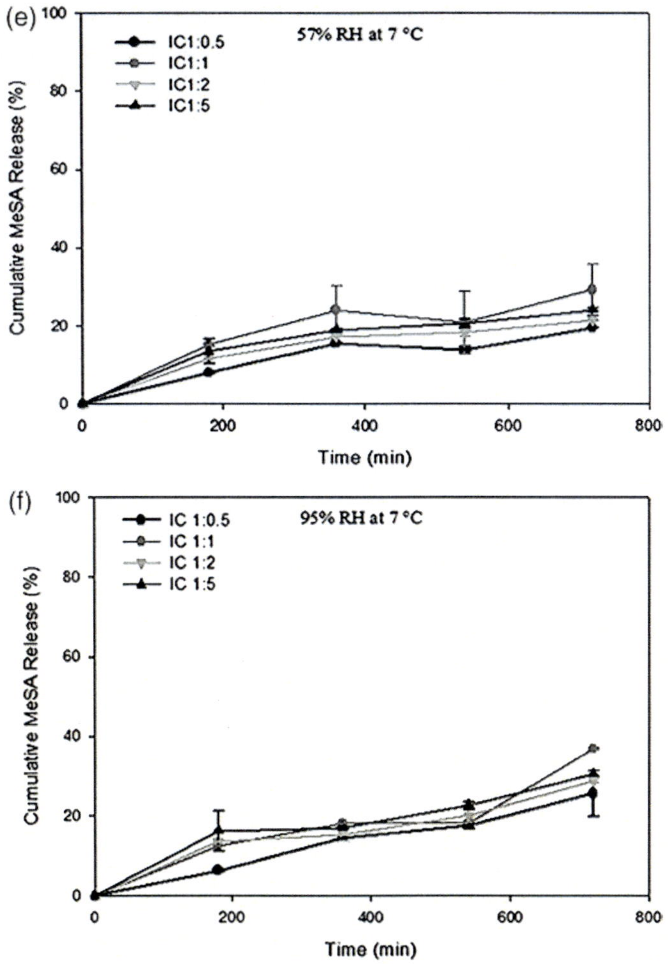

Fig. 16.4 (continued)

enhanced yellow and red color development of mangoes stored at 20 °C (González-Aguilar et al. 2001). 'Kensington Pride' mangoes were fumigated with MeJA vapors at a concentration of 30 µmol/L for 20 h at 21 ± 1 °C which promoted biosynthesis of ethylene, fatty acids, ripening and aroma volatile compounds (Lalel et al. 2003). Moreover, MeJA treatment at 230 µmol/L promoted the development and maturation of strawberries which was associated with increased fruit softening, anthocyanin and sugar content (Han et al. 2019). Applying 1 and 10 µmol/L of MeJA increased respiration rate, ethylene production and yellowing of 'Kumagai' guavas at ripe stage (Pereira da Silva et al. 2017). MeJA (50 µmol/L) increased activities of ACC synthase and ACC oxidase of 'Pajaro' strawberries (Mukkun and Singh 2009). On the other hand, González-Aguilar et al. (2003) reported that MeJA vapors at 50 µmol/L delayed fruit ripening by maintaining firmness of 'Sunrise' papayas for

14–32 days at 10 °C and 4 days of shelf life at 20 °C. MeJA at 5 μmol/L suppressed ethylene production, delayed weight loss, firmness and maintained anthocyanin content in eggplants (Fan et al. 2016). Moreover, MeJA concentration below 100 μmol/L extended the shelf life of half-red maturity of jujubes (Dong et al. 2016).

The function of exogenously MeJA on fruit ripening is dependent on fruit varieties, fruit developmental stage, concentration of MeJA and storage temperature (Buta and Moline 1998). Applying MeJA for fruit ripening is extremely complicated due to its volatile and hydrophobic character. These problems can be overcome by encapsulating MeJA in the cavity of CDs because of their ability to slow the release of this compound for a prolonged time (Alvarez-Parrilla et al. 2007). The chemical structure of MeJA contains a hydrocarbon ring with a carbonyl and a methyl ester as function groups. The interaction of MeJA with β-CD is more efficient than α-CD and γ-CD (Matencio et al. 2020).

The controlled release property of MeJA/β-CD complex in term of reduction of polyphenol oxidase (PPO) substrate is depicted in Fig. 16.5. When relative humidity increased, the amount of MeSA can be released passively and controlled from the complex. Alvarez-Parrilla et al. (2007) demonstrated that β-CD forms a 1:1 inclusion complex with MeJA inhibited PPO activity of 'Red Delicious' apples. Similarly, Sojo et al. (1999) described that β-CD inhibited PPO activity by inactivating 4-iodophenol and cinnamic acid in banana pulp. These results implied that using MeJA/β-CD inclusion complex can effectively maintain the quality of fresh produce by delaying the onset of ripening caused by enzymic activities. More studies are needed in the effectiveness of MeJA/β-CD complex on controlling the ripening of the fruits.

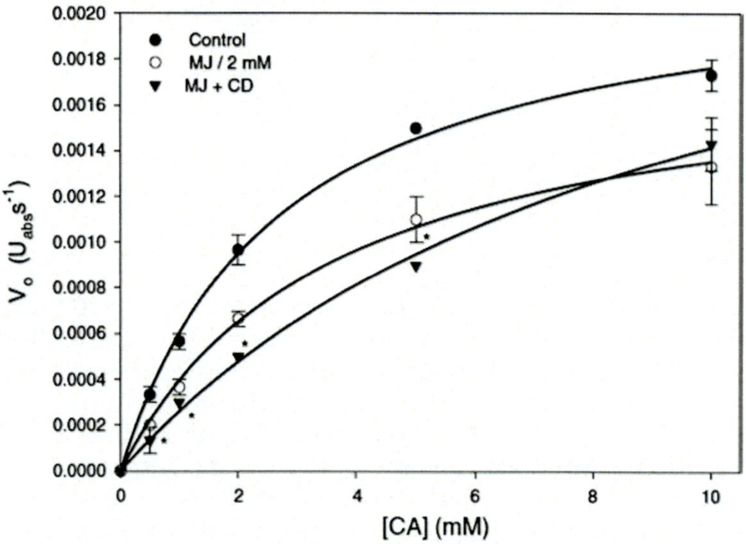

Fig. 16.5 Effects of MeJA and mixture (2 mM MeJA+5 mM β-CD), on chlorogenic acid (CA) oxidation catalyzed by Red Delicious PPO (Adapted with permission from Alvarez-Parrilla et al. 2007)

16.6 2-Chloroethylphosphonic Acid (Ethrel) Encapsulation Using Alpha and Beta Cyclodextrins

The commercial chemical 2-chloroethylphosphonic acid or ethrel has been used for fruit ripening (Alexander and Grierson 2002). When fruit immersed in an aqueous solution of 2-chloroethylphosphonic acid at pH above 3.5, the 2-chloroethylphosphonic acid penetrates in fruit cells and decomposes to ethylene along with chloride and phosphate (Kaur 2017). Generally, the fruit harvested at a mature stage can be damaged during transportation and distribution causing significant loss to farmers. For this reason, farmers harvest their fruit at a mature green but unripe stage leading to fruit ripening slowly and unevenly. The uniform ripening can be achieved by immersing fruit in diluted 2-chloroethylphosphonic acid solution which is recommended for wide varieties of climacteric fruits such as mangoes (Bhandari et al. 2017), bananas (Tan et al. 2014), tomatoes (Moniruzzaman et al. 2015), apples (Awad and de Jager 2002), and pears (Dhillon and Mahajan 2011). The recommended dosage of 2-chloroethylphosphonic acid is 500–1000 μL/L. Sabuz et al. (2019) reported that 2-chloroethylphosphonic acid can be applied at 750–1000 μL/L for more uniform ripening of mangoes at 23 ± 2 °C for 85 ± 5% relative humidity. Application of 2-chloroethylphosphonic acid (600 μL/L) improved uniform ripening of 'Dashehari' mangoes at 33 ± 2 °C for 40–60% relative humidity (Jawandha et al. 2016). Using 2-chloroethylphosphonic acid (500 μL/L) showed uniform color, acceptable flavor, appropriate firmness and good quality in 'Grand Nain' bananas after 4 days at 16–18 °C and 90–95% relative humidity (Mahajan et al. 2010). For nonclimacteric fruits, determining the harvesting time of fruits such as oranges, lemons and grapefruits depends on the color of the fruit peel. Thus, applying ethylene is also used to develop desirable fruit color. Using 2-chloroethylphosphonic acid has been reported to enhance peel color of cranberries (Murphey and Dilley 1988). Applying 2-chloroethylphosphonic acid stimulated the carotenoid accumulation in the peel of Gannan Newhall navel oranges (Hu et al. 2018).

The properties of 2-chloroethylphosphonic acid solutions are very acidic and highly corrosive (Szejtli et al. 1979). Moreover, ethylene released by 2-chloroethylphosphonic acid decomposition is quickly released leading to significant losses. Therefore, encapsulating 2-chloroethylphosphonic acid in the form of inclusion complexes with CDs is potentially an alternative technique which is safe and easy to apply.

In a process for the preparation of inclusion complexes 2-chloroethylphosphonic acid and CDs, an aqueous solution CDs is prepared in hot distilled water and an equal volume of a solution of 2-chloroethylphosphonic acid is added with vigorous agitation. The solution is cooled with ice water and kept in the refrigerator overnight to complete the complexation and crystallize the complex. The crystalline complex is filtered, washed and dried in a vacuum desiccator (Szejtli et al. 1979). Budai and Szejtli (1981) confirmed that using 2-chloroethylphosphonic acid to form an inclusion complex with α-CDs is slightly less acidic and corrosive than 2-chloroethylphosphonic acid solutions.

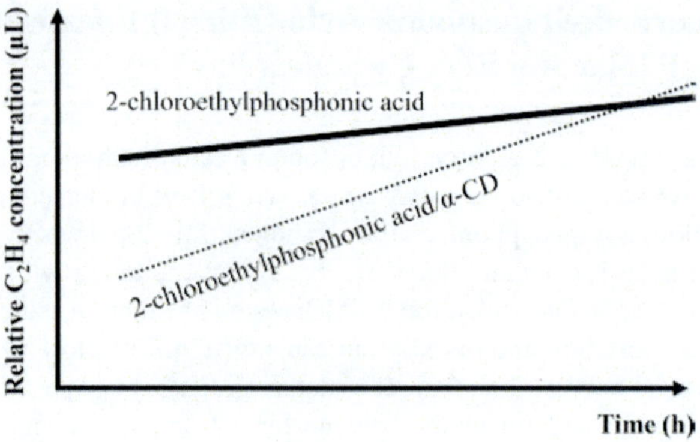

Fig. 16.6 Controlled release pattern of ethylene from 2-chloroethylphosphonic acid and 2-chloroethylphosphonic acid/α-cyclodextrin complex

The release properties of ethylene from the complex show signs of the effectiveness of the prolonged ethylene effect is depicted in Fig. 16.6 (Tétényi and Szejtli 1984). Tétényi and Szejtli (1984) reported that the release process from 2-chloroethylphosphonic acid/α-CDs complex is slower than ethylene release from free 2-chloroethylphosphonic acid. In addition, Tétényi (1982) reported that an inclusion complex of 2-chloroethylphosphonic acid with α-CD accelerated uniform ripening in sweet peppers. However, β-CD is cheaper than α-CD (Redente et al. 2001) and commercially available with high purity, so this host is of primary importance for practical applications. Szejtli et al. (1983) reported that 2-chloroethylphosphonic acid/β-CD inclusion complex can be successfully used to accelerate the ripening of red peppers by a slow release of ethylene complex. There are no recent studies on the 2-chloroethylphosphonic acid encapsulation using CD may be due to many opinions concerns about the toxicity of the 2-chloroethylphosphonic acid. Nevertheless, FAO point out the 2-chloroethylphosphonic acid residues did not affect the health of consumers at the lowest dose (0.05 mg/kg body weight/day).

16.7 Conclusion and Future Perspectives

This chapter investigated the encapsulation of fruit ripening controlling compounds such as 1-MCP, ethylene, MeSA, MeJA and 2-chloroethylphosphonic acid in various types of CDs complexes to apply postharvest. Ethylene gas has wide ranging effects on the plant; it can produce either a good effect when used to achieve uniform ripening of fruit or stimulates fruit ripening. Normally, 1-MCP gas has been widely used to inhibit ethylene action by blocking ethylene receptors to maintain the quality and extend the shelf life of fresh products. At standard temperature and pressure, ethylene and 1-MCP are gaseous forms. Moreover, 2-chloroethylphosphonic acid is a liquid formulation of exogenous ethylene, whereas MeSA and MeJA are hydrophobic volatile compounds insoluble in water. They have limitations in terms of quick release. However, encapsulation can convert from gases and liquids to powder form which led to improve its safety, handling and application. CDs constitute food grade encapsulant materials; they are nontoxic and included in the GRAS list of the FDA. The 1-MCP/α-CD, ethylene/α-CD, 2-chloroethylphosphonic acid/α-CD, MeSA/β-CD, MeJA/β-CD, 2-chloroethylphosphonic acid/β-CD and 1-MCP/Mβ-CD are successful inclusion complexes because these compounds are nonpolar, which can easily occupy the inner cavity of CD molecules. Furthermore, encapsulation of fruit ripening controlling compounds by CDs inclusion complexes can achieved controlled release of active compounds to promote uniform ripening or delay the ripening of fruit.

References

Aghdam MS, Motallebiazar A, Mostofi Y et al (2011) Methyl salicylate affects the quality of Hayward kiwifruit during storage at low temperature. J Agric Sci 3:149–156. https://doi.org/10.5539/jas.v3n2p149

Aghdam MS, Asghari M, Moradbeygi H et al (2012) Effect of postharvest salicylic acid treatment on reducing chilling injury in tomato fruit. Rom Biotechnol Lett 17:7466–7473

Alexander L, Grierson D (2002) Ethylene biosynthesis and action in tomato: a model for climacteric fruit ripening. J Exp Bot 53(377):2039–2055. https://doi.org/10.1093/jxb/erf072

Alvarez-Parrilla E, De la Rosa LA, Torres-Rivas F et al (2005) Complexation of apple antioxidants: chlorogenic acid, quercetin and rutin by β-cyclodextrin (β-CD). J Incl Phenom Macrocycl Chem 53:121–129

Alvarez-Parrilla E, De la Rosa LA, Rodrigo-Garcia J et al (2007) Dual effect of β-cyclodextrin (β-CD) on the inhibition of apple polyphenol oxidase by 4-hexylresorcinol (HR) and methyl jasmonate (MJ). Food Chem 101(4):1346–1356. https://doi.org/10.1016/j.foodchem.2006.03.040

Apollo Arquiza JMR, Hay AG, Nock JF et al (2005) 1-Methylcyclopropene interactions with diphenylamine on diphenylamine degradation, α-farnesene and conjugated trienol concentrations, and polyphenol oxidase and peroxidase activities in apple fruit. J Agric Food Chem 53:7565–7570. https://doi.org/10.1021/jf0511603

Ariyanto HD, Yoshii H (2019) Effect of stepwise humidity change on the release rate constant of 1-methylcyclopropene (1-MCP) in a cyclodextrin inclusion complex powder. Food Packag Shelf Life 21:100322. https://doi.org/10.1016/j.fpsl.2019.100322

Ariyanto HD, Berci MN, Yoshii H (2019) Effect of 1-methylcyclopropene release from inclusion complexes powder on the quality of tomato by short-term treatment. In: 2nd International conference on agriculture postharvest handling and processing (ICAPHP 2018), Kuta, Bali, Indonesia, 29–31 August 2018

Awad MA, de Jager A (2002) Formation of flavonoids, especially anthocyanin and chlorogenic acid in 'Jonagold' apple skin: influences of growth regulators and fruit maturity. Sci Hortic 93(3–4):257–266. https://doi.org/10.1016/S0304-4238(01)00333-8

Beaudry R, Kays S (1988) Application of ethylene-releasing compounds in agriculture. In: Neuman PM (ed) Plant growth and leaf-applied chemicals, 1st edn. CRC Press, Boca Raton, Florida, p 127–155

Bhandari P, Meenakshi R, Patanjali PK (2017) Effect of ethephon 39% SL on postharvest applications on fruit ripening in mango. J Nat Res Dev 12(1):58–64

Blankenship SM, Dole JM (2003) 1-Methylcyclopropene: a review. Postharvest Biol Technol 28(1):1–25. https://doi.org/10.1016/s0925-5214(02)00246-6

Brighenti AF, Würz DA, Pasa M et al (2017) Plant growth regulators to enhance fruit color of 'Gala' apples. Pesqui Agropecu Bras 52(11):1118–1122. https://doi.org/10.1590/s0100-204x2017001100018

Budai Z, Szejtli J (1981) Cyclodextrin inclusion complexes of 2-chloroethyl phosphonic acid. Acta Chim Acad Sci Hung 107(3):231–236

Buta JG, Moline HE (1998) Methyl jasmonate extends shelf life and reduces microbial contamination of fresh-cut celery and peppers. J Agric Food Chem 46(4):1253–1256. https://doi.org/10.1021/jf9707492

Candan AP, Graell J, Crisosto C et al (2006) Improvement of storability and shelf-life of 'Blackamber' plums treated with 1-methylcyclopropene. Food Sci Technol Int 12(5):437–443. https://doi.org/10.1177/1082013206069909

Chen X (2015) Development of liquid 1-methylcyclopropene delivery formulations for modifying ethylene response of fresh produce. Dissertation, Rutgers University-Graduate School-New Brunswick

Chotikakham S, Faiyue B, Uthaibutra J et al (2019, 1245) Effects of methyl salicylate on senescent spotting and hydrogen peroxide concentration in ripening 'Sucrier' bananas. Acta Hortic:115–122. https://doi.org/10.17660/actahortic.2019.1245.17

Cramer F, Henglein F (1956) Einschlußverbindungen der Cyclodextrine mit Gasen. Angew Chem 68(20):649–649. https://doi.org/10.1002/ange.19560682008

Daly J, Kourelis B (2001) Synthesis methods, complexes, and delivery methods for the safe and convenient storage, transport, and application of compounds for inhibiting the ethylene response in plants. US Patent 6017849-A, 25 Jan 2000

Del Valle EMM (2004) Cyclodextrins and their uses: a review. Process Biochem 39:1033–1046. https://doi.org/10.1016/S0032-9592(03)00258-9

Dhillon W, Mahajan B (2011) Ethylene and ethephon induced fruit ripening in pear. J Stored Prod Postharvest Res 2(3):45–51

Ding CK, Wang CY, Gross KC et al (2001) Reduction of chilling injury and transcript accumulation of heat shock proteins in tomato fruit by methyl jasmonate and methyl salicylate. Plant Sci. 161:1153–1159. https://doi.org/10.1016/S0168-9452(01)00521-0

Ding CK, Wang CY (2003) The dual effects of methyl salicylate on ripening and expression of ethylene biosynthetic genes in tomato fruit. Plant Sci 164(4):589–596. https://doi.org/10.1016/S0168-9452(03)00010-4

Ding CK, Wang CY, Gross KC et al (2002) Jasmonate and salicylate induce the expression of pathogenesis-related-protein genes and increase resistance to chilling injury in tomato fruit. Planta 214(6):895–901. https://doi.org/10.1007/s00425-001-0698-9

Dong Y, Zhi HH, Xu J et al (2016) Effect of methyl jasmonate on reactive oxygen species, antioxidant systems, and microstructure of Chinese winter jujube at two major ripening stages during shelf life. J Hortic Sci Biotechnol 91(3):316–323. https://doi.org/10.1080/14620316.2016.1159924

Fan X, Mattheis JP (2001) 1-Methylcyclopropene and storage temperature influence responses of 'Gala' apple fruit to gamma irradiation. Postharvest Biol Technol 23(2):143–151. https://doi.org/10.1016/S0925-5214(01)00119-3

Fan L, Shi J, Zuo J et al (2016) Methyl jasmonate delays postharvest ripening and senescence in the non-climacteric eggplant (*Solanum melongena* L.) fruit. Postharvest Biol Technol 120:76–83. https://doi.org/10.1016/j.postharvbio.2016.05.010

Giménez MJ, Valverdea JM, Valero D et al (2015) Methyl salicylate treatments of sweet cherry trees improve fruit quality at harvest and during storage. Sci Hortic 197:665–673. https://doi.org/10.1016/j.scienta.2015.10.033

Giménez MJ, Serrano M, Valverde JM et al (2016) Preharvest salicylic acid and acetylsalicylic acid treatments preserve quality and enhance antioxidant systems during postharvest storage of sweet cherry cultivars. J Sci Food Agric 97(4):1220–1228. https://doi.org/10.1002/jsfa.7853

González-Aguilar GA, Buta JG, Wang CY (2001) Methyl jasmonate reduces chilling injury symptoms and enhances colour development of 'Kent' mangoes. J Sci Food Agric 81(13):1244–1249. https://doi.org/10.1002/jsfa.933

González-Aguilar GA, Buta JG, Wang CY (2003) Methyl jasmonate and modified atmosphere packaging (MAP) reduce decay and maintain postharvest quality of papaya 'Sunrise'. Postharvest Biol Technol 28(3):361–370. https://doi.org/10.1016/s0925-5214(02)00200-4

Han Y, Chen C, Yan Z et al (2019) The methyl jasmonate accelerates the strawberry fruits ripening process. Sci Hortic 249:250–256. https://doi.org/10.1016/j.scienta.2019.01.061

Harris DR, Seberry JA, Wills RBH et al (2000) Effect of fruit maturity on efficiency of 1-methylcyclopropene to delay the ripening of bananas. Postharvest Biol Technol 20:303–308. https://doi.org/10.1016/S0925-5214(00)00150-2

Herrera A, Rodríguez FJ, Bruna JE et al (2019) Antifungal and physicochemical properties of inclusion complexes based on β-cyclodextrin and essential oil derivatives. Food Res Int 121:127–135. https://doi.org/10.1016/j.foodres.2019.03.026

Ho B, Bhandari B (2016) Novel solid encapsulation of ethylene gas using amorphous α-cyclodextrin and the release characteristics. J Agric Food Chem 64(17):3318–3323. https://doi.org/10.1021/acs.jafc.5b06037

Ho BT, Joyce DC, Bhandari BR (2011) Release kinetics of ethylene gas from ethylene-α-cyclodextrin inclusion complexes. Food Chem 129(2):259–266. https://doi.org/10.1016/j.foodchem.2011.04.035

Ho B, Bhandari B, Joyce D et al (2012) Ethylene gas from an ethylene-α-cyclodextrin inclusion complex powder ripens 'Calypso' mango fruit. Acta Hortic 1012(1012):331–336. https://doi.org/10.17660/ActaHortic.2013.1012.41

Ho BT, Hofman PJ, Joyce DC et al (2016) Uses of an innovative ethylene-α-cyclodextrin inclusion complex powder for ripening of mango fruit. Postharvest Biol Technol 113:77–86. https://doi.org/10.1016/j.postharvbio.2015.11.005

Hu X, Liu S, Zhou G et al (2014) Electrospinning of polymeric nanofibers for drug delivery applications. J Control Release 185(1):12–21. https://doi.org/10.1016/j.jconrel.2014.04.018

Hu Y, Wang G, Pan S et al (2018) Influence of ethylene and ethephon treatments on the peel color and carotenoids of Gannan Newhall navel orange during postharvest storage. J Food Biochem 2018:e12534. https://doi.org/10.1111/jfbc.12534

Jawandha SK, Gill PPS, Kaur N et al (2016) Influence of ethephon treatments on quality changes in Dashehari mango (*Mangifera indica* L.) during ripening. Vegetos 29(4):26. https://doi.org/10.5958/2229-4473.2016.00096.3

Jeong J, Huber DJ, Sargent SA (2003) Delay of avocado (*Persea americana*) fruit ripening by 1-methylcyclopropene and wax treatments. Postharvest Biol Technol 28:247–257. https://doi.org/10.1016/S0925-5214(02)00176-X

Jiang YM, Joyce DC (2003) Softening response of 1-methylcyclopropene-treated banana fruit to high oxygen atmospheres. Plant Growth Regul 41(3):225–229. https://doi.org/10.1023/B:GROW.0000007506.38661.c6

Jiang Y, Shiina T, Nakamura N et al (2001) Electrical conductivity evaluation of postharvest strawberry damage. J Food Sci 66(9):1392–1395. https://doi.org/10.1111/j.1365-2621.2001.tb15220.x

Jin P, Wang K, Shang H et al (2009) Low-temperature conditioning combined with methyl jasmonate treatment reduces chilling injury of peach fruit. J Sci Food Agric 89(10):1690–1696. https://doi.org/10.1002/jsfa.3642

Jinwei Z, Jiangtao F, Wei Y (2019) Effects of inclusion compounds of 1-methylcyclopropene/α-cyclodextrin or 1-methyl-3-(2-methylcyclopropyl)-1-cyclopropene/Cu-β-cyclodextrin on the preservation of sweet cherry (*Prunus avium* L.). Curr Sci 117(10):1716. https://doi.org/10.18520/cs/v117/i10/1716-1721

Kader AA (2002) Standardization and inspection of fresh fruits and vegetables. In: Kader AA (ed) Postharvest technology of horticultural crops, 3rd edn. University of California, Division of Agriculture and Natural Resources, Oakland, CA, p 353

Kaur S (2017) Effect of different treatments of ethrel on ripening behaviour and post-harvest quality of mango (*Mangifera indica* L.) during storage. J Appl Nat Sci 9(1):85–93. https://doi.org/10.31018/jans.v9i1.1155

Kluge RA, Jacomino AP (2002) Shelf life of peaches treated with 1-methylcyclopropene. Sci Agric 59(1):69–72. https://doi.org/10.1590/S0103-90162002000100010

Lalel HJD, Singh Z, Tan SC et al (2003) Maturity stage at harvest affects fruit ripening, quality and biosynthesis of aroma volatile compounds in 'Kensington Pride' mango. J Hortic Sci Biotechnol 78(2):225–233. https://doi.org/10.1080/14620316.2003.11511610

Lee M, Dey KP, Lee YS (2020) Complexation of methyl salicylate with β-cyclodextrin and its release characteristics for active food packaging. Food Sci Biotechnol 29:917–925. https://doi.org/10.1007/s10068-020-00749-z

Loftsson T, Jarho P, Másson M et al (2005) Cyclodextrins in drug delivery. Expert Opin Drug Deliv 2(2):335–351. https://doi.org/10.1517/17425247.2.1.335

Mahajan BVC, Kaur T, Gill MIS et al (2010) Studies on optimization of ripening techniques for banana. J Food Sci Technol 47(3):315–319. https://doi.org/10.1007/s13197-010-0050-0

Maksimowski P, Rumianowski T (2016) Properties of the gamma-cyclodextrin/CL-20 system. Cent Eur J Energ Mater 13(1):217–229. https://doi.org/10.22211/cejem/64973

Marques HMC (2010) A review on cyclodextrin encapsulation of essential oils and volatiles. Flavour Fragr J 25:313–326. https://doi.org/10.1002/ffj.2019

Mattheis JP, Rudell DR, Buchanan DA (2004) Ethylene intensifies but is not a requirement for methyl jasmonate enhanced anthocyanin synthesis by 'Fuji' apple fruit. Acta Hortic 636:455–460. https://doi.org/10.17660/actahortic.2004.636.55

Matencio A, Navarro-Orcajada S, García-Carmona F et al (2020) Applications of cyclodextrins in food science. A review. Trends Food Sci Technol 104:132–143. https://doi.org/10.1016/j.tifs.2020.08.009

Min D, Li F, Zhang X et al (2018) Effect of methyl salicylate in combination with 1-methylcyclopropene on postharvest quality and decay caused by Botrytis cinerea in tomato fruit. J Sci Food Agric 98(10):3815–3822. https://doi.org/10.1002/jsfa.8895

Min D, Li F, Cui X et al (2020) SlMYC2 are required for methyl jasmonate-induced tomato fruit resistance to Botrytis cinerea. Food Chem 310:125901. https://doi.org/10.1016/j.foodchem.2019.125901

Mohamed HE, Abu-Goukh AA (2003) Effect of ethrel in aqueous solution and ethylene released from ethrel on mango fruit ripening. J Hortic Sci Biotechnol 78(4):568–573. https://doi.org/10.1080/14620316.2003.11511665

Moniruzzaman M, Khatoon R, Hossain MFB et al (2015) Influence of ethephon on ripening and quality of winter tomato fruit harvested at different maturity stages. Bangladesh J Agric Res 40(4):567–580. https://doi.org/10.3329/bjar.v40i4.26932

Mukkun L, Singh Z (2009) Methyl jasmonate plays a role in fruit ripening of 'Pajaro' strawberry through stimulation of ethylene biosynthesis. Sci Hortic 123(1):5–10. https://doi.org/10.1016/j.scienta.2009.07.006

Murphey AS, Dilley DR (1988) Anthocyanin biosynthesis and maturity of 'McIntosh' apples as influenced by ethylene-releasing compounds. J Am Soc Hortic Sci (USA) 113(5):718–723

Neoh TL, Yamauchi K, Yoshii H et al (2007) Kinetics of molecular encapsulation of 1-methylcyclopropene into α-Cyclodextrin. J Agric Food Chem 55:11020–11026. https://doi.org/10.1021/jf072357t

Pereira da Silva BM, Binoti RM, Cia P et al (2017) Ripening of 'Kumagai' guavas and anthracnose control as affected by methyl jasmonate. Bragantia 76(1):167–176. https://doi.org/10.1590/1678-4499.687

Pesis E, Ackerman M, Ben-Arie R et al (2002) Ethylene involvement in chilling injury symptoms of avocado during cold storage. Postharvest Biol Technol 24:171–181. https://doi.org/10.1016/S0925-5214(01)00134-X

Pinho N, Beirão G, Patrício L et al (2014) Understanding value co-creation in complex services with many actors. J Serv Manag 25(4):470–493. https://doi.org/10.1108/JOSM-02-2014-0055

Porat R, Vinocur V, Weiss B et al (1999) Effects of various elicitors on the resistance of citrus fruit against pathogens. Phytoparasitica 27:157–158

Porat R, Feng X, Huberman M et al (2001) Gibberellic acid slows postharvest degreening of 'Oroblanco' citrus fruits. HortScience 36(5):937–940

Redente E, Szente L, Szejtli J (2001) Cyclodextrin complexes of salts of acidic drugs: thermodynamic properties, structural features, and pharmaceutical applications. J Pharm Sci 90(8):979–986. https://doi.org/10.1002/jps.1050

Reineccius GA (1995) Liposomes for controlled release in the food industry. In: Risch SJ, Reineccius GA (eds) Encapsulation and controlled release of food ingredients. American Chemical Society, Washington, DC, pp 113–131

Sabuz A, Chowdhury M, Molla MM et al (2019) Effect of ethephon on ripening and postharvest quality of mango. Bangladesh J Agric Res 44(3):453–467. https://doi.org/10.3329/bjar.v44i3.43478

Sakhale BK, Gaikwad SS, Chavan RF (2018) Application of 1-methylcyclopropene on mango fruit (Cv. Kesar): potential for shelf life enhancement and retention of quality. J Food Sci Technol 55:776–781. https://doi.org/10.1007/s13197-017-2990-0

Saltveit ME (1999) Effect of ethylene on quality of fresh fruits and vegetables. Postharvest Biol Technol 15:279–292

Seglie L, Martina K, Devecchi M, et al (2011) The effects of 1-MCP in cyclodextrin-based nanosponges to improve the vase life of Dianthus caryophyllus cut flowers. Postharvest Biol Technol 59(2):200–205. https://doi.org/10.1016/j.postharvbio.2010.08.012

Sojo MM, Nuñez-Delicado E, García-Carmona F et al (1999) Cyclodextrins as activator and inhibitor of latent banana pulp polyphenol oxidase. J Agric Food Chem 47(2):518–523. https://doi.org/10.1021/jf9808978

Szejtli J, Szente L, Banky-Elod E (1979) Molecular encapsulation of volatile, easily oxidizable labile flavor substances by cyclodexrins. Acta Chim Acad Sci Hung 101:27–46

Szejtli J, Budai Z, Tétényi M et al (1983) Hormonal plant growth regulator. US Patent 4,380,626 A, 19 Apr 1983

Tan Q, Truong A, Thanh L et al (2014) The effects of ethephon on the ripening of Vietnamese Latundan bananas (*Musa sapientum*). Emir J Food Agric 26(3):229–237. https://doi.org/10.9755/ejfa.v26i3.15853

Tassoni A, Watkins CB, Davies PJ (2006) Inhibition of the ethylene response by 1-MCP in tomato suggests that polyamines are not involved in delaying ripening, but may moderate the rate of ripening or over-ripening. J Exp Bot 57(12):3313–3325. https://doi.org/10.1093/jxb/erl092

Tester RF, Debon SJJ (2000) Annealing of starch-a review. Int J Biol Macromol 27(1):1–12. https://doi.org/10.1016/S0141-8130(99)00121-X

Tétényi M (1982) Biological effects of the cyclodextrin complexes of 2-chloromethylphosphonic acid. Paper presented at the Proceedings of the 1st international symposium on cyclodextrins, Budapest, Hungary, 30 September–2 October 1981

Tétényi M, Szejtli J (1984) Prolongation of ethylene effect by cyclodextrin complexation of 2-chloroethanephosphonic acid. Acta Chim Acad Sci Hung 33(3–4):345–348

Trinchero GD, Sozzi GO, Covatta F et al (2004) Inhibition of ethylene action by 1-methylcyclopropene extends postharvest life of 'Bartlett' pears. Postharvest Biol Technol 32:193–204. https://doi.org/10.1016/j.postharvbio.2003.11.009

Wasternack C, Stenzel I, Hause B et al (2006) The wound response in tomato-role of jasmonic acid. J Plant Physiol 163(3):297–306. https://doi.org/10.1016/j.jplph.2005.10.014

Watkins C (2015) Advances in the use of 1-MCP. In: Ron BH, Wills JG (eds) Advances in postharvest fruit and vegetable technology, 1st edn. CRC Press, Boca Raton, FL, pp 117–146

Wu Y, Liu X, Fu B et al (2020) Methyl jasmonate enhances ethylene synthesis in kiwifruit by inducing NAC genes that activate ACS1. J Agric Food Chem 68(10):3267–3276. https://doi.org/10.1021/acs.jafc.9b07379

Yang SF, Hoffman NE (1984) Ethylene biosynthesis and its regulation in higher plants. Annu Rev Plant Physiol 35(1):155–189. https://doi.org/10.1146/annurev.pp.35.060184.001103

Yin XR, Zhang Y, Zhang B et al (2013) Effects of acetylsalicylic acid on kiwifruit ethylene biosynthesis and signaling components. Postharvest Biol Technol 83:27–33. https://doi.org/10.1016/j.postharvbio.2013.03.012

Zhang H (2016) Optimization of encapsulation methods and conditions to maximize 1-MCP loading in modified beta-cyclodextrins. Dissertation, Rutgers University-Graduate School-New Brunswick

Zhang Q, Zhen Z, Jiang H et al (2011) Encapsulation of the ethylene inhibitor 1-methylcyclopropene by cucurbit[6]uril. J Agric Food Chem 59(19):10539–10545. https://doi.org/10.1021/jf2019566

Zhang Z, Tian S, Zhu Z et al (2012) Effects of 1-methylcyclopropene (1-MCP) on ripening and resistance of jujube (Zizyphus jujuba cv. Huping) fruit against postharvest disease. LWT Food Sci Technol 45(1):13–19. https://doi.org/10.1016/j.lwt.2011.07.030

Index

T. M. Ho et al. (eds.), *Functionality of Cyclodextrins in Encapsulation for Food Applications*, https://doi.org/10.1007/978-3-030-80056-7

G

H